Composite Materials

Mathematical theory and exact relations

Composite Materials

Mathematical theory and exact relations

Yury Grabovsky
Department of Mathematics, Temple University, Philadelphia, USA

IOP Publishing, Bristol, UK

ISBN 978-0-7503-1048-2 (ebook)
ISBN 978-0-7503-1049-9 (print)
ISBN 978-0-7503-1115-1 (mobi)

DOI 10.1088/978-0-7503-1048-2

Version: 20161201

IOP Expanding Physics
ISSN 2053-2563 (online)
ISSN 2054-7315 (print)

British Library Cataloguing-in-Publication Data: A catalogue record for this book is available from the British Library.

Published by IOP Publishing, wholly owned by The Institute of Physics, London

IOP Publishing, Temple Circus, Temple Way, Bristol, BS1 6HG, UK

US Office: IOP Publishing, Inc., 190 North Independence Mall West, Suite 601, Philadelphia, PA 19106, USA

To my family: Irina, David and Phillip

Contents

Part II General theory of exact relations and links

Preface

Since the appearance of the general theory of exact relations in 1998, *all* microstructure-independent formulas and *all* links between pairs of composites with the same microstructure could theoretically be obtained in a systematic and largely mechanical way. However, actual derivations involve quite a bit of effort and, in some cases, a massive amount of effort. Over the intervening 18 years the author, sometimes alone, and often with the help of a group of bright undergraduates or a graduate student, has been steadily computing complete lists of exact relations and links for conducting, elastic, piezoelectric, thermoelectric and thermoelastic composites. This book contains a compilation of all the results obtained through all these years of work. The book also includes mathematically rigorous and self-contained development of the general theory of exact relations and links, which is based on the theory of homogenization. Traditionally, homogenization theorems are proved separately in each physical context. The novelty of the approach taken in this book is that the development occurs in the general L^2 framework, common to all types of composites. This level of generality has made the theory cleaner, but also raised several interesting questions that were previously obscured by the particulars of each physical context. Therefore, this book can be read on three different levels: as a single source for all exact relations and links for effective properties of composite materials (part 3); as the definitive exposition of the general theory of exact relations (part 2); and as a new streamlined development of a homogenization-based mathematical theory of composite materials (part 1).

Notwithstanding the order in which the book is written, part 3, containing complete lists of exact relations and links, can be consulted independently, without the need to refer to parts 1 and 2. Part 3 is aimed at the broadest possible audience, requiring from the reader only the understanding of the tensorial description of properties of anisotropic materials in a given physical context. The complete calculations leading to the results listed in part 3 are often very lengthy and are not presented in the book. Instead, we refer the reader to published journal articles detailing such calculations, unless the results are new and as yet unpublished.

Part 1 of the book develops a mathematically rigorous general theory of composite materials that encompasses in a single abstract framework every material property of interest, be it conductivity, elasticity, piezoelectricity or thermoelectricity. While the abstract mathematical framework used in the book is well known among experts, it was never used as a sole basis for developing the theory. The author hopes that the more mathematically inclined reader will appreciate the main technical feature of part 1: its complete divorce from partial differential equations of classical physics, leading to elegant proofs of fundamental results. The story unfolds in scaling and translation-invariant subspaces of L^2 described by the Fourier multiplier operators encoding all relevant physics. The author also hopes that the readers will take an interest in some of the unsolved problems that arose in the process. With this said, all serious technicalities are placed into chapters marked with the ‡ sign, which can be skipped, or sought, depending on the reader's proclivities.

The author wishes to acknowledge the continued financial support of the National Science Foundation, through six grants over the course of 18 years of theory development. One of the grants—the REU site grant—made it possible to assemble groups of talented undergraduate students from all over the country over the course of three summers from 2002 to 2004. Specifically, the author thanks: Thilagavathi Murugesan and Saule Zhoshina, two graduate students at the University of Utah, who helped to compute the exact relations for three-dimensional elastic composites during the summer of 1998; three REU teams of undergraduate students from all over the country who computed all the exact relations and links for fiber-reinforced conducting composites with the Hall effect: Erin Blew, David Carchedi, Edward Corcoran, Ryan Fuoss, Joseph Galante, Jerome Hodges IV, Russell Howes, Matthew Jacobs, Matthew Macauley, John Quah, Austin Roberts, Elianna Ruppin, Steven Stewart and Peter Tom-Wolverton; Hansen To, the author's graduate student at Temple University, who used Maple™ to compute all of the exact relations for three-dimensional conducting composites with the Hall effect in 2004; Meredith Hegg, the author's graduate student at Temple University, who computed an amazing link for fiber-reinforced elastic composites in 2012; Tatyana Nuzhnaya, a graduate student at Temple University who, during the summer of 2013, used MATLAB® to compute all of the algebras and ideals for fiber-reinforced elastic composites; Mark Mikida and Andrew Schneider, undergraduate researchers who, during the 2013–14 academic year, computed all of the exact relations for fiber-reinforced elastic composites; Adam Michael Jacoby, a graduate student at Temple University who, during the summer of 2014, computed using Maple™ all the factor-algebras and their isomorphisms, and wrote the elimination program that made it feasible to compute all links for fiber-reinforced elastic composites, the links were computed by three Temple undergraduate researchers Patrick Wynne, Hansen Pei and Abraham Lyle, who worked both alone and as a team during the 2014–16 academic years; Daniel Lapsley, an undergraduate researcher who, during the summer of 2016, wrote a Maple™ program that multiplies the irreducible representations of SO(3) in the space of piezoelectric tensors, which has lead to the quick calculation of all links for three-dimensional piezoelectric composites. Last but not least, the author wishes to extend special thanks to Graeme Milton who suggested that the book be written in the first place, and whose continued encouragement sustained the author's efforts to apply the theory in more and more complicated physical problems.

Acknowledgment

This material is based upon work supported by the National Science Foundation under Grant No. DMS-1412058.

Author biography

Yury Grabovsky

Yury Grabovsky is a professor at the Department of Mathematics at Temple University, where his research interests include the calculus of variations on the mathematics side and continuum mechanics on the physics side. His latest work was on the mathematical theory of composite materials, the buckling of slender bodies, such as plates, rods, and shells, and understanding the stability of equilibrium configurations with phase boundaries in nonlinear elasticity. After graduating with a PhD from the Courant Institute of Mathematical Sciences, New York University, Yury spent a year as a postdoc at the Center for Nonlinear Analysis at Carnegie Mellon University, and four years as a Wylie Instructor at the University of Utah. In 1999 he was hired as a tenure-track Assistant Professor by Temple University. Yury enjoys teaching and performing research with students. His web-page on using continued fractions in the design of calendar systems routinely draws the attention of media every leap year. It has been translated into Portuguese and published in a journal for mathematics school teachers *Educação e Matemática*.

Chapter 1

Introduction

Composite materials play an increasingly important role in our everyday lives. The usefulness of man-made composites, such as the ones used in skis and golf clubs, comes from the ability to combine raw materials having some, but not all of the desired properties of the composite. The quest to understand mathematically the relation between properties of microscopic constituents, the microstructure and the observed macroscopic behavior, began as early as the 19th century. For example, Poisson [24], Faraday [2] and Maxwell [17] considered arrays of conducting spheres embedded in an insulating or conducting matrix. In this setting the mathematical problem is to find the limit of the electric or magnetic field values as the sizes and distances between spheres go to zero, thereby identifying the effective macroscopic properties of a microscopically heterogeneous medium. In order to describe the macroscopic properties of materials with complicated, possibly disordered microstructures, the concept of representative volume element (RVE) was introduced [10, 11]. However, from the mathematical standpoint the notion of RVE is ambiguous and ultimately mathematically unsatisfactory, since the power of its wide applicability in mechanics lies in its ambiguity. For the purposes of mathematical analysis we return to the original investigations in physics, where the heterogeneous medium is modeled by an *infinite sequence* or family of structures whose properties vary on a smaller and smaller length scales. If the early models of Poisson, Faraday and Maxwell involved simple periodic geometry, modern analytical tools permit us to prove general homogenization theorems for families of structures of arbitrary complexity. Macroscopic properties of such media are defined mathematically by means of the notion of the G-limit, introduced by Spagnolo [25], and generalized by Murat and Tartar [23], who called their more general notion of homogenization the H-limit. This method of defining composites and their effective properties is adopted in this book from the very beginning. In the mathematical theory homogenization methods are as versatile as the concept of RVE in mechanics, without suffering from the ambiguity of RVE.

doi:10.1088/978-0-7503-1048-2ch1

Among the many important questions one can ask about properties of composites we single out the *G-closure problem*, introduced by Lurie and Cherkaev [14]. This is a natural question in situations where the geometric details of the microstructure are unavailable, and only the properties of constituents and possibly their volume fractions are known. The G-closure problem asks to describe the set of all effective tensors of composites that can be made from a given set of constituent materials. Even though many G-closure sets have been computed exactly (e.g. [3, 4, 8, 14–16]), the current consensus is that the problem of an explicit analytic description of such sets is largely intractable. In parallel with this development, researchers kept discovering other, somewhat counter-intuitive, facts about effective behavior of composites. In certain special situations it could be shown that the effective tensor of a composite satisfies one or more equations that hold *regardless of the microstructure*. A typical, and probably the earliest example, would be a result discovered by Hill [11, 12], whereby any composite made of isotropic elastic materials with the same shear modulus, must necessarily be isotropic with that same shear modulus. In the language of G-closures these results meant that there are special sets of materials, whose G-closures have *empty interiors* in the space of all material tensors. Viewed geometrically, such results correspond to G-closed submanifolds of positive codimension in the space of tensors of material properties. We call such submanifolds *exact relations*.

Exact relations indicate which combinations of material moduli would be insensitive to the changes in the design of the composite structure. They can be used to guide the design of new composites, and validate testing procedures and underlying assumptions by serving as benchmarks. In a series of papers [5–7, 9] the author with his collaborators developed a general theory of exact relations that provides effective methods for explicit description of exact relations in every physical context. Such a broad scope of the theory was achieved by means of the application of a unified functional analytic framework developed for random and periodic composites in [1, 13, 19–21]. In this book this unified framework is expanded further to serve as a basis for the theory of effective behavior of composites based on H-convergence.

The abstraction of the theory of exact relations from any specific physical context permits us to go beyond the original concept of an exact relation and obtain deeper and more general properties of composites, which we call *links*. A link specifies the relations between effective tensors of a pair of composite materials with the same microstructure, that differ only in the values of the material properties of their constituents. Mathematically, links are transformation rules that are stable under homogenization. The simplest, and probably earliest example of a non-obvious link is in the context of two-dimensional conducting composites. It says that if the conductivities of every constituent in an arbitrary composite are transformed according to the rule $\sigma \mapsto \sigma/\det \sigma$, then the effective conductivity of the 'transformed' composite will be $\sigma^*/\det \sigma^*$, where σ^* is the effective conductivity tensor of the original composite[1]. This result is due to Mendelson [18].

[1] The effective conductivity tensor σ^* does not have to be constant throughout the composite material, since we do not assume that the microstructure in the vicinity of different macroscopic points is statistically the same.

Part 1 of the book develops the general mathematical framework of homogenization, based on the concept of H-convergence. The main tool for proving key homogenization theorems is compensated compactness, due to Murat and Tartar [22, 26, 27]. Our approach raises new questions (unanswered in full generality) of the relation of the Fourier multiplier symbol determined by the physics of the composite and key properties of local spaces, such a compensated compactness, for example. In this part of the book we derive a new formula for the effective tensor of a periodic composite and give a transparent proof of the locality of H-closure.

As the theory of exact relations and links was developing, it produced new tools capable of harvesting *all* exact relations and links in virtually any physical context, including coupled properties, such as piezoelectricity or thermoelasticity, and even fibrous, or fiber-reinforced composites. The general theory and all the new tools are described in part 2 of the book. The complete lists of all the exact relations and links for conducting, elastic, piezoelectric, thermoelastic and thermoelectric composites are collected in part 3 and represent years of work by the author and his students.

References

[1] Dell'Antonio G F, Figari R and Orlandi E 1986 An approach through orthogonal projections to the study of inhomogeneous random media with linear response *Ann. Inst. Henri Poincaré* **44** 1–28

[2] Faraday M 1838 Experimental researches in electricity *Phil. Trans. R. Soc. Lond.* **128** 265–82

[3] Francfort G A and Milton G W 1987 Optimal bounds for conduction in two-dimensional, multiphase, polycrystalline media *J. Stat. Phys.* **46** 161–77

[4] Grabovsky Y 1993 The G-closure of two well-ordered anisotropic conductors *Proc. R. Soc. Edin.* A **123** 423–32

[5] Grabovsky Y 1998 Exact relations for effective tensors of polycrystals. I: Necessary conditions *Arch. Ration. Mech. Anal.* **143** 309–30

[6] Grabovsky Y 2004 Algebra, geometry and computations of exact relations for effective moduli of composites *Advances in Multifield Theories of Continua with Substructure* ed G Capriz and P M Mariano (*Modeling and Simulation in Science, Engineering and Technology*) (Boston, MA: Birkhäuser) pp 167–97

[7] Grabovsky Y 2009 Exact relations for effective conductivity of fiber-reinforced conducting composites with the Hall effect via a general theory *SIAM J. Math. Anal.* **41** 973–1024

[8] Grabovsky Y and Milton G W 1998 Rank one plus a null-Lagrangian is an inherited property of two-dimensional compliance tensors under homogenization *Proc. R. Soc. Edin.* A **128** 283–99

[9] Grabovsky Y, Milton G W and Sage D S 2000 Exact relations for effective tensors of polycrystals: necessary conditions and sufficient conditions *Commun. Pure Appl. Math.* **53** 300–53

[10] Hashin Z 1964 Theory of mechanical behavior of heterogeneous media *Appl. Mech. Rev.* **17** 1–9

[11] Hill R 1963 Elastic properties of reinforced solids: some theoretical principles *J. Mech. Phys. Solids.* **11** 357–72

[12] Hill R 1964 Theory of mechanical properties of fibre-strengthened materials: I. Elastic behaviour *J. Mech. Phys. Solids.* **12** 199–212

[13] Kohler W and Papanicolaou G C 1982 Bounds for effective conductivity of random media *Macroscopic Properties of Disordered Media* ed R Burridge, S Childress and G Papanicolaou (Berlin: Springer) pp 111–30

[14] Lurie K A and Cherkaev A V 1981 *G*-closure of a set of anisotropic conducting media in the case of two dimensions *Dokl. Akad. Nauk. SSSR* **259** 328–31 (in Russian)

[15] Lurie K A and Cherkaev A V 1984 Exact estimates of conductivity of composites formed by two isotropically conducting media taken in prescribed proportion *Proc. R. Soc. Edin.* A **99** 71–87

[16] Lurie K A and Cherkaev A V 1986 Exact estimates of a binary mixture of isotropic components *Proc. R. Soc. Edin.* A **104** 21–38

[17] Maxwell J C 1881 *A Treatise on Electricity and Magnetism* vol 1 2nd edn (Oxford: Clarendon)

[18] Mendelson K S 1975 A theorem on the conductivity of two-dimensional heterogeneous medium *J. Appl. Phys.* **46** 4740–1

[19] Milton G W 1987 Multicomponent composites, electrical networks and new types of continued fraction. I *Commun. Math. Phys.* **111** 281–327

[20] Milton G W 1987 Multicomponent composites, electrical networks and new types of continued fraction I: *Commun. Math. Phys.* **111** 329–72

[21] Milton G W and Kohn R V 1988 Variational bounds on the effective moduli of anisotropic composites *J. Mech. Phys. Solids* **36** 597–629

[22] Murat F 1981 Compacité par compensation: condition nécessaire et suffisante de continuité faible sous une hypothèse de rang constant *Ann. Scuola Norm. Sup. Pisa Cl. Sci.* **8** 69–102

[23] Murat F and Tartar L 1997 H-convergence *Topics in the Mathematical Modelling of Composite Materials* ed A Cherkaev and R V Kohn (*Progress in Non-linear Differential equations and Their Applications* vol 31) (Boston, MA: Birkhäuser) ch 3 pp 21–43

[24] Poisson S D 1826 Mémoire sur la théorie du magnétisme *Mém. Acad. R. Sci. Inst. France* **5** 247–338

[25] Spagnolo S 1968 Sulla convergenza di soluzioni di equazioni paraboliche ed ellittiche *Ann. Scuola Norm. Sup. Pisa* **22** 571–97

Spagnolo S 1968 Sulla convergenza di soluzioni di equazioni paraboliche ed ellittiche *Ann. Scuola Norm. Sup. Pisa* **22** 673 (erratum)

[26] Tartar L 1979 Estimation de coefficients homogénéisés *Computing Methods in Applied Sciences and Engineering (Lecture Notes in Mathematics* vol 704*)* (Berlin: Springer) pp 364–73

[27] Tartar L 1983 Compacité par compensation: résultats et perspectives *Nonlinear Partial Differential Equations and their Applications, Collège de France Seminar* vol 4 (London: Pitman) pp 350–69

Part I

Mathematical theory of composite materials

Chapter 2

Material properties and governing equations

2.1 Introduction

The fundamental physical laws of classical physics are often insufficient when the outcome depends on the properties of matter involved in the observed phenomena. A typical example is the original system of Maxwell's equations, requiring a relation between a magnetic field and a magnetic induction, and the electric field and the dielectric displacement. Postulating these relations to be linear covers a vast, but by no means entire spectrum of physical phenomena. A closer look at ordinary materials reveals a complicated picture at the microscopic level, from the fibrillar structure of wood to the polycrystalline structure of metals. Man-made composites, widely used in industry, represent another class of materials with microstructure. The mathematical tools used to analyze the relation between microstructure, the properties of constituent materials and their macroscopic properties are referred to collectively as *homogenization*.

When we talk about composites we may refer to conducting, elastic, or piezoelectric composites. It is well understood that mathematical equations governing the underlying physical phenomena, such as stationary Maxwell's equations, or equations of continuum mechanics, or both, share a common mathematical structure. For this reason many technical results in the existing literature are often established in the simpler context of conductivity with the implicit understanding that the same approach would also apply in other contexts. This understanding has lead to the development of an abstract formalism in the works of Kohler and Papanicolaou [4], Dell-Antonio *et al* [2], Milton and Kohn [9] and Milton [5, 6], where the governing systems of partial differential equations were replaced with scale and translation (and often, rotation) invariant subspaces of L^2. Our exposition here follows closely the development in [7] (see also [8]). While this abstract approach is natural for the analysis of periodic cell problems, it has never been adapted to homogenization problems in the general context of H-convergence, as is done in this book.

Typically, homogenization theorems are proved for boundary value problems for partial differential equations or systems of partial differential equations defined on the set $\Omega \subset \mathbb{R}^3$ occupied by the material. The boundary value problem for conductivity is usually formulated in terms of the electric potential ϕ. Similarly, the boundary value problem for elasticity is formulated in terms of the displacement vector $\boldsymbol{u}$. In the abstract Hilbert space approach we operate on the level of physical fields, such as the current and electric fields for conductivity, and the stress and strain fields for elasticity. The price we have to pay for this generality is the loss of access to potentials. This difficulty has already been encountered in [3]. However, in our case, several fundamental questions that did not arise in [3] remain unanswered.

To motivate the formal axioms of the abstract theory we examine equations of conductivity and elasticity and emphasize common features that form the basis of the Hilbert space formalism described in this chapter.

2.2 Conductivity and elasticity

In conductivity, the electric field $\boldsymbol{e}$ is a gradient of the potential ϕ, while the current density $\boldsymbol{j}$ is divergence-free:

$$\boldsymbol{e} = \nabla\phi, \qquad \nabla \cdot \boldsymbol{j} = 0. \tag{2.1}$$

If we think of $\boldsymbol{e}$ and $\boldsymbol{j}$ as elements of $L^2(\mathbb{R}^3; \mathbb{R}^3)$ then we can write equations (2.1) in Fourier space as follows

$$\hat{e}(\boldsymbol{\xi}) = 2\pi \mathrm{i}\boldsymbol{\xi}\hat{\phi}(\boldsymbol{\xi}), \qquad \boldsymbol{\xi} \cdot \hat{j}(\boldsymbol{\xi}) = 0, \tag{2.2}$$

where

$$\hat{f}(\boldsymbol{\xi}) = \int_{\mathbb{R}^3} f(\boldsymbol{x})\mathrm{e}^{-2\pi \mathrm{i}\boldsymbol{x}\cdot\boldsymbol{\xi}}\,\mathrm{d}\boldsymbol{x}$$

is the Fourier transform. We observe that for every $\boldsymbol{\xi} \in \mathbb{R}^3 \backslash \{\boldsymbol{0}\}$ equations (2.2) restrict vectors $\hat{e}(\boldsymbol{\xi})$ and $\hat{j}(\boldsymbol{\xi})$ to lie in certain subspaces of $\mathbb{C}^3$, namely

$$\hat{e}(\boldsymbol{\xi}) \in \mathbb{C}\boldsymbol{\xi} = \{c\boldsymbol{\xi}: c \in \mathbb{C}\} = \mathcal{E}_{\boldsymbol{\xi}} \otimes \mathbb{C}, \qquad \hat{j}(\boldsymbol{\xi}) \in \left\{\boldsymbol{a} \in \mathbb{C}^3: \boldsymbol{a} \cdot \boldsymbol{\xi} = 0\right\} = \mathcal{J}_{\boldsymbol{\xi}} \otimes \mathbb{C},$$

where $\mathcal{E}_{\boldsymbol{\xi}}$ and $\mathcal{J}_{\boldsymbol{\xi}}$ are the real versions of these subspaces

$$\mathcal{E}_{\boldsymbol{\xi}} = \{c\boldsymbol{\xi} : c \in \mathbb{R}\}, \quad \mathcal{J}_{\boldsymbol{\xi}} = \left\{\boldsymbol{a} \in \mathbb{R}^3 : \boldsymbol{a} \cdot \boldsymbol{\xi} = 0\right\}.$$

We also observe that $\mathbb{R}^3 = \mathcal{E}_{\boldsymbol{\xi}} \oplus \mathcal{J}_{\boldsymbol{\xi}}$, where the sum is orthogonal. This orthogonality in Fourier space defines the orthogonal decomposition $L^2(\mathbb{R}^3; \mathbb{R}^3) = \mathcal{E} \oplus \mathcal{J}$, where

$$\mathcal{E} = \left\{\boldsymbol{E} \in L^2\left(\mathbb{R}^3; \mathbb{R}^3\right) : \hat{\boldsymbol{E}}(\boldsymbol{\xi}) \in \mathcal{E}_{\boldsymbol{\xi}} \otimes \mathbb{C}\right\},$$

$$\mathcal{J} = \left\{ \boldsymbol{J} \in L^2\left(\mathbb{R}^3; \mathbb{R}^3\right) : \hat{\boldsymbol{J}}(\boldsymbol{\xi}) \in \mathcal{J}_{\xi} \otimes \mathbb{C} \right\}.$$

The operator Γ of the orthogonal projection onto the subspace $\mathcal{E}$ is a Fourier multiplier:

$$\left(\widehat{\Gamma f}\right)(\boldsymbol{\xi}) = \frac{\left(\hat{f} \cdot \boldsymbol{\xi}\right)\boldsymbol{\xi}}{|\boldsymbol{\xi}|^2} = \boldsymbol{\Gamma}\left(\frac{\boldsymbol{\xi}}{|\boldsymbol{\xi}|}\right)\hat{f}(\boldsymbol{\xi}),$$

where

$$\boldsymbol{\Gamma}(\boldsymbol{\eta}) = \boldsymbol{\eta} \otimes \boldsymbol{\eta}. \tag{2.3}$$

The Fourier multiplier operators commute with translations, expressing the translation-invariant nature of physical laws. The homogeneity, i.e. the dependence of the Fourier multiplier only on $\boldsymbol{\xi}/|\boldsymbol{\xi}|$, expresses scaling-invariance of physical laws. The rotational invariance of physical laws can also be expressed in terms of the Fourier multiplier symbol $\boldsymbol{\Gamma}(\boldsymbol{\eta})$:

$$\boldsymbol{R}\boldsymbol{\Gamma}(\boldsymbol{\eta})\boldsymbol{R}^T = \boldsymbol{\Gamma}(\boldsymbol{R}\boldsymbol{\eta}), \qquad \boldsymbol{R} \in \mathrm{SO}(3).$$

Next we examine the equations of linear elasticity in all-space. The strain field $\boldsymbol{\varepsilon}$ is a symmetrized gradient, while the stress field $\boldsymbol{\sigma}$ is symmetric and divergence free:

$$\boldsymbol{\varepsilon} = e(\boldsymbol{u}) = \frac{1}{2}\left(\nabla \boldsymbol{u} + (\nabla \boldsymbol{u})^T\right), \qquad \nabla \cdot \boldsymbol{\sigma} = \boldsymbol{0}, \quad \boldsymbol{\sigma}^T = \boldsymbol{\sigma}. \tag{2.4}$$

In Fourier space $\hat{\boldsymbol{\varepsilon}}(\boldsymbol{\xi}) \in \mathcal{E}_{\xi} \otimes \mathbb{C}$ and $\hat{\boldsymbol{\sigma}}(\boldsymbol{\xi}) \in \mathcal{J}_{\xi} \otimes \mathbb{C}$

$$\mathcal{E}_{\xi} = \left\{ \boldsymbol{a} \otimes \boldsymbol{\xi} + \boldsymbol{\xi} \otimes \boldsymbol{a} : \boldsymbol{a} \in \mathbb{R}^3 \right\}, \qquad \mathcal{J}_{\xi} = \left\{ \boldsymbol{S} \in \mathrm{Sym}\left(\mathbb{R}^3\right) : \boldsymbol{S}\boldsymbol{\xi} = \boldsymbol{0} \right\},$$

where $\mathrm{Sym}(\mathbb{R}^3)$ denotes a vector space of all real symmetric 3×3 matrices. We can easily verify the orthogonal decomposition

$$\mathrm{Sym}\left(\mathbb{R}^3\right) = \mathcal{E}_{\xi} \oplus \mathcal{J}_{\xi}, \qquad \boldsymbol{\xi} \in \mathbb{R}^3 \setminus \{\boldsymbol{0}\},$$

with respect to the Frobenius inner product

$$(\boldsymbol{A}, \boldsymbol{B})_{\mathrm{F}} = \mathrm{Tr}\,(\boldsymbol{A}\boldsymbol{B}) \tag{2.5}$$

on $\mathrm{Sym}(\mathbb{R}^3)$. This leads to the orthogonal decomposition $L^2(\mathbb{R}^3; \mathrm{Sym}(\mathbb{R}^3)) = \mathcal{E} \oplus \mathcal{J}$,

$$\mathcal{E} = \left\{ \boldsymbol{E} \in L^2\left(\mathbb{R}^3; \mathrm{Sym}\left(\mathbb{R}^3\right)\right) : \hat{\boldsymbol{E}}(\boldsymbol{\xi}) \in \mathcal{E}_{\xi} \otimes \mathbb{C} \right\},$$

$$\mathcal{J} = \left\{ \boldsymbol{J} \in L^2\left(\mathbb{R}^3; \mathrm{Sym}\left(\mathbb{R}^3\right)\right) : \hat{\boldsymbol{J}}(\boldsymbol{\xi}) \in \mathcal{J}_{\xi} \otimes \mathbb{C} \right\}.$$

The orthogonal projection Γ onto the space $\mathcal{E}$ is a homogeneous of degree zero Fourier multiplier: $(\widehat{\Gamma f})(\boldsymbol{\xi}) = \boldsymbol{\Gamma}(\boldsymbol{\xi}/|\boldsymbol{\xi}|)\hat{f}(\boldsymbol{\xi})$, where

$$\Gamma(\eta)S = S\eta \otimes \eta + \eta \otimes S\eta - (S\eta \cdot \eta)\eta \otimes \eta. \tag{2.6}$$

The rotational invariance of elasticity equations can be expressed by the corresponding property of the projections $\Gamma(\eta)$:

$$R(\Gamma(\eta)\varepsilon)R^T = \Gamma(R\eta)\left(R\varepsilon R^T\right).$$

2.3 Abstract Hilbert space framework

We now part with explicitly stated differential equations such as (2.1) or (2.4) and work instead with the abstract fields $E(x)$ and $J(x)$ taking their values in an abstract finite-dimensional inner product space $\mathcal{T}$. (For conductivity $\mathcal{T} = \mathbb{R}^3$ with the standard dot product and for elasticity $\mathcal{T} = \mathrm{Sym}(\mathbb{R}^3)$ with the Frobenius inner product (2.5).) The field E is called the *intensity field* and J—the *flux field*. Instead of the differential constraints on E and J we postulate the following geometric structure of these fields in the Fourier space.

Postulate 2.1 (Orthogonality). *Assume that $\Gamma(\eta)$ is an orthogonal projection onto a proper subspace $\mathcal{E}_\eta \subset \mathcal{T}$, for every unit vector $\eta \in \mathbb{R}^d$. Furthermore, $\Gamma(\eta)$ is assumed to be a continuous even function on the unit sphere $\mathbb{S}^{d-1}$ in $\mathbb{R}^d$, in other words,*

$$\mathcal{T} = \mathcal{E}_\eta \oplus \mathcal{J}_\eta, \qquad |\eta| = 1, \qquad \mathcal{E}_{-\eta} = \mathcal{E}_\eta, \qquad \mathcal{J}_{-\eta} = \mathcal{J}_\eta, \tag{2.7}$$

where $\mathcal{J}_\eta = \mathcal{E}_\eta^\perp$ is the orthogonal complement of $\mathcal{E}_\eta$ in $\mathcal{T}$.

The continuity of $\Gamma(\eta)$ implies, in particular, that all subspaces $\mathcal{E}_\eta$ are of the same dimension (equal to $\mathrm{Tr}\,\Gamma(\eta)$). The spaces $\mathcal{E}$ and $\mathcal{J} = \mathcal{E}^\perp$ are then *defined by*

$$\begin{cases} \mathcal{E} = \left\{ E \in L^2\left(\mathbb{R}^d; \mathcal{T}\right) : \hat{E}(\xi) \in \mathcal{E}_{\frac{\xi}{|\xi|}} \otimes \mathbb{C} \text{ for almost all } \xi \in \mathbb{R}^d \right\}, \\ \mathcal{J} = \left\{ J \in L^2\left(\mathbb{R}^d; \mathcal{T}\right) : \hat{J}(\xi) \in \mathcal{J}_{\frac{\xi}{|\xi|}} \otimes \mathbb{C} \text{ for almost all } \xi \in \mathbb{R}^d \right\}. \end{cases} \tag{2.8}$$

The operator Γ of the orthogonal projection onto $\mathcal{E}$ is a Fourier multiplier

$$\widehat{\Gamma E}(\xi) = \Gamma\left(\frac{\xi}{|\xi|}\right)\hat{E}(\xi). \tag{2.9}$$

While we do not need rotational invariance of physical laws in order to develop the general abstract theory, it becomes essential for the analysis of polycrystalline composites. It can also be destroyed, for example, by the presence of a magnetic field.

Postulate 2.2 (Rotational equivariance). *We assume that rotations SO(d) act by orthogonal transformations on the inner product space $\mathcal{T}$. Moreover,*

$$\boldsymbol{R}\cdot(\boldsymbol{\Gamma}(\boldsymbol{\eta})\boldsymbol{e}) = \boldsymbol{\Gamma}(\boldsymbol{R\eta})(\boldsymbol{R}\cdot\boldsymbol{e}), \qquad \boldsymbol{R}\in \mathrm{SO}(d),\ \boldsymbol{\eta}\in\mathbb{S}^{d-1},\ \boldsymbol{e}\in\mathcal{T}, \tag{2.10}$$

where $\boldsymbol{R}\cdot\boldsymbol{e}$ denotes the action of the rotation $\boldsymbol{R}\in\mathrm{SO}(d)$ on the vector $\boldsymbol{e}\in\mathcal{T}$[1].

We remark that if the rotational equivariance[2] postulate holds, then the function $\boldsymbol{\Gamma}(\boldsymbol{\eta})$ is determined by a single finite-dimensional orthogonal projection operator $\boldsymbol{\Gamma}(\boldsymbol{\eta}_0)$ and the property $\boldsymbol{R}\cdot\boldsymbol{\Gamma}(\boldsymbol{\eta}_0)=\boldsymbol{\Gamma}(\boldsymbol{\eta}_0)$ for all rotations $\boldsymbol{R}$ for which $\boldsymbol{R\eta}_0=\pm\boldsymbol{\eta}_0$. In that case the function $\boldsymbol{\Gamma}(\boldsymbol{\eta})$ will necessarily agree with an even homogeneous polynomial on the unit sphere, which means, in particular, that our abstract Hilbert space setting can also be described in terms of partial differential equations.

In order to treat the homogenization of boundary value problems on specific domains Ω we need *local versions* of the spaces $\mathcal{E}$ and $\mathcal{J}$. To describe them we need to define the extension and restriction operators. The extension operator $X_0(\Omega)$ extends an L^2 field from Ω to $\mathbb{R}^d$ by zero:

$$X_0(\Omega): L^2(\Omega;\mathcal{T})\to L^2\left(\mathbb{R}^d;\mathcal{T}\right), \qquad (X_0(\Omega)\boldsymbol{f})(\boldsymbol{x}) = \begin{cases}\boldsymbol{f}(\boldsymbol{x}), & \boldsymbol{x}\in\Omega,\\ \boldsymbol{0}, & \boldsymbol{x}\in\Omega^c,\end{cases} \tag{2.11}$$

where $\Omega^c=\mathbb{R}^d\backslash\Omega$ denotes the complement of Ω. The restriction operator maps $L^2(\mathbb{R}^d;\mathcal{T})$ onto $L^2(\Omega;\mathcal{T})$ by restriction to Ω. Its action is denoted by $\boldsymbol{f}|\Omega$ for $\boldsymbol{f}\in L^2(\mathbb{R}^d;\mathcal{T})$. We now define the local versions of the spaces $\mathcal{E}$ and $\mathcal{J}$:

$$\mathcal{E}(\Omega)=\{\boldsymbol{E}|\Omega:\boldsymbol{E}\in\mathcal{E}\}, \qquad \mathcal{E}_0(\Omega)=\left\{\boldsymbol{f}\in L^2(\Omega;\mathcal{T}): X_0(\Omega)\boldsymbol{f}\in\mathcal{E}\right\}, \tag{2.12}$$

and

$$\mathcal{J}(\Omega)=\{\boldsymbol{J}|\Omega:\boldsymbol{J}\in\mathcal{J}\}, \qquad \mathcal{J}_0(\Omega)=\left\{\boldsymbol{f}\in L^2(\Omega;\mathcal{T}): X_0(\Omega)\boldsymbol{f}\in\mathcal{J}\right\}. \tag{2.13}$$

The orthogonality of the global spaces $\mathcal{E}$ and $\mathcal{J}$ has a local version.

Lemma 2.3. *The Hilbert space $L^2(\Omega;\mathcal{T})$ can be decomposed into the orthogonal sums as follows*

$$L^2(\Omega;\mathcal{T}) = \mathcal{J}_0(\Omega)\oplus\overline{\mathcal{E}(\Omega)} = \mathcal{E}_0(\Omega)\oplus\overline{\mathcal{J}(\Omega)}, \tag{2.14}$$

where $\bar{V}$ denotes the closure in $L^2(\Omega;\mathcal{T})$ of the subspace $V\subset L^2(\Omega;\mathcal{T})$.

---

[1] Rotations may act in a different way depending on the nature of the space $\mathcal{T}$. For example, $\boldsymbol{R}\cdot\boldsymbol{e}=\boldsymbol{Re}$, when $\mathcal{T}=\mathbb{R}^d$ in the case of conductivity and $\boldsymbol{R}\cdot\boldsymbol{\varepsilon}=\boldsymbol{R\varepsilon R}^T$, when $\mathcal{T}=\mathrm{Sym}(\mathbb{R}^d)$ in the case of elasticity.

[2] Rotational equivariance refers to operators whose action commutes with all rotations, while rotational invariance refers to vectors or sets of vectors that are unchanged by rotations. In this postulate $\boldsymbol{\Gamma}$ operates on pairs of vectors $(\boldsymbol{n},\boldsymbol{e})$. The postulate says that applying $\boldsymbol{\Gamma}$ to the pair $(\boldsymbol{n},\boldsymbol{e})$ and then rotating the result is the same as rotating $\boldsymbol{n}$ and $\boldsymbol{e}$ and then applying $\boldsymbol{\Gamma}$ to the pair $(\boldsymbol{Rn},\boldsymbol{R}\cdot\boldsymbol{e})$.

Proof. Let $f \in \mathcal{E}(\Omega)^{\perp}$. Then for any $E \in \mathcal{E}$ we have

$$\int_{\mathbb{R}^d} \left(E, X_0(\Omega)f\right) \mathrm{d}x = \int_{\Omega} \left((E|\Omega)(x), f(x)\right) \mathrm{d}x = 0,$$

since $E|\Omega \in \mathcal{E}(\Omega)$. Thus, $X_0(\Omega)f \in \mathcal{E}^{\perp} = \mathcal{J}$. It follows that $f \in \mathcal{J}_0(\Omega)$ by definition of $\mathcal{J}_0(\Omega)$. It is obvious that if $f \in \mathcal{J}_0(\Omega)$ then $f \in \mathcal{E}(\Omega)^{\perp}$. Hence, $\mathcal{J}_0(\Omega) = \mathcal{E}(\Omega)^{\perp}$. The decomposition (2.14) follows from the fact that $\mathcal{J}_0(\Omega)^{\perp} = (\mathcal{E}(\Omega)^{\perp})^{\perp} = \overline{\mathcal{E}(\Omega)}$. The second decomposition in (2.14) is proved similarly. □

It is easy to see that the spaces $\mathcal{E}_0(\Omega)$ and $\mathcal{J}_0(\Omega)$ are closed for any measurable subset Ω of $\mathbb{R}^d$. What properties of $\mathbf{\Gamma}(\boldsymbol{\eta})$ guarantee that the spaces $\mathcal{E}(\Omega)$ and $\mathcal{J}(\Omega)$ are closed, at least when Ω is a unit ball, is a major unresolved problem in this general theory. In appendix A we show that for $\mathbf{\Gamma}(\boldsymbol{\eta})$ given by (2.3) and (2.6) the spaces $\mathcal{E}(\Omega)$ and $\mathcal{J}(\Omega)$ are closed, when Ω is a unit ball.

2.4 Boundary value problems

The fields $E(x)$ and $J(x)$ in the physical material can be found from the solution of a boundary value problem, if one postulates a linear constitutive relation $J = \mathsf{L}E$, where $\mathsf{L} \in \mathrm{End}^{+}(\mathcal{T})$ is a positive definite linear operator on $\mathcal{T}$. In almost all examples the tensor L is symmetric.[3] The tensor L is called the *tensor of material properties.* If L is a constant operator we refer to a *homogeneous* material and if L depends on $x \in \Omega$, then we refer to an *inhomogeneous* material. In this book we do not treat high contrast composites, and hence we always assume that $\mathsf{L}(x)$ takes its values in a compact subset of $\mathrm{End}^{+}(\mathcal{T})$—the set of all positive definite linear operators on $\mathcal{T}$. For $0 < \alpha < \beta$ we define

$$\mathcal{M}(\alpha, \beta) = \left\{\mathsf{L} \in \mathrm{End}^{+}(\mathcal{T}) : (\mathsf{L}E, E) \geqslant \alpha|E|^2, \ \left(\mathsf{L}^{-1}E, E\right) \geqslant \beta^{-1}|E|^2\right\}. \tag{2.15}$$

We remark that $\mathsf{L} \in \mathcal{M}(\alpha, \beta)$ is equivalent to $\mathsf{L}^{-1} \in \mathcal{M}(\beta^{-1}, \alpha^{-1})$.

Our goal is not a solution of specific boundary value problems, but rather an understanding of macroscopic properties of composite materials. For that purpose it is sufficient to test the response of a composite to prescribed Dirichlet boundary conditions. However, since we do not have a direct access to potentials, the boundary value problem needs to be stated in a somewhat round about way. Let Γ_{Ω} be an orthogonal projection onto $\mathcal{E}_0(\Omega)$. The governing equations can therefore be written in the following form

$$E \in \mathcal{E}_0(\Omega), \qquad J = \mathsf{L}(x)E, \qquad \Gamma_{\Omega}J = f, \qquad f \in \mathcal{E}_0(\Omega). \tag{2.16}$$

[3] The presence of the magnetic field can break the symmetry, as in the Hall effect.

Theorem 2.4. *For every $f \in \mathcal{E}_0(\Omega)$ the boundary value problem (2.16) has a unique solution $E \in \mathcal{E}_0(\Omega)$.*

Proof. Indeed, the linear operator

$$\mathsf{T}_{\mathsf{L}} : \mathcal{E}_0(\Omega) \to \mathcal{E}_0(\Omega), \qquad \mathsf{T}_{\mathsf{L}} \boldsymbol{E} = \Gamma_\Omega(\mathsf{L}(\boldsymbol{x})\boldsymbol{E}) \tag{2.17}$$

is bounded and coercive:

$$\|\mathsf{T}_{\mathsf{L}}\| \leqslant \beta, \qquad (\mathsf{T}_{\mathsf{L}}\boldsymbol{E}, \boldsymbol{E})_{L^2(\Omega)} = (\mathsf{L}(\boldsymbol{x})\boldsymbol{E}, \boldsymbol{E})_{L^2(\Omega)} \geqslant \alpha \, \|\boldsymbol{E}\|^2_{L^2(\Omega)}.$$

Therefore, it is invertible by the Lax–Milgram lemma. In fact,

$$\|T_{\mathsf{L}}^{-1}\| \leqslant \frac{1}{\alpha}. \tag{2.18}$$

□

Equations (2.16) do not mean that the analysis is restricted to the zero boundary conditions on $\partial\Omega$, because the case of arbitrary Dirichlet boundary conditions can be reduced to (2.16). Indeed, we may choose a field $\boldsymbol{g}_0 \in \mathcal{E}(\Omega)$ satisfying the desired Dirichlet boundary conditions, and then solve $\Gamma_\Omega(\mathsf{L}(\boldsymbol{x})(\boldsymbol{E} + \boldsymbol{g}_0)) = \boldsymbol{f}$, which is equivalent to $\mathsf{T}_{\mathsf{L}}\boldsymbol{E} = \boldsymbol{f} - \Gamma_\Omega(\mathsf{L}(\boldsymbol{x})\boldsymbol{g}_0)$.

In the context of conductivity, for example,

$$\mathcal{E}_0(\Omega) = \left\{ \nabla\phi : \phi \in H_0^1(\Omega) \right\},$$

if the domain Ω is diffeomorphic to a ball.[4] In this case the operator Γ_Ω is the Helmholtz projection onto the space of gradients:

$$\Gamma_\Omega \boldsymbol{f} = \nabla\psi, \qquad \Delta\psi = \nabla \cdot \boldsymbol{f}, \qquad \psi \in H_0^1(\Omega).$$

There is an isomorphism between $\mathcal{E}_0(\Omega)$ and $H^{-1}(\Omega)$, whereby every $\boldsymbol{f} \in \mathcal{E}_0(\Omega)$, given by $\boldsymbol{f} = \nabla g$, $g \in H_0^1(\Omega)$ corresponds to $F = \Delta g \in H^{-1}(\Omega)$. Then, the system of equations (2.16) can be rewritten as

$$\nabla \cdot (\mathsf{L}(\boldsymbol{x})\nabla\phi) = F, \qquad \phi \in H_0^1(\Omega).$$

From the theory of elliptic partial differential equations it is well known that this boundary value problem has a unique solution.

2.5 Geometry of local spaces†

In this section we build some of the technical tools needed for rigorous analysis of homogenization problems in the next chapter. It is hoped that mathematically inclined readers will enjoy this section, since it raises novel mathematical questions

[4] If Ω has a non-trivial homology, such as a spherical shell, then the space $\mathcal{E}_0(\Omega)$ can be larger than $\{\nabla\phi : \phi \in H_0^1(\Omega)\}$.

that, in the classical partial-differential-equation-based approach, have been obviated by direct access to potentials.

2.5.1 Additional structural constraints on the symbol

The basic assumptions imposed on the symbol $\mathbf{\Gamma}(\boldsymbol{\eta})$ (i.e. Postulate 2.1) are sufficient for analysis of periodic cell problems [8]. However, for the purposes of the general homogenization theory developed in the next chapter, more delicate conditions that are not apparent at this moment will be needed.

Definition 2.5. *The symbol $\mathbf{\Gamma}(\boldsymbol{\eta})$ will be called $\mathcal{E}$-regular, if $\mathcal{E}(B_1)$ is a closed subspace of $L^2(B_1; \mathcal{T})$, where B_1 denotes the unit ball in $\mathbb{R}^d$. Similarly, the symbol $\mathbf{\Gamma}(\boldsymbol{\eta})$ is $\mathcal{J}$-regular,* if *$\mathcal{J}(B_1)$ is a closed subspace of $L^2(B_1; \mathcal{T})$.*

Lemma 2.6. *Let Ω be a measurable subset of $\mathbb{R}^d$. Then, $\mathcal{E}(\Omega)$ is a closed subspace of $L^2(\Omega; \mathcal{T})$ if and only if there exists a bounded linear extension operator $X_{\mathcal{E}}(\Omega)$: $\mathcal{E}(\Omega) \to \mathcal{E}$.*

Proof. To prove sufficiency we assume that there exists a bounded linear extension operator $X_{\mathcal{E}}(\Omega) : \mathcal{E}(\Omega) \to \mathcal{E}$. For any $\boldsymbol{E} \in \overline{\mathcal{E}(\Omega)}$ there exists $\boldsymbol{E}_n \in \mathcal{E}(\Omega)$, $\boldsymbol{E}_n \to \boldsymbol{E}$, as $n \to \infty$ in $L^2(\Omega; \mathcal{T})$. Then, by continuity of the extension operator, $X_{\mathcal{E}}(\Omega)\boldsymbol{E}_n \to \tilde{\boldsymbol{E}} \in \mathcal{E}$, as $n \to \infty$ in $L^2(\mathbb{R}^d; \mathcal{T})$. Thus, $\boldsymbol{E} = \tilde{\boldsymbol{E}}|\Omega \in \mathcal{E}(\Omega)$ and $\mathcal{E}(\Omega)$ is closed.

To prove necessity we assume that $\mathcal{E}(\Omega)$ is a closed subspace of $L^2(\Omega; \mathcal{T})$. Let $D \subset \mathbb{R}^d$ be measurable and

$$\mathcal{E}_0^0(D) = X_0(D)\mathcal{E}_0(D) = \left\{X_0(D)\boldsymbol{E} : \boldsymbol{E} \in \mathcal{E}_0(D)\right\}. \tag{2.19}$$

In other words, $\mathcal{E}_0^0(D)$ consists of functions in $\mathcal{E}_0(D)$ extended by 0 to all of $\mathbb{R}^d$. It is obvious that $\mathcal{E}_0^0(D)$ is a closed subspace of $\mathcal{E}$. Let V be the orthogonal complement of $\mathcal{E}_0^0(\Omega^c)$ in $\mathcal{E}$. Let $R : V \to \mathcal{E}(\Omega)$ be the restriction map $V \ni \boldsymbol{v} \mapsto \boldsymbol{v}|\Omega$. Then R is a bounded linear map. Let us show that it is also a bijection. If $R\boldsymbol{v} = 0$, then $\boldsymbol{v} \in \mathcal{E}_0^0(\Omega^c)$. Hence, $\boldsymbol{v} = 0$, since $\boldsymbol{v} \in V$. It follows that R is an injective map. Let $\boldsymbol{E} \in \mathcal{E}$ be arbitrary. To prove the surjectivity of R we need to show that there exists $\boldsymbol{v} \in V$, such that $R\boldsymbol{v} = \boldsymbol{E}|\Omega$. Let $\boldsymbol{E}_0 \in \mathcal{E}_0^0(\Omega^c)$ be the orthogonal projection of $\boldsymbol{E}$ onto $\mathcal{E}_0^0(\Omega^c)$. Then $\boldsymbol{v} = \boldsymbol{E} - \boldsymbol{E}_0 \in V$ and $R\boldsymbol{v} = \boldsymbol{E}|\Omega$ because $\boldsymbol{E}_0|\Omega = 0$. We have proved that R is a bijection. The subspace V of $\mathcal{E}$ is closed by construction and $\mathcal{E}(\Omega)$ is closed by assumption. Therefore, by the Banach inverse operator theorem $R^{-1} : \mathcal{E}(\Omega) \to V \subset \mathcal{E}$ is a bounded extension operator. □

In fact, regardless of whether the spaces $\mathcal{E}(\Omega)$ and $\mathcal{J}(\Omega)$ are closed, there is a canonical choice of the extension operators $X_{\mathcal{E}}(\Omega) : \mathcal{E}(\Omega) \to \mathcal{E}$ and $X_{\mathcal{J}}(\Omega)$: $\mathcal{J}(\Omega) \to \mathcal{J}$.

Lemma 2.7. *Let Ω be a measurable subset of $\mathbb{R}^d$. Let $\boldsymbol{E} \in \mathcal{E}(\Omega)$ and let $\boldsymbol{E}' \in \mathcal{E}$ be an extension of $\boldsymbol{E}$. Then*

$$X_{\mathcal{E}}(\Omega)\boldsymbol{E} = \boldsymbol{E}' - X_0(\Omega^c)\Gamma_{\Omega^c}(\boldsymbol{E}'|\Omega^c) \tag{2.20}$$

is also an extension $\boldsymbol{E}$. Moreover, it is independent of the choice of $\boldsymbol{E}'$.

Proof. For $\boldsymbol{E} \in \mathcal{E}(\Omega)$ let $\{\boldsymbol{E}', \boldsymbol{E}''\} \subset \mathcal{E}$ be two possible extensions of $\boldsymbol{E}$. Let

$$\boldsymbol{E}_1 = \boldsymbol{E}' - X_0(\Omega^c)\Gamma_{\Omega^c}(\boldsymbol{E}'|\Omega^c), \qquad \boldsymbol{E}_2 = \boldsymbol{E}'' - X_0(\Omega^c)\Gamma_{\Omega^c}(\boldsymbol{E}''|\Omega^c).$$

Let us show that $\boldsymbol{E}_1 = \boldsymbol{E}_2$. Clearly $\boldsymbol{E}_1|\Omega = \boldsymbol{E}'|\Omega = \boldsymbol{E} = \boldsymbol{E}_2|\Omega$. We also observe that

$$\{\boldsymbol{E}_1|\Omega^c, \boldsymbol{E}_2|\Omega^c\} \subset \overline{\mathcal{J}(\Omega^c)}.$$

Therefore, on the one hand $\boldsymbol{E}_1|\Omega^c - \boldsymbol{E}_2|\Omega^c \in \overline{\mathcal{J}(\Omega^c)}$, on the other hand $\boldsymbol{E}_1 - \boldsymbol{E}_2 \in \mathcal{E}$ and $(\boldsymbol{E}_1 - \boldsymbol{E}_2)|\Omega = 0$. Thus, $(\boldsymbol{E}_1 - \boldsymbol{E}_2)|\Omega^c \in \mathcal{E}_0(\Omega^c)$. The orthogonality of $\overline{\mathcal{J}(\Omega^c)}$ and $\mathcal{E}_0(\Omega^c)$ implies that $(\boldsymbol{E}_1 - \boldsymbol{E}_2)|\Omega^c = 0$. Thus, $\boldsymbol{E}_1 = \boldsymbol{E}_2$. ☐

The same argument establishes the existence of the canonical extension for $\mathcal{J}(\Omega)$. The canonical extensions $X_{\mathcal{E}}(\Omega)$ and $X_{\mathcal{J}}(\Omega)$ are also called the *harmonic extensions*, since they are the unique extensions that are 'harmonic' on Ω^c.

Definition 2.8. *The space $\mathcal{H}(\Omega) = \overline{\mathcal{E}(\Omega)} \cap \overline{\mathcal{J}(\Omega)}$ is called the space of harmonic functions on Ω.*

The harmonic extensions $X_{\mathcal{E}}(\Omega)$ and $X_{\mathcal{J}}(\Omega)$ have the minimal norm property: among all extensions $\tilde{\boldsymbol{E}}$ of $\boldsymbol{E} \in \mathcal{E}(\Omega)$ to $\mathcal{E}$ the canonical extension $X_{\mathcal{E}}(\Omega)\boldsymbol{E}$ has the smallest L^2 norm. Indeed, by (2.20)

$$\|\tilde{\boldsymbol{E}}\|^2_{L^2(\Omega^c)} = \|X_{\varepsilon}(\Omega)\boldsymbol{E} + \Gamma_{\Omega^c}\tilde{\boldsymbol{E}}\|^2_{L^2(\Omega^c)} = \|X_{\mathcal{E}}(\Omega)\boldsymbol{E}\|^2_{L^2(\Omega^c)} + \|\Gamma_{\Omega^c}\tilde{\boldsymbol{E}}\|^2_{L^2(\Omega^c)},$$

since $(X_{\mathcal{E}}(\Omega)\boldsymbol{E})|\Omega^c \in \overline{\mathcal{J}(\Omega^c)}$ and is therefore orthogonal to $\Gamma_{\Omega^c}\tilde{\boldsymbol{E}} \in \mathcal{E}_0(\Omega^c)$. This observation can be used to prove several equivalent reformulations of $\mathcal{E}$-regularity of the symbol $\boldsymbol{\Gamma}(\boldsymbol{\eta})$ stated as a properties of the operator Γ.

Theorem 2.9. *Let Ω be a measurable subset of $\mathbb{R}^d$. Then the following statements are equivalent.*

(a) *The space $\mathcal{E}(\Omega)$ is closed*

(b) *There exists a constant $C > 0$, such that*

$$\|\boldsymbol{e}\|_{L^2(\Omega)} \leqslant C\,\|\Gamma X_0(\Omega)\boldsymbol{e}\|_{L^2(\Omega)}, \tag{2.21}$$

for every $\boldsymbol{e} \in \mathcal{E}(\Omega)$.

(c) *There exists a constant* $C > 0$, *such that*

$$\|\boldsymbol{e}\|_{L^2(\Omega)} \leqslant C\, \|\Gamma X_0(\Omega)\boldsymbol{e}\|_{L^2(\mathbb{R}^d)}, \tag{2.22}$$

for every $\boldsymbol{e} \in \mathcal{E}(\Omega)$.

(d) *There exists a constant* $C > 0$, *so that for every* $\boldsymbol{h} \in L^2(\Omega; \mathcal{T})$

$$\|\Gamma X_0(\Omega)\boldsymbol{h}\|_{L^2(\mathbb{R}^d)} \leqslant C\, \|\Gamma X_0(\Omega)\boldsymbol{h}\|_{L^2(\Omega)}. \tag{2.23}$$

(e) *The space* $\mathcal{J}(\Omega^c)$ *is closed.*

Proof. First we will prove the circle of implications $(a) \Rightarrow (b) \Rightarrow (c) \Rightarrow (d) \Rightarrow (a)$. Then we will prove $(e) \Rightarrow (a)$. The reverse implication follows from the fact that all statements remain true if we interchange $\mathcal{E}$ and $\mathcal{J}$. Hence, interchanging $\mathcal{E}$ and $\mathcal{J}$ in the implication $(e) \Rightarrow (a)$ and replacing Ω with Ω^c we obtain the reverse implication $(a) \Rightarrow (e)$.

$(a) \Rightarrow (b)$. Suppose (2.21) is false. Then there exists a sequence $\boldsymbol{e}_n \in \mathcal{E}(\Omega)$, such that

$$\|\boldsymbol{e}_n\|_{L^2(\Omega)} = 1, \qquad \lim_{n\to\infty} \|\Gamma X_0(\Omega)\boldsymbol{e}_n\|_{L^2(\Omega)} = 0.$$

We note that $X_0(\Omega)\boldsymbol{e}_n - \Gamma X_0(\Omega)\boldsymbol{e}_n \in \mathcal{J}$. Thus,

$$\left(\Gamma X_0(\Omega)\boldsymbol{e}_n\right)\Big|\Omega^c \in \mathcal{J}(\Omega^c) \cap \mathcal{E}(\Omega^c).$$

It follows that $\Gamma X_0(\Omega)\boldsymbol{e}_n = X_{\mathcal{E}}(\Omega)((\Gamma X_0(\Omega)\boldsymbol{e}_n)|\Omega)$. By assumption and lemma 2.6 the operator $X_{\mathcal{E}}(\Omega)$ is bounded. Therefore, there exists a constant $C > 0$, such that

$$\|\Gamma X_0(\Omega)\boldsymbol{e}_n\|_{L^2(\mathbb{R}^d)} \leqslant C\, \|\Gamma X_0(\Omega)\boldsymbol{e}_n\|_{L^2(\Omega)}.$$

Thus,

$$\lim_{n\to\infty} \|\Gamma X_0(\Omega)\boldsymbol{e}_n\|_{L^2(\mathbb{R}^d)} = 0. \tag{2.24}$$

Let $\boldsymbol{E}_n = X_{\mathcal{E}}(\Omega)\boldsymbol{e}_n$. Then $\|\boldsymbol{E}_n\|_{L^2(\mathbb{R}^d)} \leqslant C$. However,

$$0 = \left(X_0(\Omega)\boldsymbol{e}_n - \Gamma X_0(\Omega)\boldsymbol{e}_n,\, \boldsymbol{E}_n\right)_{L^2(\mathbb{R}^d)} = \|\boldsymbol{e}_n\|^2_{L^2(\Omega)} - \left(\Gamma X_0(\Omega)\boldsymbol{e}_n,\, \boldsymbol{E}_n\right)_{L^2(\mathbb{R}^d)}.$$

This is a contradiction, since $\|\boldsymbol{e}_n\|^2_{L^2(\Omega)} = 1$ and

$$\lim_{n\to\infty} \left(\Gamma X_0(\Omega)\boldsymbol{e}_n,\, \boldsymbol{E}_n\right)_{L^2(\mathbb{R}^d)} = 0$$

due to (2.24) and boundedness of $\|\boldsymbol{E}_n\|_{L^2(\mathbb{R}^d)}$. Hence, we have proved that inequality (2.21) must hold.

$(b) \Rightarrow (c)$ is trivial:

$$\|\boldsymbol{e}\|_{L^2(\Omega)} \leqslant C\, \|\Gamma X_0(\Omega)\boldsymbol{e}\|_{L^2(\Omega)} \leqslant C\, \|\Gamma X_0(\Omega)\boldsymbol{e}\|_{L^2(\mathbb{R}^d)}.$$

$(c) \Rightarrow (d)$. Using the decomposition (2.14) we write $\boldsymbol{h} = \boldsymbol{e} + \boldsymbol{j}$, where $\boldsymbol{e} \in \overline{\mathcal{E}(\Omega)}$ and $\boldsymbol{j} \in \mathcal{J}_0(\Omega)$. In particular $X_0(\Omega)\boldsymbol{j} \in \mathcal{J}$. Hence,

$$\Gamma X_0(\Omega)\boldsymbol{h} = \Gamma X_0(\Omega)\boldsymbol{e}. \tag{2.25}$$

Since Γ is a projection operator we have

$$\|\Gamma X_0(\Omega)\boldsymbol{e}\|^2_{L^2(\mathbb{R}^d)} = \left(\boldsymbol{e}, \Gamma X_0(\Omega)\boldsymbol{e}\right)_{L^2(\Omega)} \leqslant \|\boldsymbol{e}\|_{L^2(\Omega)} \|\Gamma X_0(\Omega)\boldsymbol{e}\|_{L^2(\Omega)}.$$

Hence, inequality (2.22) implies $\|\Gamma X_0(\Omega)\boldsymbol{e}\|_{L^2(\mathbb{R}^d)} \leqslant C\,\|\Gamma X_0(\Omega)\boldsymbol{e}\|_{L^2(\Omega)}$. Inequality (2.23) now follows from (2.25).

$(d) \Rightarrow (a)$. Now let us assume that (2.23) holds. Let $\boldsymbol{e} \in \mathcal{E}(\Omega)$. Since $X_{\mathcal{E}}(\Omega)$ is the harmonic extension operator, there exists $\boldsymbol{j}_n \in \mathcal{J}$, such that $\boldsymbol{j}_n|\Omega^c \to (X_{\mathcal{E}}(\Omega)\boldsymbol{e})|\Omega^c$, as $n \to \infty$, in $L^2(\Omega^c)$. Let $\Delta_n = \boldsymbol{j}_n - X_{\mathcal{E}}(\Omega)\boldsymbol{e}$. Then,

$$X_{\mathcal{E}}(\Omega)\boldsymbol{e} = \boldsymbol{j}_n - \Delta_n = \boldsymbol{j}_n + X_0(\Omega)\left(\boldsymbol{e} - \boldsymbol{j}_n|\Omega\right) - \chi_{\Omega^c}(\boldsymbol{x})\Delta_n.$$

Thus,

$$X_{\mathcal{E}}(\Omega)\boldsymbol{e} = \Gamma X_{\mathcal{E}}(\Omega)\boldsymbol{e} = \Gamma X_0(\Omega)\left(\boldsymbol{e} - \boldsymbol{j}_n|\Omega\right) - \Gamma\left(\chi_{\Omega^c}(\boldsymbol{x})\Delta_n\right). \tag{2.26}$$

By inequality (2.23)

$$\|X_{\mathcal{E}}(\Omega)\boldsymbol{e}\|_{L^2(\mathbb{R}^d)} \leqslant C\,\|\Gamma X_0(\Omega)(\boldsymbol{e} - \boldsymbol{j}_n|\Omega)\|_{L^2(\Omega)} + \|\chi_{\Omega^c}(\boldsymbol{x})\Delta_n\|_{L^2(\mathbb{R}^d)}.$$

Restricting (2.26) to Ω we obtain

$$\left(\Gamma X_0(\Omega)\left(\boldsymbol{e} - \boldsymbol{j}_n|\Omega\right)\right)\Big|\Omega = \boldsymbol{e} + \Gamma\left(\chi_{\Omega^c}(\boldsymbol{x})\Delta_n\right)\Big|\Omega.$$

Recalling that $\chi_{\Omega^c}(\boldsymbol{x})\Delta_n \to 0$ in $L^2(\mathbb{R}^d)$ we obtain

$$\|X_{\mathcal{E}}(\Omega)\boldsymbol{e}\|_{L^2(\mathbb{R}^d)} \leqslant C\,\|\boldsymbol{e}\|_{L^2(\Omega)},$$

which proves that $X_{\mathcal{E}}(\Omega)$ is a continuous extension.

$(e) \Rightarrow (a)$. Let us show that we get a contradiction if we assume that $\mathcal{E}(\Omega)$ is not closed. According to part (c), inequality (2.22) is false, and hence, there exists a sequence $\boldsymbol{e}_n \in \mathcal{E}(\Omega)$, such that $\|\boldsymbol{e}_n\|_{L^2(\Omega)} = 1$, while $\Gamma X_0(\Omega)\boldsymbol{e}_n \to 0$ in $L^2(\mathbb{R}^d)$. Let

$$\boldsymbol{j}_n = X_0(\Omega)\boldsymbol{e}_n - \Gamma X_0(\Omega)\boldsymbol{e}_n.$$

Then $\boldsymbol{j}_n \in \mathcal{J}$, $\boldsymbol{j}_n|\Omega^c = -\,(\Gamma X_0(\Omega)\boldsymbol{e}_n)|\Omega^c \to 0$ in $L^2(\Omega^c)$. At the same time $\boldsymbol{j}_n|\Omega = \boldsymbol{e}_n - (\Gamma X_0(\Omega)\boldsymbol{e}_n)|\Omega \in \mathcal{E}(\Omega)$. We conclude that $\boldsymbol{j}_n$ is harmonic on Ω. In particular, $\boldsymbol{j}_n$ is the harmonic extension of $\boldsymbol{j}_n|\Omega^c$. By assumption $\mathcal{J}(\Omega^c)$ is closed and therefore, the harmonic extension $X_{\mathcal{J}}(\Omega^c)$ is a bounded operator. But then $\boldsymbol{j}_n = X_{\mathcal{J}}(\Omega^c)(\boldsymbol{j}_n|\Omega^c) \to 0$ in $L^2(\mathbb{R}^d)$. This contradicts the fact that $\boldsymbol{j}_n|\Omega = \boldsymbol{e}_n - (\Gamma X_0(\Omega)\boldsymbol{e}_n)|\,\Omega \not\to 0$ in $L^2(\Omega)$. □

The question of $\mathcal{E}$ and $\mathcal{J}$-regularity of symbols, is surprisingly hard. Indeed, the existence of an extension operator for Sobolev functions is a highly non-trivial fact [1, 10]. The existence of an extension operator in our setting, where only certain combinations of partial derivatives are known to be in L^2 is substantially more delicate. At the moment there is no understanding of $\mathcal{E}$- and $\mathcal{J}$-regularity in terms of the symbol $\boldsymbol{\Gamma}(\boldsymbol{\eta})$. Nevertheless, we achieve a certain degree of unity by singling out this property of physical laws, which enables us to generalize the classical homogenization results in the abstract Hilbert space framework. In appendix A we show that conductivity and elasticity have $\mathcal{E}$ and $\mathcal{J}$-regular symbols. This immediately implies that any other coupled field physics, such as piezoelectricity and thermoelectricity, also has $\mathcal{E}$ and $\mathcal{J}$-regular symbols.

Another useful property of physical symbols is nondegeneracy.

Definition 2.10. *We say that the symbol* $\boldsymbol{\Gamma}(\boldsymbol{\eta})$ *is* $\mathcal{E}$*-nondegenerate if* $\boldsymbol{\Gamma}(\boldsymbol{\eta})\boldsymbol{e}_0 = 0$ *for all* $\boldsymbol{\eta} \in \mathbb{S}^{d-1}$ *implies that* $\boldsymbol{e}_0 = 0$. *Equivalently,*

$$\mathcal{T}_{\mathcal{E}} = \mathcal{T}, \qquad \mathcal{T}_{\mathcal{E}} = \sum_{\boldsymbol{\eta}\in\mathbb{S}^{d-1}} \mathcal{E}_{\boldsymbol{\eta}}. \tag{2.27}$$

Similarly, the symbol $\boldsymbol{\Gamma}(\boldsymbol{\eta})$ is $\mathcal{J}$*-nondegenerate if* $\boldsymbol{\Gamma}(\boldsymbol{\eta})\boldsymbol{e}_0 = \boldsymbol{e}_0$ *for all* $\boldsymbol{\eta} \in \mathbb{S}^{d-1}$ *implies that* $\boldsymbol{e}_0 = 0$, *or*

$$\mathcal{T}_{\mathcal{J}} = \mathcal{T}, \qquad \mathcal{T}_{\mathcal{J}} = \sum_{\boldsymbol{\eta}\in\mathbb{S}^{d-1}} \mathcal{J}_{\boldsymbol{\eta}}. \tag{2.28}$$

One can easily verify that the symbols $\boldsymbol{\Gamma}(\boldsymbol{\eta})$ for conductivity and elasticity are $\mathcal{E}$ and $\mathcal{J}$-nondegenerate.

2.5.2 Density theorems

Theorem 2.11. *Let* Ω *be a measurable subset of* $\mathbb{R}^d$. *Then the spaces* $C^\infty(\overline{\Omega}; \mathcal{T}) \cap \mathcal{E}(\Omega)$ *and* $C^\infty(\overline{\Omega}; \mathcal{T}) \cap \mathcal{J}(\Omega)$ *are dense in* $\mathcal{E}(\Omega)$ *and* $\mathcal{J}(\Omega)$, *respectively.*

Proof. Let ρ_ϵ be a standard convolution kernel. Then $\boldsymbol{E}_\epsilon = \rho_\epsilon * \boldsymbol{E} \in C^\infty(\mathbb{R}^d; \mathcal{T}) \cap \mathcal{E}$ for any $\boldsymbol{E} \in \mathcal{E}$. Thus, $\boldsymbol{E}_\epsilon|\Omega \in C^\infty(\overline{\Omega}; \mathcal{T}) \cap \mathcal{E}(\Omega)$. Moreover, $\boldsymbol{E}_\epsilon \to \boldsymbol{E}$ in $L^2(\mathbb{R}^d; \mathcal{T})$. Hence, $\boldsymbol{E}_\epsilon|\Omega \to \boldsymbol{E}|\Omega$ in $L^2(\Omega; \mathcal{T})$. The statement about $\mathcal{J}(\Omega)$ is proved in the same way. □

Theorem 2.12. *Let us assume that* $\boldsymbol{\Gamma}(\boldsymbol{\eta})$ *agrees with an even homogeneous polynomial* $P(\boldsymbol{\eta})$ *of degree* 2m *on the unit sphere* $\mathbb{S}^{d-1}$. *Then the sets of compactly supported functions*

$$\mathcal{E}_c = \left\{ P(\mathrm{i}\nabla)\boldsymbol{\Phi} : \boldsymbol{\Phi} \in C_0^\infty\left(\mathbb{R}^d; \mathcal{T}\right) \right\},$$

$$\mathcal{J}_c = \left\{ \left(\Delta^m - P(\mathrm{i}\nabla) \right) \mathbf{\Phi} : \mathbf{\Phi} \in C_0^\infty\left(\mathbb{R}^d; \mathcal{T} \right) \right\}$$

are dense in $\mathcal{E}$ and $\mathcal{J}$, *respectively.*

Proof. By assumption there is a homogeneous polynomial $P(\boldsymbol{\xi})$ of degree $2m$ such that $\mathbf{\Gamma}(\boldsymbol{\eta}) = P(\boldsymbol{\eta})$ for all $\boldsymbol{\eta} \in \mathbb{S}^{d-1}$. Let $\boldsymbol{E} \in \mathcal{E}$ be arbitrary. Then

$$\hat{\boldsymbol{E}}(\boldsymbol{\xi}) = P(\boldsymbol{\xi}) \left(\frac{\hat{\boldsymbol{E}}(\boldsymbol{\xi})}{|\boldsymbol{\xi}|^{2m}} \right).$$

Let $\boldsymbol{\phi}_\epsilon(\boldsymbol{x})$ be given by its Fourier transform

$$\hat{\boldsymbol{\phi}}_\epsilon(\boldsymbol{\xi}) = \chi_{B_\epsilon^c}(\boldsymbol{\xi}) \frac{\hat{\boldsymbol{E}}(\boldsymbol{\xi})}{|\boldsymbol{\xi}|^{2m}}.$$

Clearly, $(1 + |\boldsymbol{\xi}|^{2m}) \hat{\boldsymbol{\phi}}_\epsilon(\boldsymbol{\xi}) \in L^2(\mathbb{R}^d; \mathcal{T})$ for any $\epsilon > 0$. Hence (see e.g. [1]) $\boldsymbol{\phi}_\epsilon \in H^{2m}(\mathbb{R}^d; \mathcal{T})$. We also see that

$$\lim_{\epsilon \to 0} P(\boldsymbol{\xi}) \hat{\boldsymbol{\phi}}_\epsilon(\boldsymbol{\xi}) = \lim_{\epsilon \to 0} \mathbf{\Gamma}\left(\frac{\boldsymbol{\xi}}{|\boldsymbol{\xi}|} \right) \left(\chi_{B_\epsilon^c}(\boldsymbol{\xi}) \hat{\boldsymbol{E}}(\boldsymbol{\xi}) \right) = \mathbf{\Gamma}\left(\frac{\boldsymbol{\xi}}{|\boldsymbol{\xi}|} \right) \hat{\boldsymbol{E}}(\boldsymbol{\xi}) = \hat{\boldsymbol{E}}(\boldsymbol{\xi}),$$

where the limit as $\epsilon \to 0$ is understood in the sense of $L^2(\mathbb{R}^d)$. This shows that the space

$$\mathcal{C}'(\mathcal{E}) = \left\{ P(\mathrm{i}\nabla)\boldsymbol{\phi} : \boldsymbol{\phi} \in H^{2m}\left(\mathbb{R}^d; \mathcal{T} \right) \right\}$$

is dense in $\mathcal{E}$. It is well known that $C_0^\infty(\mathbb{R}^d)$ is dense in $H^{2m}(\mathbb{R}^d)$. Furthermore, if $C_0^\infty(\mathbb{R}^d; \mathcal{T}) \ni \boldsymbol{\psi}_n \to \boldsymbol{\phi} \in H^{2m}(\mathbb{R}^d; \mathcal{T})$, as $n \to \infty$ in $H^{2m}(\mathbb{R}^d; \mathcal{T})$, then $P(\mathrm{i}\nabla)\boldsymbol{\psi}_n \to P(\mathrm{i}\nabla)\boldsymbol{\phi}$, as $n \to \infty$ in $L^2(\mathbb{R}^d; \mathcal{T})$. It remains to observe that $\boldsymbol{E}_n = P(\mathrm{i}\nabla)\boldsymbol{\psi}_n \in C_0^\infty(\mathbb{R}^d; \mathcal{T})$, since $P(\mathrm{i}\nabla)$ is a differential operator. Also, $\boldsymbol{E}_n \in \mathcal{E}$, since

$$\hat{\boldsymbol{E}}_n(\boldsymbol{\xi}) = (2\pi)^{2m} P(\boldsymbol{\xi}) \hat{\boldsymbol{\psi}}_n(\boldsymbol{\xi}) = \mathbf{\Gamma}\left(\frac{\boldsymbol{\xi}}{|\boldsymbol{\xi}|} \right) \left((2\pi)^{2m} |\boldsymbol{\xi}|^{2m} \hat{\boldsymbol{\psi}}_n(\boldsymbol{\xi}) \right),$$

and $|\boldsymbol{\xi}|^{2m} \hat{\boldsymbol{\psi}}_n(\boldsymbol{\xi}) \in L^2(\mathbb{R}^d; \mathcal{T})$. Thus, $\mathcal{E}_c$ is dense in $\mathcal{C}'(\mathcal{E})$. The L^2 topology comes from a norm. Therefore, $\mathcal{E}_c$ is dense in $\mathcal{E}$. The statement about $\mathcal{J}$ is proved by replacing $\mathbf{\Gamma}(\boldsymbol{\eta})$ with $\boldsymbol{I} - \mathbf{\Gamma}(\boldsymbol{\eta})$ and $P(\boldsymbol{\xi})$ with $|\boldsymbol{\xi}|^{2m} - P(\boldsymbol{\xi})$. □

Corollary 2.13. *Under the assumptions of theorem 2.12, let* Ω *and* Ω' *be open subsets of* $\mathbb{R}^d$, *such that* $\overline{\Omega} \subset \Omega'$. *Then the sets of functions*

$$\mathcal{E}_0^\infty\left(\Omega, \Omega' \right) = \left\{ P(\mathrm{i}\nabla)\mathbf{\Phi}|\Omega : \mathbf{\Phi} \in C_0^\infty\left(\Omega'; \mathcal{T} \right) \right\},$$

$$\mathcal{J}_0^\infty\left(\Omega, \Omega'\right) = \left\{\left(\Delta^m - P(\mathrm{i}\nabla)\right)\boldsymbol{\Phi}\middle|\Omega : \boldsymbol{\Phi} \in C_0^\infty\left(\Omega'; \mathcal{T}\right)\right\}$$

are dense in $\mathcal{E}(\Omega)$ *and* $\mathcal{J}(\Omega)$, *respectively.*

Proof. For any $\boldsymbol{E} \in \mathcal{E}(\Omega)$ there exists $\tilde{\boldsymbol{E}} \in \mathcal{E}$, such that $\tilde{\boldsymbol{E}}|\Omega = \boldsymbol{E}$. By theorem 2.12 there exists a sequence $\boldsymbol{\Phi}_n \in C_0^\infty(\mathbb{R}^d; \mathcal{T})$, such that $P(\mathrm{i}\nabla)\boldsymbol{\Phi}_n \to \tilde{\boldsymbol{E}}$ in $L^2(\mathbb{R}^d; \mathcal{T})$. Let $\eta \in C_0^\infty(\Omega')$ be such that $\eta(\boldsymbol{x}) = 1$ for all $\boldsymbol{x} \in \Omega$. Then

$$\boldsymbol{E}_n = P(\mathrm{i}\nabla)\boldsymbol{\Phi}_n|\Omega = P(\mathrm{i}\nabla)(\eta\boldsymbol{\Phi}_n)|\Omega \in \mathcal{E}_0^\infty(\Omega, \Omega'),$$

and $\boldsymbol{E}_n \to \boldsymbol{E}$ in $L^2(\Omega; \mathcal{T})$. The statement for $\mathcal{J}(\Omega)$ is proved in the same way. □

Corollary 2.14. *Under the assumptions of corollary 2.13 the sets of functions*

$$\{\boldsymbol{E}|\Omega : \boldsymbol{E} \in \mathcal{E}_0(\Omega')\}, \qquad \{\boldsymbol{J}|\Omega : \boldsymbol{E} \in \mathcal{J}_0(\Omega')\}$$

are dense in $\mathcal{E}(\Omega)$ *and* $\mathcal{J}(\Omega)$, *respectively.*

Theorem 2.15. *Suppose that the symbol* $\boldsymbol{\Gamma}$ *is* $\mathcal{J}$*-regular. Then the set of compactly supported functions*

$$\mathcal{C}(\mathcal{E}) = \bigcup_{n=1}^{\infty} \mathcal{E}_0^0(B_n) \tag{2.29}$$

is dense in $\mathcal{E}$, *where the space* $\mathcal{E}_0^0(B_n)$ *is defined in* (2.19), *and* B_r *is a ball of radius r, centered at* $\mathbf{0} \in \mathbb{R}^n$. *The statement with* $\mathcal{E}$ *and* $\mathcal{J}$ *interchanged is also true.*

Proof. The proof of the theorem is based on lemma 2.6. To prove density we consider a field $\boldsymbol{E}_0 \in \mathcal{E}$ that is orthogonal to all functions in $\mathcal{C}(\mathcal{E})$. By lemma 2.3, $\boldsymbol{E}_0|B_n \in \mathcal{J}(B_n)$ for any $n \geqslant 1$. Let us construct extensions $\boldsymbol{J}_n \in \mathcal{J}$ of $\boldsymbol{E}_0|B_n \in \mathcal{J}(B_n)$ using the scale-invariance of the spaces $\mathcal{E}$ and $\mathcal{J}$. The functions $B_1 \ni z \mapsto \boldsymbol{E}_n(z) = \boldsymbol{E}_0(nz)$ are in $\mathcal{J}(B_1)$. By lemma 2.6 there exists a bounded extension operator $X_{\mathcal{J}}(B_1) : \mathcal{J}(B_1) \to \mathcal{J}$. We define $\boldsymbol{J}_n(\boldsymbol{x}) = (X_{\mathcal{J}}(B_1)\boldsymbol{E}_n)(\boldsymbol{x}/n) \in \mathcal{J}$. Then, for any $\boldsymbol{x} \in B_n$

$$\boldsymbol{J}_n(\boldsymbol{x}) = \left(X_{\mathcal{J}}(B_1)\boldsymbol{E}_n\right)\left(\frac{\boldsymbol{x}}{n}\right) = \boldsymbol{E}_n\left(\frac{\boldsymbol{x}}{n}\right) = \boldsymbol{E}_0(\boldsymbol{x}).$$

Hence, $\boldsymbol{J}_n$ is an extension of $\boldsymbol{E}_0|B_n$ to $\mathcal{J}$. Therefore, for any $\boldsymbol{\phi} \in C_0^\infty(\mathbb{R}^d, \mathcal{T})$

$$\lim_{n\to\infty} \left(\boldsymbol{J}_n, \boldsymbol{\phi}\right)_{L^2(\mathbb{R}^d)} = \int_{\mathbb{R}^d} \left(\boldsymbol{E}_0, \boldsymbol{\phi}\right) \mathrm{d}\boldsymbol{x},$$

since B_n contains support of $\boldsymbol{\phi}$ for sufficiently large n. We will be able to conclude that $\boldsymbol{J}_n \rightharpoonup \boldsymbol{E}_0$ in $L^2(\mathbb{R}^d; \mathcal{T})$ if we prove that the sequence $\{\boldsymbol{J}_n : n \geqslant 1\}$ is bounded in $L^2(\mathbb{R}^d; \mathcal{T})$, since $C_0^\infty(\mathbb{R}^d, \mathcal{T})$ is dense in $L^2(\mathbb{R}^d; \mathcal{T})$. We compute

$$\|J_n\|_{L^2(\mathbb{R}^d)} = n^{d/2}\left\|\left(X_{\mathcal{J}}(B_1)E_n\right)\right\|_{L^2(\mathbb{R}^d)}$$
$$\leqslant n^{d/2}C(B_1)\|E_n\|_{L^2(B_1)} = C(B_1)\|E_0\|_{L^2(B_n)}$$
$$\leqslant C(B_1)\|E_0\|_{L^2(\mathbb{R}^d)}.$$

Thus, $J_n \rightharpoonup E_0$ in $L^2(\mathbb{R}^d; \mathcal{T})$. But $J_n \in \mathcal{J}$ for all $n \geqslant 1$ and hence its weak limit must be in $\mathcal{J}$. It follows that $E_0 = \mathbf{0}$, since $E_0 \in \mathcal{E}$. The density of $\mathcal{C}(\mathcal{E})$ is proved. The result for $\mathcal{C}(\mathcal{J})$ is proved in the same way. □

2.5.3 Pointwise behavior of the fields

We now proceed to establish pointwise properties of the space $\mathcal{E}$.

Theorem 2.16. *Assuming postulate 2.1, the following statements are true.*

(a) *If $E \in \mathcal{E}$ then $E(x) \in \mathcal{T}_{\mathcal{E}}$ for a.e. $x \in \mathbb{R}^d$, where $\mathcal{T}_{\mathcal{E}}$ is defined in (2.27).*

(b) *Assume that Ω is bounded. Then for any $e_0 \in \mathcal{T}_{\mathcal{E}}$ the function $E_0(x) = e_0$ for all $x \in \Omega$ is in $\overline{\mathcal{E}(\Omega)}$.*

(c) *Assume that Ω is an open and bounded subset of $\mathbb{R}^d$ and $x_0 \in \Omega$ is fixed. Then*

$$\operatorname{Span}\left\{E(x_0) : E \in \mathcal{E}_0(\Omega),\ \text{such that } x_0 \text{ is a Lebesgue point of } E\right\}$$

is the same subspace $V_{\mathcal{E}} \subset \mathcal{T}_{\mathcal{E}}$ for all open and bounded subsets $\Omega \subset \mathbb{R}^d$ and all $x_0 \in \Omega$.

(d) *If $C_0^\infty(\mathbb{R}^d; \mathcal{T}) \cap \mathcal{E}$ is dense in $\mathcal{E}$ then $V_{\mathcal{E}} = \mathcal{T}_{\mathcal{E}}$.*

Proof. Part (a).

$$E(x) = \int_{\mathbb{R}^d} \hat{E}(\xi)e^{2\pi i x\cdot\xi}\,d\xi \in \left(\mathcal{T}_{\mathcal{E}} \otimes \mathbb{C}\right) \cap \mathcal{T} = \mathcal{T}_{\mathcal{E}}.$$

Part (b). For any $J \in \mathcal{J}_0(\Omega)$ we have $J_0 = X_0(\Omega)J \in L^1(\mathbb{R}^d; \mathcal{T})$, since Ω is a bounded set. Then $\hat{J}_0(\xi)$ is continuous on $\mathbb{R}^d$ and for any $\epsilon > 0$ and $\xi \neq \mathbf{0}$ we have

$$\mathbf{0} = \lim_{\epsilon\to 0}\Gamma\left(\frac{\epsilon\xi}{|\epsilon\xi|}\right)\hat{J}_0(\epsilon\xi) = \Gamma\left(\frac{\xi}{|\xi|}\right)\hat{J}_0(\mathbf{0}),$$

since $J_0 \in \mathcal{J}$. Thus,

$$\hat{J}_0(\mathbf{0}) \in \bigcap_{|\eta|=1} \mathcal{E}_\eta^\perp = \mathcal{T}_{\mathcal{E}}^\perp.$$

Thus, for all $e_0 \in \mathcal{T}_{\mathcal{E}}$ and any $J \in \mathcal{J}_0(\Omega)$ we have

$$(J, E_0)_{L^2(\Omega)} = \left(e_0, \int_{\mathbb{R}^d} J_0(x)\,dx\right) = \left(e_0, \hat{J}_0(\mathbf{0})\right) = 0,$$

where $E_0(x) = e_0$ for all $x \in \Omega$. Therefore, $E_0 \in (\mathcal{J}_0(\Omega))^\perp = \overline{\mathcal{E}(\Omega)}$.

Part (c). Before the statement of part (c) is proved let us temporarily denote[5]

$$V_x(\Omega) = \mathrm{Span}\{\boldsymbol{E}(\boldsymbol{x}) : \boldsymbol{E} \in \mathcal{E}_0(\Omega),\ \boldsymbol{x} \text{ is a Lebesgue point of } \boldsymbol{E}\}.$$

Let

$$\mathcal{N}(\Omega) = \{\boldsymbol{f} \in L^2(\Omega; \mathcal{T}) : (\boldsymbol{f}(\boldsymbol{x}), \boldsymbol{E}(\boldsymbol{x})) = 0 \text{ for a.e. } \boldsymbol{x} \in \Omega \text{ and all } \boldsymbol{E} \in \mathcal{E}_0(\Omega)\}. \tag{2.30}$$

It is easy to see that $\mathcal{N}(\Omega)$ is a closed subspace of $L^2(\Omega; \mathcal{T})$. The subspace $\mathcal{N}(\Omega)$ has the following properties: $\chi_D(\boldsymbol{x})\boldsymbol{f} \in \mathcal{N}(\Omega)$, for any $D \subset \Omega$, provided $\boldsymbol{f} \in \mathcal{N}(\Omega)$. Also, $\boldsymbol{f}(\alpha\boldsymbol{x} + \beta) \in \mathcal{N}(\Omega)$, when $\boldsymbol{f} \in \mathcal{N}(\Omega)$, provided $\alpha\Omega + \beta \subset \Omega$. Indeed, for any $\boldsymbol{E} \in \mathcal{E}_0^0(\Omega)$ the function $\tilde{\boldsymbol{E}}(\boldsymbol{x}) = \boldsymbol{E}((\boldsymbol{x} - \beta)/\alpha) \in \mathcal{E}_0^0(\Omega)$. Therefore, $(\boldsymbol{f}(\boldsymbol{x}), \tilde{\boldsymbol{E}}(\boldsymbol{x})) = 0$ for a.e. $\boldsymbol{x} \in \Omega$. But then for a.e. $\boldsymbol{x} \in \Omega$ we also have

$$0 = \big(\boldsymbol{f}(\alpha\boldsymbol{x} + \beta), \tilde{\boldsymbol{E}}(\alpha\boldsymbol{x} + \beta)\big) = \big(\boldsymbol{f}(\alpha\boldsymbol{x} + \beta), \boldsymbol{E}(\boldsymbol{x})\big).$$

We conclude that $\boldsymbol{f}(\alpha\boldsymbol{x} + \beta) \in \mathcal{N}(\Omega)$.

Let $\{\boldsymbol{E}_1, \boldsymbol{E}_2, \ldots\}$ be a countable dense subset of $\mathcal{E}_0(\Omega)$. Let $\Omega_0 \subset \Omega$ be the set of all common Lebesgue points of $\{\boldsymbol{E}_1, \boldsymbol{E}_2, \ldots\}$. Let

$$\tilde{V}_x = \mathrm{Span}\{\boldsymbol{E}_n(\boldsymbol{x}) : n \geqslant 1\}, \qquad \boldsymbol{x} \in \Omega_0.$$

We now define the map $\mathcal{P} : \Omega_0 \to \mathrm{Sym}(\mathcal{T})$ such that $\mathcal{P}(\boldsymbol{x})$ is the orthogonal projection onto $\tilde{V}_x^{\perp}$. Let $\Omega_0^* \subset \Omega_0$ be the set of all Lebesgue points of $\mathcal{P}(\boldsymbol{x})$.

Let $\{\boldsymbol{x}_0, \boldsymbol{x}_1\} \subset \Omega_0^*$. Let $\boldsymbol{v} \in \tilde{V}_{x_0}^{\perp}$. Let $\boldsymbol{p}_v(\boldsymbol{x}) = \mathcal{P}(\boldsymbol{x})\boldsymbol{v}$. Then $\boldsymbol{p}_v(\boldsymbol{x})$ is measurable and bounded, $\boldsymbol{p}_v(\boldsymbol{x}_0) = \boldsymbol{v}$, and $\boldsymbol{x}_0$ is a Lebesgue point of $\boldsymbol{p}_v(\boldsymbol{x})$, since $\boldsymbol{x}_0$ is a Lebesgue point of $\mathcal{P}(\boldsymbol{x})$. We claim that $\boldsymbol{p}_v \in \mathcal{N}(\Omega)$. Indeed, by construction $\boldsymbol{p}_v(\boldsymbol{x}) \in \tilde{V}_x^{\perp}$ for all $\boldsymbol{x} \in \Omega_0^*$. Then $(\boldsymbol{p}_v(\boldsymbol{x}), \boldsymbol{E}_n(\boldsymbol{x})) = 0$ for all $n \geqslant 1$ and $\boldsymbol{x} \in \Omega_0^*$. Let $\boldsymbol{E} \in \mathcal{E}_0(\Omega)$ and $\boldsymbol{E}_{n_k} \to \boldsymbol{E}$ in $L^2(\Omega; \mathcal{T})$. But then in $L^1(\Omega)$ we have

$$0 = \big(\boldsymbol{p}_v(\boldsymbol{x}), \boldsymbol{E}_{n_k}(\boldsymbol{x})\big) \to \big(\boldsymbol{p}_v(\boldsymbol{x}), \boldsymbol{E}(\boldsymbol{x})\big) \text{ in } L^1(\Omega; \mathcal{T}).$$

Thus $\boldsymbol{p}_v \in \mathcal{N}(\Omega)$. Now for sufficiently small α we have $\alpha(\Omega - \boldsymbol{x}_1) + \boldsymbol{x}_0 \subset \Omega$. Therefore,

$$\tilde{\boldsymbol{p}}(\boldsymbol{x}) = \boldsymbol{p}_v\big(\alpha(\boldsymbol{x} - \boldsymbol{x}_1) + \boldsymbol{x}_0\big) \in \mathcal{N}(\Omega).$$

Also, since the affine function $\boldsymbol{x} \mapsto \alpha(\boldsymbol{x} - \boldsymbol{x}_1) + \boldsymbol{x}_0$ maps $\boldsymbol{x}_1$ into $\boldsymbol{x}_0$, we conclude that $\boldsymbol{x}_1$ is a Lebesgue point of $\tilde{\boldsymbol{p}}$. We also have $\tilde{\boldsymbol{p}}(\boldsymbol{x}_1) = \boldsymbol{p}_v(\boldsymbol{x}_0) = \boldsymbol{v}$. Recalling that $\boldsymbol{x}_1$ is a Lebesgue point of both $\boldsymbol{E}_n(\boldsymbol{x})$ and $\tilde{\boldsymbol{p}}(\boldsymbol{x})$ and that $\tilde{\boldsymbol{p}}(\boldsymbol{x})$ is bounded, we conclude that $\boldsymbol{x}_1$ is also a Lebesgue point of $(\tilde{\boldsymbol{p}}(\boldsymbol{x}), \boldsymbol{E}_n(\boldsymbol{x}))$. However, $(\tilde{\boldsymbol{p}}(\boldsymbol{x}), \boldsymbol{E}_n(\boldsymbol{x})) = 0$ for a.e. $\boldsymbol{x} \in \Omega$, since $\tilde{\boldsymbol{p}}(\boldsymbol{x}) \in \mathcal{N}(\Omega)$. Thus,

---

[5] Our goal is to prove that $V_x(\Omega)$ depends neither on $\boldsymbol{x}$, nor on Ω.

$$0 = \lim_{r\to 0} \fint_{B_r(x_0)} \Big(\tilde{p}(x),\, E_n(x)\Big)\,\mathrm{d}x = \Big(\tilde{p}(x_1),\, E_n(x_1)\Big) = \Big(v,\, E_n(x_1)\Big).$$

Hence, $v \in \tilde{V}_{x_1}^{\perp}$. We have proved that $\tilde{V}_{x_0}^{\perp} \subset \tilde{V}_{x_1}^{\perp}$ for any $\{x_0, x_1\} \subset \Omega_0^*$. Switching the order of x_0 and x_1 we obtain the reverse inclusion. Thus, the subspaces $\tilde{V}_x$ are independent of x. Let us denote this common subspace by $\tilde{V}_\Omega$.

We have just proved that $(v, E_n(x)) = 0$ for all $v \in \tilde{V}_\Omega^{\perp}$ and all $x \in \Omega_0^*$. Therefore, any function $f \in L^2(\Omega) \otimes \tilde{V}_\Omega^{\perp}$ is in $\mathcal{N}(\Omega)$, since $(f(x), E_n(x)) = 0$ for a.e. $x \in \Omega$ and all $n \geqslant 1$. Conversely, if $f \in \mathcal{N}(\Omega)$, but $f \notin L^2(\Omega) \otimes \tilde{V}_\Omega^{\perp}$, then there exists a vector $e \in \tilde{V}_\Omega$ such that $(f(x), e) \neq 0$ in $L^2(\Omega)$. Then there exists $x_0 \in \Omega_0^*$, such that x_0 is a Lebesgue point of $f(x)$ and $(f(x), E_n(x))$, for all $n \geqslant 1$, and such that $(f(x_0), e) \neq 0$. Recalling that $f \in \mathcal{N}(\Omega)$, we conclude that

$$0 = \lim_{r\to 0} \fint_{B_r(x_0)} \Big(f(x),\, E_n(x)\Big)\,\mathrm{d}x = \Big(f(x_0),\, E_n(x_0)\Big).$$

But $e \in \tilde{V}_\Omega$ is a finite linear combination of vectors $\{E_1(x_0), E_2(x_0), \dots\}$. It follows that $(f(x_0), e) = 0$. This is a contradiction. We have thus proved that

$$\mathcal{N}(\Omega) = L^2(\Omega) \otimes \tilde{V}_\Omega^{\perp}.$$

Observe that for any $E \in \mathcal{E}_0(\Omega)$ and any of its Lebesgue points $x \in \Omega$ we have $(v, E(x)) = 0$ for any $v \in \tilde{V}_\Omega^{\perp}$, since the constant function $f(x) = v$ is in $L^2(\Omega) \otimes \tilde{V}_\Omega^{\perp} = \mathcal{N}(\Omega)$. Therefore, $E(x) \in \tilde{V}_\Omega$. In other words, $V_x(\Omega) \subset \tilde{V}_\Omega$. But also $\tilde{V}_\Omega \subset V_x(\Omega)$ for any $x \in \Omega$. Indeed, let $x_0 \in \Omega$ and $x_1 \in \Omega_0^*$ be arbitrary, and let α be so small that $\alpha(\Omega - x_1) + x_0 \subset \Omega$. Then the functions

$$\tilde{E}_j(x) = X_0(\Omega) E_j\!\left(x_1 + \frac{x - x_0}{\alpha}\right), \qquad j \geqslant 1$$

are in $\mathcal{E}_0^0(\Omega)$ and x_0 is a common Lebesgue point of all of them. Thus

$$\tilde{V}_\Omega = \tilde{V}_{x_1} = \mathrm{Span}\Big\{E_1(x_1),\, E_2(x_1),\, \dots\Big\} = \mathrm{Span}\Big\{\tilde{E}_1(x_0),\, \tilde{E}_2(x_0),\, \dots\Big\} \subset V_{x_0}(\Omega).$$

This proves that the spaces $V_x(\Omega)$ are independent of $x \in \Omega$.

Let us prove now that the spaces $V_x(\Omega) = \tilde{V}_\Omega$ are also independent of Ω. Let Ω' be another open and bounded subset of $\mathbb{R}^d$. Let us fix $x_0' \in \Omega'$ and $x_0 \in \Omega$ arbitrarily. For sufficiently small α we have $\alpha(\Omega - x_0) + x_0' \subset \Omega'$. Let $\mathcal{C}(x_0)$ be the set of all $E \in \mathcal{E}_0^0(\Omega)$ for which x_0 is a Lebesgue point. Then for any $E \in \mathcal{C}(x_0)$ we define

$$\tilde{E}(x') = E\!\left(x_0 + \frac{x' - x_0'}{\alpha}\right).$$

Then $\tilde{E} \in \mathcal{E}_0^0(\Omega')$ and x_0' is its Lebesgue point. Therefore,

$$\tilde{V}_\Omega = \mathrm{Span}\left\{ \boldsymbol{E}(\boldsymbol{x}_0) : \boldsymbol{E} \in \mathcal{C}(\boldsymbol{x}_0) \right\} = \mathrm{Span}\left\{ \tilde{\boldsymbol{E}}\left(\boldsymbol{x}_0'\right) : \boldsymbol{E} \in \mathcal{C}(\boldsymbol{x}_0) \right\} \subset \tilde{V}_{\Omega'}.$$

Switching the order of Ω and Ω' we obtain the reverse inclusion. Hence, we have proved that $V_x(\Omega) = \tilde{V}_\Omega = V_{\mathcal{E}}$.

Part (d). Suppose $\boldsymbol{e}_0 \in \mathcal{T}_{\mathcal{E}}$. By part (b) of this theorem there exists $\boldsymbol{E}_n \in \mathcal{E}$ such that $\boldsymbol{E}_n|B_1 \to \boldsymbol{e}_0$, as $n \to \infty$ in $L^2(B_1; \mathcal{T})$. The topology of $L^2(B_1; \mathcal{T})$ comes from a norm. Hence, by the density assumption there exist $\boldsymbol{E}_n' \in \mathcal{E}_0(B_{R_n})$, such that $\boldsymbol{E}_n'|B_1 \to \boldsymbol{e}_0$ in $L^2(B_1, \mathcal{T})$. But $\boldsymbol{E}_n'|B_1 \in L^2(B_1, V_{\mathcal{E}})$, which is a closed subspace of $L^2(B_1, \mathcal{T})$. Therefore, $\boldsymbol{e}_0 \in L^2(B_1, V_{\mathcal{E}})$. In particular, $\boldsymbol{e}_0 \in V_{\mathcal{E}}$. Hence, $\mathcal{T}_{\mathcal{E}} \subset V_{\mathcal{E}}$. The reverse inclusion is a direct corollary of part (a). The theorem is proved. □

Corollary 2.17. *Let* $\mathcal{N}(\Omega)$ *be given by* (2.30). *We observe that the proof of theorem* 2.16 *establishes the formula for* $\mathcal{N}(\Omega)$:

$$\mathcal{N}(\Omega) = L^2\left(\Omega;\, V_{\mathcal{E}}^{\perp}\right). \tag{2.31}$$

Bibliography

[1] Calderón A 1961 Lebesgue spaces of differentiable functions *Proc. Symp. Pure Math.* **4** 33–49

[2] Dell'Antonio G F, Figari R and Orlandi E 1986 An approach through orthogonal projections to the study of inhomogeneous random media with linear response *Ann. Inst. Henri Poincaré* **44** 1–28

[3] Fonseca I and Müller S 1999 A-quasiconvexity, lower semicontinuity and young measures *Siam J. Math. Anal.* **30** 1355–90

[4] Kohler W and Papanicolaou G C 1982 Bounds for effective conductivity of random media *Macroscopic Properties of Disordered Media* (Berlin: Springer) pp 111–30

[5] Milton G W 1987 Multicomponent composites, electrical networks and new types of continued fraction. I. *Commun. Math. Phys.* **111** 281–327

[6] Milton G W 1987 Multicomponent composites, electrical networks and new types of continued fraction. II. *Commun. Math. Phys.* **111** 329–72

[7] Milton G W 1990 On characterizing the set of possible effective tensors of composites: the variational method and the translation method *Commun. Pure Appl. Math.* **43** 63–125

[8] Milton G W 2002 *The Theory of Composites* (*Cambridge Monographs on Applied and Computational Mathematics* vol. 6) (Cambridge: Cambridge University Press)

[9] Milton G W and Kohn R V 1988 Variational bounds on the effective moduli of anisotropic composites *J. Mech. Phys. Solids* **36** 597–629

[10] Stein E M 1970 *Singular Integrals and Differentiability Properties of Functions* (*Princeton Mathematical Series* vol. 30) (Princeton, NJ: Princeton University Press)

IOP Publishing

Composite Materials
Mathematical theory and exact relations
Yury Grabovsky

Chapter 3

Composite materials

3.1 Mathematical definition of a composite

The intuitive notion of a composite material and its effective properties is clear to most people. However, a rigorous mathematical analysis requires a definition. Let us consider an example of a composite domain $\Omega \subset \mathbb{R}^3$ occupied by two materials A and B. To describe such a medium one needs to specify a subset of Ω occupied by the material A. (The complement will be occupied by the material B.) However, such a description does not draw a clear line between a true composite and a compound object, such as a wooden door with metal door knob. This is the reason why in the mathematical theory of composite materials we define a composite as a *limit of a sequence of structures* [5, 17, 19, 20]. With such a definition we imagine that a specific composite material is part of a family of heterogeneous media labeled by a small parameter ϵ that may (but does not have to) represent a typical length scale. The effective properties of the 'composite' will then be defined to be the properties of the 'limit material', as $\epsilon \to 0$. In this book we discuss questions of existence and properties of such limit materials with full mathematical rigor. In practice, a composite material is identified with a finite (but 'small') value of ϵ; see, e.g. [10] for a numerical comparison between the finite ϵ and limit materials for periodic composites, where ϵ is the ratio of the size of the period cell to the size of the sample.

In this book by a 'composite material' we mean a family of heterogeneous media, whose tensors of local material properties will be denoted by $\mathsf{L}_\epsilon(\boldsymbol{x})$, $\epsilon > 0$, and that converge to a limit material in the sense of definition 3.1. below. For a subset $U \subset \mathrm{End}^+(\mathcal{T})$ of material properties we define

$$U^\Omega = \left\{\mathsf{L} \in L^\infty(\Omega; \mathrm{End}(\mathcal{T})) : \mathsf{L}(\boldsymbol{x}) \in U \text{ for a.e. } \boldsymbol{x} \in \Omega\right\}. \tag{3.1}$$

In other words, U^Ω denotes the set of all possible heterogeneous media made with materials from a given set U, and occupying the domain Ω. In this book we assume

doi:10.1088/978-0-7503-1048-2ch3

that $\mathsf{L}_\epsilon \in \mathcal{M}(\alpha, \beta)^\Omega$, for some $\beta > \alpha > 0$, all $\epsilon > 0$, where the sets $\mathcal{M}(\alpha, \beta)$ are defined in (2.15). We can now make rigorous the intuitive picture of a composite. On the microscale a composite is described by the family $\mathsf{L}_\epsilon(\boldsymbol{x})$ that oscillates on smaller and smaller length scales as $\epsilon \to 0$. Yet, the oscillations of $\mathsf{L}_\epsilon(\boldsymbol{x})$ describing a specific composite should not be completely random, in the sense that for small ϵ the *response* of the composite should be indistinguishable from the response of a material, which is locally homogeneous, i.e. its pointwise properties may vary only on the macroscale. The idea is then to look at the limiting behavior of the solutions $\boldsymbol{E}_\epsilon$ of (2.16) for each ϵ, as $\epsilon \to 0$.

Definition 3.1. *We say that the family* $\{\mathsf{L}_\epsilon : \epsilon > 0\}$ H-converges to $\mathsf{L}_*(\boldsymbol{x})$ *(denoted by* $\mathsf{L}_\epsilon \overset{H}{\rightharpoonup} \mathsf{L}_*$*) if the family of solutions* $\boldsymbol{E}_\epsilon$ *of*

$$\boldsymbol{E}_\epsilon \in \mathcal{E}_0(\Omega), \qquad \boldsymbol{J}_\epsilon = \mathsf{L}_\epsilon(\boldsymbol{x})\boldsymbol{E}_\epsilon, \qquad \Gamma_\Omega \boldsymbol{J}_\epsilon = \boldsymbol{f} \tag{3.2}$$

converges weakly in $L^2(\Omega; \mathcal{T})$ *to the solution* $\boldsymbol{E}_*$ *of*

$$\boldsymbol{E}_* \in \mathcal{E}_0(\Omega), \qquad \boldsymbol{J}_* = \mathsf{L}_*(\boldsymbol{x})\boldsymbol{E}_*, \qquad \Gamma_\Omega \boldsymbol{J}_* = \boldsymbol{f}, \tag{3.3}$$

for every $\boldsymbol{f} \in \mathcal{E}_0(\Omega)$, *and additionally* $\boldsymbol{J}_\epsilon = \mathsf{L}_\epsilon \boldsymbol{E}_\epsilon$ *converges weakly in* $L^2(\Omega; \mathcal{T})$ *to* $\boldsymbol{J}_* = \mathsf{L}_*(\boldsymbol{x})\boldsymbol{E}_*$. *The H-limit* $\mathsf{L}_*(\boldsymbol{x})$ *is called the effective tensor of the composite* $\mathsf{L}_\epsilon(\boldsymbol{x})$.

The notion of H-convergence was first studied by Spagnolo [20] for the conductivity problem. In that specific context it was shown that what we defined as H-convergence can be described in terms of the convergence of Green's functions for the operators $\phi \mapsto \nabla \cdot (\boldsymbol{\sigma}_\epsilon(\boldsymbol{x})\nabla\phi)$. For that reason the name 'G-convergence' was given to this kind of limit. Later, Murat and Tartar [17] generalized this notion to the case of non-symmetric conductivity tensors $\boldsymbol{\sigma}_\epsilon$ by adding the requirement of convergence of fluxes. However, as our approach reveals, the need for convergence of fluxes is related not to the lack of symmetry of material tensors $\mathsf{L}_\epsilon(\boldsymbol{x})$ per se, but to deeper geometric properties of the spaces $\mathcal{E}$ and $\mathcal{J}$. Since our approach to homogenization in this book does not deal with specific partial differential equations and their Green's functions explicitly, and is based on Murat and Tartar's compensated compactness approach [16, 21], we adopt the H-convergence terminology, where 'H' stands for homogenization. It is important to remark that for a physical composite material there is no mathematically sharp boundary between the microstructure and the macroscale inhomogeneity. In modeling a specific composite material it is a judgment call to identify the oscillatory regions (microstructure) and non-oscillatory regions (macroscale inhomogeneity). The latter can be described mathematically as pointwise convergent family $\mathsf{L}_\epsilon(\boldsymbol{x})$, in which case the H-limit reduces to the pointwise limit.

Lemma 3.2. *Suppose* $\mathsf{L}_\epsilon(\boldsymbol{x}) \in \mathcal{M}(\alpha, \beta)$ *for a.e.* $\boldsymbol{x} \in \Omega$ *and* $\mathsf{L}_\epsilon(\boldsymbol{x}) \to \mathsf{L}_*(\boldsymbol{x})$ *in measure (strong convergence) then* $\mathsf{L}_\epsilon \overset{H}{\rightharpoonup} \mathsf{L}_*$ *in* Ω.

Proof. The sequence of solutions $\boldsymbol{E}_\epsilon$ of (2.16) is uniformly bounded in $L^2(\Omega; \mathcal{T})$. Therefore, there is a weakly convergent subsequence $\boldsymbol{E}_{\epsilon_k} \rightharpoonup \boldsymbol{E}_0$ in $L^2(\Omega; \mathcal{T})$. The uniform boundedness of L_ϵ and convergence in measure implies that $\mathsf{L}_{\epsilon_k}\boldsymbol{E}_{\epsilon_k} \rightharpoonup \mathsf{L}_*\boldsymbol{E}_0$ in $L^2(\Omega; \mathcal{T})$. Thus, the weak limit $\boldsymbol{E}_0 \in \mathcal{E}_0(\Omega)$ of $\boldsymbol{E}_{\epsilon_k}$ is uniquely determined by the equation $\Gamma_\Omega(\mathsf{L}_*\boldsymbol{E}_0) = \boldsymbol{f}$, and is independent of the chosen weakly convergent subsequence. It follows that the entire family $\boldsymbol{E}_\epsilon$ weakly converges to $\boldsymbol{E}_0$ in $L^2(\Omega; \mathcal{T})$. Hence, $\mathsf{L}_\epsilon \overset{H}{\rightharpoonup} \mathsf{L}_*$. □

H-convergence has a remarkable compactness property [17].

Theorem 3.3. *Assume that the Fourier multiplier symbol* $\boldsymbol{\Gamma}(\boldsymbol{\eta})$ *is* $\mathcal{E}$ and $\mathcal{J}$*-regular in the sense of definition 2.5 and* $\mathcal{E}$*-nondegenerate in the sense of definition 2.10. Suppose* $\mathsf{L}_\epsilon \in \mathcal{M}(\alpha, \beta)^\Omega$ *is an arbitrary family of microstructures. Then there exists a subsequence* $\epsilon_k \to 0$ *and* $\mathsf{L}_* \in \mathcal{M}(\alpha, \beta)^\Omega$, *such that* $\mathsf{L}_{\epsilon_k} \overset{H}{\rightharpoonup} \mathsf{L}_*$.

The proof is technical and can be found in section 3.3.4. This theorem tells us that up to a subsequence *any* infinite family of microstructures can be homogenized.

3.2 Periodic composites

Periodic composites are the most important (and, as we will see, quite general) examples of H-convergent families of heterogeneous media. Let $\{\boldsymbol{p}_1, \ldots, \boldsymbol{p}_d\}$ be a basis of $\mathbb{R}^d$ and let

$$Q = \left\{ \sum_{i=1}^{d} \lambda_i \boldsymbol{p}_i : 0 < \lambda_i < 1,\ i = 1, .., d \right\}$$

be the parallelepiped of periods. For $\mathsf{L} \in \mathcal{M}(\alpha, \beta)^Q$ and $0 < \alpha < \beta$ we define

$$\mathsf{L}_\epsilon(\boldsymbol{x}) = \mathsf{L}_{\mathrm{per}}\left(\frac{\boldsymbol{x}}{\epsilon}\right), \quad \boldsymbol{x} \in \mathbb{R}^d, \tag{3.4}$$

where $\mathsf{L}_{\mathrm{per}}(\boldsymbol{z})$, $\boldsymbol{z} \in \mathbb{R}^d$ is the Q-periodic extension of $\mathsf{L}(\boldsymbol{z})$ from the period cell Q to all of $\mathbb{R}^d$. We observe that tensor fields $\mathsf{L}_\epsilon(\boldsymbol{x})$ are ϵQ-periodic. For example, if $A \cap B = \varnothing$ are complementary subsets of Q occupied by materials L_A, L_B then

$$\mathsf{L}(\boldsymbol{z}) = \chi_A(\boldsymbol{z})\mathsf{L}_A + \chi_B(\boldsymbol{z})\mathsf{L}_B, \quad \boldsymbol{z} \in Q$$

defines a tensor field of local material properties of a two-phase composite material. The family $\{\mathsf{L}_\epsilon(\boldsymbol{x}) : \boldsymbol{x} \in \Omega\}$, where $\mathsf{L}_\epsilon(\boldsymbol{x})$ is given by (3.4), is then interpreted as a periodic two-phase composite occupying the domain $\Omega \subset \mathbb{R}^d$.

In general, we will call the function $\mathsf{L}(\boldsymbol{z})$ *the local tensor* of material properties. Here and throughout this entire section we assume that $\mathsf{L} \in \mathcal{M}(\alpha, \beta)^Q$. We will prove that L_ϵ H-converges, in the sense of definition 3.1, to a constant effective tensor $\mathsf{L}_* \in \mathrm{End}^+(\mathcal{T})$, which is intuitively clear, since the microstructure in the vicinity of any point of Ω is exactly the same periodic microstructure, as in the vicinity of any other point of Ω. In order to describe the H-limit L_* we first need to define periodic versions of subspaces $\mathcal{E}$ and $\mathcal{J}$ of $L^2_{\mathrm{per}}(Q; \mathcal{T})$—the set of Q-periodic locally L^2 vector fields.

Let $\boldsymbol{Q}$ be the invertible $d \times d$ matrix, such that $\boldsymbol{Q}Q = [0, 1]^d$. (The basis vectors $\{\boldsymbol{p}_1, \ldots, \boldsymbol{p}_d\}$ are the columns of $\boldsymbol{Q}^{-1}$.) We define

$$\mathcal{E}_{\mathrm{per}} = \left\{ \boldsymbol{E} \in L^2_{\mathrm{per}}(Q; \mathcal{T}) : \widehat{\boldsymbol{E}}(\boldsymbol{0}) = \boldsymbol{0},\ \widehat{\boldsymbol{E}}(\boldsymbol{k}) \in \mathcal{E}_{\frac{\boldsymbol{Q}^T\boldsymbol{k}}{|\boldsymbol{Q}^T\boldsymbol{k}|}} \otimes \mathbb{C},\ \boldsymbol{k} \neq \boldsymbol{0} \right\}, \tag{3.5}$$

$$\mathcal{J}_{\mathrm{per}} = \left\{ \boldsymbol{J} \in L^2_{\mathrm{per}}(Q; \mathcal{T}) : \widehat{\boldsymbol{J}}(\boldsymbol{0}) = \boldsymbol{0},\ \widehat{\boldsymbol{J}}(\boldsymbol{k}) \in \mathcal{J}_{\frac{\boldsymbol{Q}^T\boldsymbol{k}}{|\boldsymbol{Q}^T\boldsymbol{k}|}} \otimes \mathbb{C},\ \boldsymbol{k} \neq \boldsymbol{0} \right\}, \tag{3.6}$$

where $\hat{\boldsymbol{J}}(\boldsymbol{k})$ (and $\hat{\boldsymbol{E}}(\boldsymbol{k})$), $\boldsymbol{k} \in \mathbb{Z}^d$, are the Fourier coefficients in the expansion

$$\boldsymbol{J}(z) = \sum_{\boldsymbol{k} \in \mathbb{Z}^d} \widehat{\boldsymbol{J}}(\boldsymbol{k}) \mathrm{e}^{2\pi \mathrm{i} \boldsymbol{Q}z \cdot \boldsymbol{k}}, \tag{3.7}$$

and are given by

$$\widehat{\boldsymbol{J}}(\boldsymbol{k}) = \fint_Q \boldsymbol{J}(z) \mathrm{e}^{-2\pi \mathrm{i} \boldsymbol{Q}z \cdot \boldsymbol{k}}\, \mathrm{d}z, \quad \boldsymbol{k} \in \mathbb{Z}^d, \tag{3.8}$$

where

$$\fint_Q = \frac{1}{|Q|} \int_Q$$

denotes the average over a period cell.

Let $\mathcal{U}$ be the space of constant vector fields in $L^2_{\mathrm{per}}(Q; \mathcal{T})$. Then we have the orthogonal decomposition

$$L^2_{\mathrm{per}}(Q; \mathcal{T}) = \mathcal{E}_{\mathrm{per}} \oplus \mathcal{J}_{\mathrm{per}} \oplus \mathcal{U}.$$

In order to compute the effective tensor of the periodic composite with period cell Q and local tensor $\mathsf{L}(z)$ we need to solve the *periodic cell problem*

$$\boldsymbol{E} \in \mathcal{E}_{\mathrm{per}} \oplus \mathcal{U}, \quad \boldsymbol{J}(z) = \mathsf{L}(z)\boldsymbol{E}(z), \quad \boldsymbol{J} \in \mathcal{J}_{\mathrm{per}} \oplus \mathcal{U}, \quad \langle \boldsymbol{E} \rangle = \boldsymbol{e}_0, \tag{3.9}$$

where

$$\langle \boldsymbol{E} \rangle = \fint_Q \boldsymbol{E}(z)\, \mathrm{d}z$$

is a compact notation for the average over a period cell.

Theorem 3.4. *Assume that the symbol $\boldsymbol{\Gamma}(\boldsymbol{\eta})$ is $\mathcal{E}$- and $\mathcal{J}$-regular in the sense of definition 2.5. Then the effective tensor L_* of the periodic composite (3.4) is given by its action on the arbitrary field $\boldsymbol{e}_0 \in \mathcal{T}$:*

$$\mathsf{L}_* \boldsymbol{e}_0 = \fint_Q \mathsf{L}(z)\boldsymbol{E}(z)\, dz, \tag{3.10}$$

where $\boldsymbol{E}(z)$ solves the cell problem (3.9).

The proof of this theorem requires the development of a sophisticated machinery of homogenization and will be given in section 3.3.

3.2.1 G-closure

In this book it will generally be assumed that properties of the constituent materials and their volume fractions are known exactly. However, properties of composites can also be strongly dependent upon the microstructure, which is rarely known exactly. This gives rise to the so-called G-closure problem [11] of describing the set of all possible properties that a composite can have, provided that it is made of a given set of materials. In this book we deal exclusively with the situation where the microstructure is not constrained in any way at all, except possibly by the specified volume fractions of the phases. Our methods are also applicable to fiber-reinforced or fibrous composites, since from the mathematical point of view they can be regarded as two-dimensional composites, with unconstrained two-dimensional microstructure, modified governing equations and constitutive laws fitting perfectly into our unified theoretical framework.

Suppose that have a closed subset $U \subset \mathcal{M}(\alpha, \beta)$ of available materials and we are free to construct composites made with materials from U in any way we wish. The set U does not have to be finite. For example, if we are allowed to use an anisotropic material in arbitrary orientations, then U will be a union of $SO(d)$ orbits of a set of anisotropic tensors.

Definition 3.5. *The G-closure $G(U)$ of a subset $U \subset \mathcal{M}(\alpha, \beta)$ of tensors of material properties is the closure of the set of all effective tensors L_* of periodic composites $\mathsf{L}(\boldsymbol{x}/\epsilon)$, where $\mathsf{L}(z)$ can be any measurable function on Q with values in U.*

Remark 3.6. *Let $U \subset \mathcal{M}(\alpha, \beta)$ be a subset. Then the set $G(U)$ from definition 3.5 is independent of the period cell Q. Indeed, let $\boldsymbol{Q}$ map Q into $[0, 1]^d$ and $\mathsf{L}_0 : [0, 1]^d \to U$. Then $\mathsf{L}(z) = \mathsf{L}_0(\boldsymbol{Q}z)$ maps Q into U. Conversely, every measurable map $\mathsf{L} : Q \to U$ can be represented in this way, since $\mathsf{L}_0(z) = \mathsf{L}(\boldsymbol{Q}^{-1}z)$ maps $[0, 1]^d$ into U. Obviously, $\boldsymbol{E}_0(z)$ solves the $[0, 1]^d$-periodic cell problem* if and only if $\boldsymbol{E}(z) = \boldsymbol{E}_0(\boldsymbol{Q}z)$ *solves the Q-periodic cell problem. But then formula (3.10) shows that $\mathsf{L}_0^* = \mathsf{L}^*$.*

So far we have presented periodic composites as a class of examples of H-convergent sequences of microstructures. We now present a remarkable theorem formulated by Kohn and Dal Maso and proved by Tartar [24] and Raitums [18] in the context of conductivity. See also [2] for the analogous result in elasticity.

Theorem 3.7. *Assume that the Fourier multiplier symbol* $\boldsymbol{\Gamma}(\eta)$ *is* $\mathcal{E}$*-and* $\mathcal{J}$*-regular in the sense of definition 2.5 and* $\mathcal{E}$*-nondegenerate in the sense of definition 2.10. Let U be any closed subset of* $\mathcal{M}(\alpha, \beta)$:

(i) *Assume that* $\mathsf{L}_\epsilon \in U^\Omega$ *is such that* $\mathsf{L}_\epsilon \overset{H}{\rightharpoonup} \mathsf{L}_*$. *Then* $\mathsf{L}_* \in G(U)^\Omega$.

(ii) *Assume that* $\mathsf{L}_* \in G(U)^\Omega$. *Then there exists* $\mathsf{L}_\epsilon \in U^\Omega$, *such that* $\mathsf{L}_\epsilon \overset{H}{\rightharpoonup} \mathsf{L}_*$.

The proof of this result is in section 3.3.5. In essence it says that there is no loss of generality in studying periodic composites if we are interested in questions about G-closure, such as the ones studied in this book.

It is natural to expect that the G-closure of a set U would be G-closed. This is indeed the case. However, the mathematical proof of this simple statement requires going rather deep into the relationship between G-closures and H-convergence. See corollary 3.47 in section 3.3.5

We now distinguish two special cases. In the first one the set U is a finite set $U = \{\mathsf{L}_1, \dots \mathsf{L}_r\}$. The tensors in $G(U)$ will be referred to as the effective tensors of r-material composites.

In the second case U is the union of finitely many SO(d) orbits

$$U = \bigcup_{j=1}^{r} \{\boldsymbol{R} \cdot \mathsf{L}_j : \boldsymbol{R} \in \mathrm{SO}(d)\}. \tag{3.11}$$

In this case the composites made from the materials in U are called polycrystals. Since single crystals $\mathsf{L}_1, \dots, \mathsf{L}_r$ determine the set U we will talk about the polycrystalline G-closure of materials $\mathsf{L}_1, \dots, \mathsf{L}_r$ and write $G^{\mathrm{pc}}(\{\mathsf{L}_1, \dots, \mathsf{L}_r\})$ to denote the G-closure of the set U in (3.11).

In the two special cases above we can restrict the set of admissible microstructures so that each material or each crystal L_j is used in specified volume fraction θ_j.

Definition 3.8. *Let* $\vartheta = (\theta_1, \dots, \theta_r)$ *be such that* $0 \leqslant \theta_j \leqslant 1$ *and* $\theta_1 + \cdots \theta_r = 1$. *The partition of the period cell Q into r disjoint measurable subsets* $A_1, \dots, A_r$ *so that* $|A_j|/|Q| = \theta_j$, $j = 1, \dots, r$ *is called a* ϑ*-partition of Q.*

Definition 3.9. *The set of all effective tensors of periodic composites with*

$$\mathsf{L}(z) = \sum_{j=1}^{r} \mathsf{L}_j \chi_{A_j}(z),$$

where $A_1, \dots, A_r$ *is a* ϑ*-partition of Q, is called* G_ϑ*-closure of the set* $U = \{\mathsf{L}_1, \dots, \mathsf{L}_r\}$.

Definition 3.10. *The polycrystalline* G_ϑ*-closure of the set* $U = \{\mathsf{L}_1, \dots, \mathsf{L}_r\}$ *is the set of all effective tensors of periodic composites with*

$$\mathsf{L}(z) = \sum_{j=1}^{r} \boldsymbol{R}_j(z) \cdot \mathsf{L}_j \chi_{A_j}(z),$$

where $A_1, \ldots, A_r$ is a ϑ-partition of Q, and $\boldsymbol{R}_j : Q \to \mathrm{SO}(d)$ are arbitrary measurable functions, that collectively are called 'the texture' of the polycrystal.

The knowledge of the G-closure of a set of given materials provides bounds on the possible properties of a specific composite made of these materials. It can also tell us whether or not a desirable combination of properties can theoretically be achieved by mixing materials that possess only a partial set of the required properties. Very few G-closures have been computed exactly, e.g. [6, 8, 9, 11–13]. In general it is an almost intractable problem [4].

3.2.2 The formula for the effective tensor

So far the effective tensor of a periodic composite is defined in terms of a solution of a periodic cell problem (3.9). In this section we will derive a formula for L_*, which will be central for the theory of exact relations. Moreover, we will show that this formula has other remarkable features, such as superior convergence properties for effective tensors of high contrast composites.

If we approximate local material properties by a given homogeneous reference medium $\mathsf{L}_0 \in \mathrm{End}^+(\mathcal{T})$, then the resulting flux field $\boldsymbol{J}$ will be approximated by $\mathsf{L}_0\boldsymbol{E}$. The discrepancy $\boldsymbol{P} = \boldsymbol{J} - \mathsf{L}_0\boldsymbol{E}$ is called the polarization field. It turns out that the system of three equations (3.9) can be written as a single equation in terms of $\boldsymbol{P}$, if we introduce a scale-invariant Fourier multiplier operator

$$\widehat{\Gamma' f}(\boldsymbol{k}) = \begin{cases} \boldsymbol{\Gamma}'\left(\dfrac{Q^T\boldsymbol{k}}{|Q^T\boldsymbol{k}|}\right)\widehat{f}(\boldsymbol{k}), & \boldsymbol{k} \in \mathbb{Z}^d\setminus\{\boldsymbol{0}\}, \\ \boldsymbol{0}, \ \boldsymbol{k} = \boldsymbol{0}, \end{cases} \qquad Q\mathcal{Q} = [0, 1]^d,$$

where $\boldsymbol{\Gamma}'(\boldsymbol{n})$ is the non-orthogonal projection operator onto the subspace $\mathsf{L}_0\mathcal{E}_{\boldsymbol{n}}$ along the subspace $\mathcal{J}_{\boldsymbol{n}}$. In fact, Γ' is the non-orthogonal projection onto $\mathsf{L}_0\mathcal{E}_{\mathrm{per}}$ along $\mathcal{J}_{\mathrm{per}} \oplus \mathcal{U}$. We note that the projection $\boldsymbol{\Gamma}'(\boldsymbol{n})$ is well defined as long as L_0 is positive definite, since in that case $\mathsf{L}_0\mathcal{E}_{\boldsymbol{n}}$ and $\mathcal{J}_{\boldsymbol{n}}$ have trivial intersection. Indeed, if $\mathsf{L}_0\boldsymbol{e} = \boldsymbol{j}$ for some $\boldsymbol{e} \in \mathcal{E}_{\boldsymbol{n}}$ and $\boldsymbol{j} \in \mathcal{J}_{\boldsymbol{n}}$, then taking the inner product with $\boldsymbol{e}$ we obtain

$$0 = (\boldsymbol{j}, \boldsymbol{e}) = (\mathsf{L}_0\boldsymbol{e}, \boldsymbol{e}),$$

and thus, $\boldsymbol{e} = \boldsymbol{0}$. Noting that a non-orthogonal projection operator is necessarily non-symmetric, it will be convenient to introduce a related symmetric operator

$$\Gamma_0 = \mathsf{L}_0^{-1}\Gamma'. \tag{3.12}$$

Lemma 3.11. *Let*

$$\Gamma_0(\boldsymbol{n}) = \mathsf{L}_0^{-1}\Gamma'(\boldsymbol{n}). \tag{3.13}$$

Then $\Gamma_0(\boldsymbol{n}) \in \mathrm{Sym}(\mathcal{T})$, *provided* $\mathsf{L}_0 \in \mathrm{Sym}^+(\mathcal{T})$.

Proof. For $\boldsymbol{u}_1, \boldsymbol{u}_2 \subset \mathcal{T}$ we write

$$\boldsymbol{u}_1 = \boldsymbol{j}_1 + \mathsf{L}_0\boldsymbol{e}_1, \qquad \boldsymbol{u}_2 = \boldsymbol{j}_2 + \mathsf{L}_0\boldsymbol{e}_2, \qquad \{\boldsymbol{e}_1, \boldsymbol{e}_2\} \subset \mathcal{E}_{\boldsymbol{n}}, \quad \{\boldsymbol{j}_1, \boldsymbol{j}_2\} \subset \mathcal{J}_{\boldsymbol{n}}.$$

Then $\Gamma_0\boldsymbol{u}_1 = \boldsymbol{e}_1$, $\Gamma_0\boldsymbol{u}_2 = \boldsymbol{e}_2$, so that

$$\left(\Gamma_0(\boldsymbol{n})\boldsymbol{u}_1, \boldsymbol{u}_2\right) = \left(\boldsymbol{e}_1, \mathsf{L}_0\boldsymbol{e}_2\right) = \left(\mathsf{L}_0\boldsymbol{e}_1, \boldsymbol{e}_2\right) = \left(\boldsymbol{u}_1, \Gamma_0(\boldsymbol{n})\boldsymbol{u}_2\right).$$

□

We can now restate the cell problem (3.9) as a single equation.

Lemma 3.12. *Let* $\boldsymbol{P}(z) = \mathsf{L}(z)\boldsymbol{E}(z) - \mathsf{L}_0\boldsymbol{E}(z)$ *be the polarization field. Then the cell problem* (3.9) *is equivalent to a single equation*

$$\Gamma_0\boldsymbol{P} = \boldsymbol{e}_0 - \boldsymbol{E}, \qquad \boldsymbol{E} \in L^2_{\mathrm{per}}(Q; \mathcal{T}). \tag{3.14}$$

Proof. Let us first assume that $\boldsymbol{E} \in L^2_{\mathrm{per}}(Q; \mathcal{T})$ solves (3.14). Let us prove that all equations in (3.9) are satisfied. We note that

$$\left\langle \Gamma_0\boldsymbol{P}\right\rangle = \left(\widehat{\Gamma_0\boldsymbol{P}}\right)(\boldsymbol{0}) = 0.$$

Thus, $\boldsymbol{e}_0 = \langle \boldsymbol{E}\rangle$. Also, $\Gamma_0\boldsymbol{P} \in \mathcal{E}_{\mathrm{per}}$, by definition of the operator Γ_0. Hence, $\boldsymbol{E} = \boldsymbol{e}_0 - \Gamma_0\boldsymbol{P} \in \mathcal{E}_{\mathrm{per}} \oplus \mathcal{U}$. Rewriting (3.14) as

$$\Gamma'\boldsymbol{P} = \mathsf{L}_0(\boldsymbol{e}_0 - \boldsymbol{E}) \tag{3.15}$$

and using the fact that Γ' is the projection operator, we conclude that

$$\boldsymbol{P} - \mathsf{L}_0(\boldsymbol{e}_0 - \boldsymbol{E}) \in \mathcal{J}_{\mathrm{per}} \oplus \mathcal{U}. \tag{3.16}$$

Finally, defining $\boldsymbol{J} = \mathsf{L}(z)\boldsymbol{E}(z)$ we can rewrite (3.16) as $\boldsymbol{J} \in \mathcal{J}_{\mathrm{per}} \oplus \mathcal{U}$.

Now let us assume that equations (3.9) hold. Then $\boldsymbol{J} = \mathsf{L}(z)\boldsymbol{E}(z) \in \mathcal{J}_{\mathrm{per}} \oplus \mathcal{U}$ implies

$$\Gamma'\boldsymbol{J} = \Gamma'(\mathsf{L}(z)\boldsymbol{E}(z)) = 0. \tag{3.17}$$

Also,

$$\mathsf{L}_0\boldsymbol{E} - \langle \mathsf{L}_0\boldsymbol{E}\rangle \in \mathsf{L}_0\mathcal{E}_{\text{per}}.$$

Therefore,

$$\Gamma'(\mathsf{L}_0\boldsymbol{E} - \langle \mathsf{L}_0\boldsymbol{E}\rangle) = \mathsf{L}_0\boldsymbol{E} - \langle \mathsf{L}_0\boldsymbol{E}\rangle = \mathsf{L}_0\big(\boldsymbol{E} - e_0\big).$$

Using the fact that Γ' annihilates constant fields we obtain

$$\Gamma'(\mathsf{L}_0\boldsymbol{E}) = \mathsf{L}_0(\boldsymbol{E} - e_0). \tag{3.18}$$

Subtracting (3.18) from (3.17) we obtain (3.15), which is equivalent to (3.14). □

We note that the relation between the intensity field $\boldsymbol{E}$ and the polarization field $\boldsymbol{P}$ may become singular, if $\mathsf{L}(z) - \mathsf{L}_0$ is not invertible. This problem can be resolved using the idea of preconditioning, i.e. rewriting the same linear equation in an equivalent form which behaves in a more stable manner with respect to its parameters. For this reason we will work with a generalized polarization field, which is as yet undermined, linear function of the intensity and flux fields:

$$\boldsymbol{S} = \mathsf{A}\boldsymbol{E} + \mathsf{M}\boldsymbol{J} = (\mathsf{A} + \mathsf{ML})\boldsymbol{E}, \tag{3.19}$$

where possible values of constant operators $\{\mathsf{A}, \mathsf{M}\} \subset \text{End}(\mathcal{T})$ will be determined in the course of our analysis. The idea is to chose operators A and M, so that $\mathsf{A} + \mathsf{ML}_*$ and $\mathsf{A} + \mathsf{ML}(z)$ are stably invertible on Q.

Definition 3.13. *We say that* $\mathsf{F} \in L^\infty(Q; \text{End}(\mathcal{T}))$ *is* stably invertible *on* Q *if* $\mathsf{F}(z)^{-1}$ *exists for almost all* $z \in Q$ *and* $\mathsf{F}^{-1} \in L^\infty(Q; \text{End}(\mathcal{T}))$.

Taking the average of (3.19) over a period cell we obtain

$$\langle \boldsymbol{S}\rangle = (\mathsf{A} + \mathsf{ML}_*)e_0. \tag{3.20}$$

Using (3.19), we can express the polarization field in terms of $\boldsymbol{S}$:

$$\boldsymbol{P} = (\mathsf{L} - \mathsf{L}_0)(\mathsf{A} + \mathsf{ML})^{-1}\boldsymbol{S}.$$

We can also express the average of the polarization field in terms of $\langle \boldsymbol{S}\rangle$, using (3.20):

$$\langle \boldsymbol{P}\rangle = (\mathsf{L}_* - \mathsf{L}_0)e_0 = (\mathsf{L}_* - \mathsf{L}_0)(\mathsf{A} + \mathsf{ML}_*)^{-1}\langle \boldsymbol{S}\rangle.$$

This suggests defining a transformation

$$W(\mathsf{L}) = (\mathsf{L} - \mathsf{L}_0)(\mathsf{A} + \mathsf{ML})^{-1},$$

relating the classical and the generalized polarization fields, so that in terms of $W(\mathsf{L})$ we have

$$P = W(\mathsf{L})S, \qquad \langle P\rangle = W(\mathsf{L}_*)\langle S\rangle. \tag{3.21}$$

We would like the W-transformation to preserve symmetry. In other words, if L and L_0 are symmetric, we want $W(\mathsf{L})$ to be symmetric as well. This means that we must have

$$(\mathsf{L} - \mathsf{L}_0)(\mathsf{A} + \mathsf{ML})^{-1} = (\mathsf{A}^T + \mathsf{LM}^T)^{-1}(\mathsf{L} - \mathsf{L}_0).$$

for every $\mathsf{L} \in \mathrm{Sym}(\mathcal{T})$. Equivalently,

$$(\mathsf{A}^T + \mathsf{LM}^T)(\mathsf{L} - \mathsf{L}_0) = (\mathsf{L} - \mathsf{L}_0)(\mathsf{A} + \mathsf{ML}).$$

It follows that $\mathsf{M}^T = \mathsf{M}$ and

$$\mathsf{L}(\mathsf{A} + \mathsf{ML}_0) = \left(\mathsf{A} + \mathsf{ML}_0\right)^T\mathsf{L}, \quad \forall\, \mathsf{L} \in \mathrm{Sym}(\mathcal{T}).$$

It follows that $\mathsf{A} + \mathsf{ML}_0 = \alpha\mathsf{I}$ for some $\alpha \in \mathbb{R}$. Conversely, if $\mathsf{M} \in \mathrm{Sym}(\mathcal{T})$ and

$$\mathsf{A} = \alpha\mathsf{I} - \mathsf{ML}_0, \tag{3.22}$$

then

$$W(\mathsf{L}) = (\mathsf{L} - \mathsf{L}_0)\left(\alpha\mathsf{I} + \mathsf{M}(\mathsf{L} - \mathsf{L}_0)\right)^{-1} \in \mathrm{Sym}(\mathcal{T})$$

for every $\mathsf{L} \in \mathrm{Sym}(\mathcal{T})$. We will set the value of A according to (3.22), even if L is not necessarily symmetric, in which case M will not be assumed to be symmetric either.

We now proceed to eliminate $\boldsymbol{E}$ from the cell problem in the form (3.14), taking (3.22) into account. Replacing $\boldsymbol{E}$ and e_0 in (3.14) by their expressions in terms of $\boldsymbol{S}$ and $\langle \boldsymbol{S}\rangle$, found from (3.19) and (3.20), respectively, we obtain

$$\Gamma_0(W(\mathsf{L})\boldsymbol{S}) = \left(\alpha\mathsf{I} + \mathsf{M}(\mathsf{L}_* - \mathsf{L}_0)\right)^{-1}\langle \boldsymbol{S}\rangle - \left(\alpha\mathsf{I} + \mathsf{M}(\mathsf{L} - \mathsf{L}_0)\right)^{-1}\boldsymbol{S}.$$

It is easy to see that

$$\left(\alpha\mathsf{I} + \mathsf{M}(\mathsf{L} - \mathsf{L}_0)\right)^{-1} = \frac{1}{\alpha}(\mathsf{I} - \mathsf{M}W(\mathsf{L})).$$

The cell problem (3.14) now takes the form

$$\Gamma_0(W(\mathsf{L})\boldsymbol{S}) = \frac{1}{\alpha}\left(\langle \boldsymbol{S}\rangle - \boldsymbol{S} + \mathsf{M}(W(\mathsf{L})\boldsymbol{S} - W(\mathsf{L}_*)\langle \boldsymbol{S}\rangle)\right). \tag{3.23}$$

Recalling that Γ_0 annihilates uniform fields and using (3.21) we can rewrite (3.23) as

$$\left(\Gamma_0 - \frac{1}{\alpha}\mathsf{M}\right)\left(W(\mathsf{L})\boldsymbol{S} - W(\mathsf{L}_*)\langle \boldsymbol{S}\rangle\right) = \frac{1}{\alpha}\left(\langle \boldsymbol{S}\rangle - \boldsymbol{S}\right). \tag{3.24}$$

Noting that

$$W(\mathsf{L}) = \frac{1}{\alpha}(\mathsf{L} - \mathsf{L}_0)\big(\mathsf{I} + \mathsf{M}(\mathsf{L} - \mathsf{L}_0)/\alpha\big)^{-1},$$

we find it convenient to rename M/α into M and $\alpha W(\mathsf{L})$ into $W_{\mathsf{M}}(\mathsf{L})$, obtaining

$$(\Gamma_0 - \mathsf{M})\Big(W_{\mathsf{M}}(\mathsf{L})\boldsymbol{S} - W_{\mathsf{M}}(\mathsf{L}_*)\langle \boldsymbol{S}\rangle\Big) = \langle \boldsymbol{S}\rangle - \boldsymbol{S}, \tag{3.25}$$

where

$$W_{\mathsf{M}}(\mathsf{L}) = (\mathsf{L} - \mathsf{L}_0)\big(\mathsf{I} + \mathsf{M}(\mathsf{L} - \mathsf{L}_0)\big)^{-1}. \tag{3.26}$$

Choosing M in such a way that the operators $\mathsf{I} + \mathsf{M}(\mathsf{L}_* - \mathsf{L}_0)$ and $\mathsf{I} + \mathsf{M}(\mathsf{L}(z) - \mathsf{L}_0)$ are stably invertible on Q, the derivation of (3.25) becomes rigorous. The new form (3.25) of the cell problem makes it convenient to introduce a Fourier multiplier operator Λ_{M}, given by

$$\widehat{\Lambda_{\mathsf{M}}\boldsymbol{f}}(\boldsymbol{k}) = \begin{cases} \left(\mathsf{M} - \boldsymbol{\Gamma}_0\left(\dfrac{Q^T\boldsymbol{k}}{|Q^T\boldsymbol{k}|}\right)\right)\widehat{\boldsymbol{f}}(\boldsymbol{k}), & \boldsymbol{k} \in \mathbb{Z}^d\backslash\{\boldsymbol{0}\}, \\ \boldsymbol{0}, \ \boldsymbol{k} = \boldsymbol{0}, \end{cases} \qquad QQ = [0, 1]^d. \tag{3.27}$$

With this notation (3.25) simplifies:

$$\big(\mathsf{I} - \Lambda_{\mathsf{M}} W_{\mathsf{M}}(\mathsf{L})\big)\boldsymbol{S} = \langle \boldsymbol{S}\rangle. \tag{3.28}$$

In order to solve (3.28) for $\boldsymbol{S}$ we need to know that the operator $\mathsf{I} - \Lambda_{\mathsf{M}} W_{\mathsf{M}}(\mathsf{L})$ is invertible on $L^2_{\text{per}}(Q, \mathcal{T})$.

Lemma 3.14. *Suppose that the operators* $\mathsf{I} + \mathsf{M}(\mathsf{L}_* - \mathsf{L}_0)$ *and* $\mathsf{I} + \mathsf{M}(\mathsf{L}(z) - \mathsf{L}_0)$ *are stably invertible on* Q. *Then the operator* $\mathfrak{T} = \mathsf{I} - \Lambda_{\mathsf{M}} W_{\mathsf{M}}(\mathsf{L})$ *is invertible on* $L^2_{\text{per}}(Q, \mathcal{T})$.

Proof. By our assumptions the operator $\mathfrak{T}$ is bounded. It remains to prove that for every $\boldsymbol{f} \in L^2_{\text{per}}(Q, \mathcal{T})$ there exists unique $\boldsymbol{u} \in L^2_{\text{per}}(Q, \mathcal{T})$ such that $\mathfrak{T}\boldsymbol{u} = \boldsymbol{f}$. Then, by the Banach invertibility theorem this would imply that $\mathfrak{T}^{-1}$ is a bounded operator on $L^2_{\text{per}}(Q, \mathcal{T})$. Let us examine the equation $\mathfrak{T}\boldsymbol{u} = \boldsymbol{f}$. Using the definition of Λ_{M} and W_{M} we can rewrite it as

$$\Gamma_0(W_{\mathsf{M}}(\mathsf{L})\boldsymbol{u}) = \boldsymbol{f} - \mathsf{M}\langle W_{\mathsf{M}}(\mathsf{L})\boldsymbol{u}\rangle - \big(\mathsf{I} + \mathsf{M}(\mathsf{L} - \mathsf{L}_0)\big)^{-1}\boldsymbol{u} \tag{3.29}$$

It follows that the right-hand side of (3.29) belongs to $\mathcal{E}_{\text{per}}$. If we denote it by $\boldsymbol{e}$ and the constant field $\mathsf{M}\langle W_{\mathsf{M}}(\mathsf{L})\boldsymbol{u}\rangle$ by $\boldsymbol{e}_0$, then we obtain the relation between $\boldsymbol{u}$ and $\boldsymbol{e}$

$$\boldsymbol{u} = \big(\mathsf{I} + \mathsf{M}(\mathsf{L} - \mathsf{L}_0)\big)(\boldsymbol{f} - \boldsymbol{e} - \boldsymbol{e}_0). \tag{3.30}$$

Multiplying equation (3.29) on the left by L_0 we obtain

$$\Gamma'\big(W_{\mathsf{M}}(\mathsf{L})\boldsymbol{u}\big) = \mathsf{L}_0(\boldsymbol{f} - \boldsymbol{e}_0) - \mathsf{L}_0\big(\mathsf{I} + \mathsf{M}(\mathsf{L} - \mathsf{L}_0)\big)^{-1}\boldsymbol{u}.$$

Thus,

$$W_{\mathsf{M}}(\mathsf{L})\boldsymbol{u} - \mathsf{L}_0(\boldsymbol{f} - \boldsymbol{e}_0) + \mathsf{L}_0\big(\mathsf{I} + \mathsf{M}(\mathsf{L} - \mathsf{L}_0)\big)^{-1}\boldsymbol{u} \in \mathcal{J}_{\text{per}} \oplus \mathcal{U}.$$

If Γ_{per} denotes the orthogonal projection onto $\mathcal{E}_{\text{per}}$ in $L^2_{\text{per}}(Q, \mathcal{T})$, then we have

$$\Gamma_{\text{per}}\Big(\mathsf{L}\big(\mathsf{I} + \mathsf{M}(\mathsf{L} - \mathsf{L}_0)\big)^{-1}\boldsymbol{u}\Big) = \Gamma_{\text{per}}(\mathsf{L}_0\boldsymbol{f}). \tag{3.31}$$

Eliminating $\boldsymbol{u}$, using (3.30) we rewrite (3.31) as

$$\Gamma_{\text{per}}\Big(\mathsf{L}(\boldsymbol{e} + \boldsymbol{e}_0)\Big) = \Gamma_{\text{per}}\Big((\mathsf{L} - \mathsf{L}_0)\boldsymbol{f}\Big), \qquad \boldsymbol{e} \in \mathcal{E}_{\text{per}}. \tag{3.32}$$

Conversely, it is easy verify by reversing the arguments that the system of equations (3.32), (3.30), together with

$$\boldsymbol{e}_0 = \mathsf{M}\langle W_{\mathsf{M}}(\mathsf{L})\boldsymbol{u}\rangle \tag{3.33}$$

is equivalent to (3.29).

We note that by analogy with the proof of theorem 2.4 the Lax–Milgram lemma guarantees that the operator $\Gamma_{\text{per}}(\mathsf{L}\boldsymbol{e})$ is invertible on $\mathcal{E}_{\text{per}}$. Hence, the solution $\boldsymbol{e}$ of (3.32) can be represented as $\boldsymbol{e} + \boldsymbol{e}_0 = \tilde{\boldsymbol{e}} + \boldsymbol{E}$, where $\tilde{\boldsymbol{e}} \in \mathcal{E}_{\text{per}}$ is the unique solution of

$$\Gamma_{\text{per}}(\mathsf{L}\tilde{\boldsymbol{e}}) = \Gamma_{\text{per}}\big((\mathsf{L} - \mathsf{L}_0)\boldsymbol{f}\big),$$

while $\boldsymbol{E} \in \mathcal{E}_{\text{per}} \oplus \mathcal{U}$ is the unique solution of the periodic cell problem (3.9). The solution $\boldsymbol{u}$ of $\mathfrak{T}\boldsymbol{u} = \boldsymbol{f}$ is then given by (3.30), establishing existence and uniqueness, if we show that $\boldsymbol{e}_0 \in \mathcal{U}$ can be determined uniquely.

In order to determine $\boldsymbol{e}_0$ we use (3.30) to compute

$$W_{\mathsf{M}}(\mathsf{L})\boldsymbol{u} = (\mathsf{L} - \mathsf{L}_0)(\boldsymbol{f} - \boldsymbol{e} - \boldsymbol{e}_0). \tag{3.34}$$

Averaging (3.34) over the period cell, multiplying by M, and using (3.33) we obtain

$$\boldsymbol{e}_0 = \mathsf{M}\Big\langle(\mathsf{L} - \mathsf{L}_0)(\boldsymbol{f} - \boldsymbol{e} - \boldsymbol{e}_0)\Big\rangle.$$

Replacing $\boldsymbol{e} + \boldsymbol{e}_0$ with $\tilde{\boldsymbol{e}}+\boldsymbol{E}$ we obtain

$$\boldsymbol{e}_0 = \mathsf{M}\Big\langle(\mathsf{L} - \mathsf{L}_0)(\boldsymbol{f} - \tilde{\boldsymbol{e}})\Big\rangle - \mathsf{M}(\mathsf{L}_* - \mathsf{L}_0)\boldsymbol{e}_0,$$

where we took into account that $\boldsymbol{E}$ solves (3.9). This gives

$$\boldsymbol{e}_0 = \big(\mathsf{I} + \mathsf{M}(\mathsf{L}_* - \mathsf{L}_0)\big)^{-1}\mathsf{M}\Big\langle(\mathsf{L} - \mathsf{L}_0)(\boldsymbol{f} - \tilde{\boldsymbol{e}})\Big\rangle. \tag{3.35}$$

Conversely, it is easy to verify that (3.35) implies (3.33). Thus, for every $f \in L^2_{\mathrm{per}}(Q, \mathcal{T})$ we have exhibited the unique solution $\boldsymbol{u}$ of $\mathfrak{T}\boldsymbol{u} = \boldsymbol{f}$. The lemma is proved now. □

We can now return to (3.28) and write

$$\boldsymbol{S} = \left(\mathsf{I} - \Lambda_{\mathsf{M}} W_{\mathsf{M}}(\mathsf{L})\right)^{-1} \langle \boldsymbol{S} \rangle.$$

Multiplying this equation by $W_{\mathsf{M}}(\mathsf{L})$ and taking averages (taking (3.21) into account) we obtain

$$W_{\mathsf{M}}(\mathsf{L}_*)\langle \boldsymbol{S} \rangle = \left\langle W_{\mathsf{M}}(\mathsf{L})(\mathsf{I} - \Lambda_{\mathsf{M}} W_{\mathsf{M}}(\mathsf{L}))^{-1} \right\rangle \langle \boldsymbol{S} \rangle. \tag{3.36}$$

Finally, we note that there is a one-to-one correspondence between $\langle \boldsymbol{S} \rangle$ and $\boldsymbol{e}_0$

$$\langle \boldsymbol{S} \rangle = \left(\mathsf{I} + \mathsf{M}(\mathsf{L}_* - \mathsf{L}_0)\right)\boldsymbol{e}_0.$$

Thus, we obtain the general formula for the W-transform of the effective tensor of the composite in terms of the W-transform of the local tensor:

$$W_{\mathsf{M}}(\mathsf{L}_*) = \left\langle W_{\mathsf{M}}(\mathsf{L})(\mathsf{I} - \Lambda_{\mathsf{M}} W_{\mathsf{M}}(\mathsf{L}))^{-1} \right\rangle. \tag{3.37}$$

We summarize our results in the form of a theorem.

Theorem 3.15. *Let* L_* *be the effective tensor of the composite with the local tensor* $\mathsf{L}(z)$, $z \in Q$. *Let* $\mathsf{L}_0 \in \mathrm{End}^+(\mathcal{T})$ *and* $\mathsf{M} \in \mathrm{End}(\mathcal{T})$ *be such that the operators* $\mathsf{I} + \mathsf{M}(\mathsf{L}_* - \mathsf{L}_0)$ *and* $\mathsf{I} + \mathsf{M}(\mathsf{L}(z) - \mathsf{L}_0)$ *are stably invertible on* Q *in the sense of definition 3.13. Then the effective tensor* L_* *can be determined by (3.37).*

Remark 3.16. *Substituting (3.22) into (3.19) with* $\alpha = 1$, *we obtain* $\boldsymbol{S} = \boldsymbol{E} + \mathsf{M}\boldsymbol{P}$. *Hence,* $\boldsymbol{S}$ can be interpreted as a modified total field, which is the sum of the applied field $\boldsymbol{E}$ and the modified polarization field $\mathsf{M}\boldsymbol{P}$. *Formula (3.21) gives the physical meaning of* $W_{\mathsf{M}}(\mathsf{L})$ *as the tensor in the linear constitutive relation between the polarization field* $\boldsymbol{P}$ *and the modified total field* $\boldsymbol{S}$.

We reiterate that the effective tensor L_* of the periodic composite does not depend on reference medium L_0 and the translation tensor M. These objects can be regarded as preconditioners for the solution of the cell problem (3.9). In particular, it was shown in [15], section 14.9 that for $\mathsf{L} \in \mathcal{M}(\alpha, \beta)^{\Omega}$ choosing $\mathsf{L}_0 = \sqrt{\alpha\beta}\,\mathsf{I}$ and $\mathsf{M} = \mathsf{I}/2\sqrt{\alpha\beta}$, we can make the operator norm of $\Lambda_{\mathsf{M}} W_{\mathsf{M}}(\mathsf{L})$ no larger than $(\sqrt{c} - 1)/(\sqrt{c} + 1)$, where $c = \beta/\alpha$ is the contrast ratio. Hence, formula (3.37) can be written in the form of the geometrically fast converging Neumann series

$$W_{\mathsf{M}}(\mathsf{L}_*) = \sum_{n=0}^{\infty} \left\langle W_{\mathsf{M}}(\mathsf{L})\left(\Lambda_{\mathsf{M}} W_{\mathsf{M}}(\mathsf{L})\right)^n \right\rangle. \tag{3.38}$$

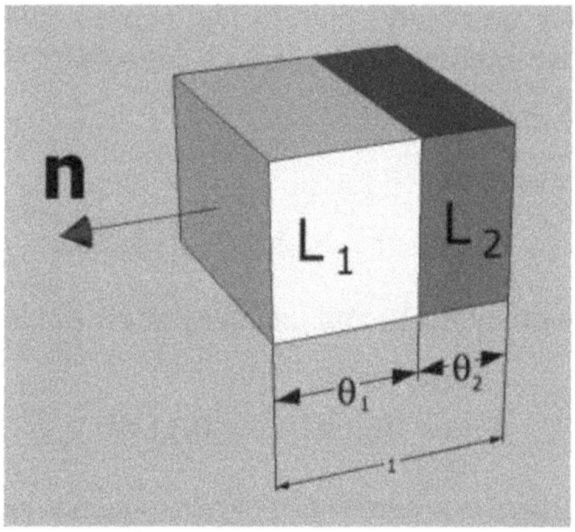

Figure 3.1. The period cell of the laminate.

We note that it is possible to avoid having to check stable invertibility of $\mathsf{I} + \mathsf{M}(\mathsf{L}(z) - \mathsf{L}_0)$ and $\mathsf{I} + \mathsf{M}(\mathsf{L}_* - \mathsf{L}_0)$ for each specific periodic composite by choosing M with the *universal invertibility* property.

Definition 3.17. *We say that* $\mathsf{M} \in \mathrm{End}(\mathcal{T})$ *has the universal invertibility property if* $\mathsf{F} = \mathsf{I} + \mathsf{M}(\mathsf{L} - \mathsf{L}_0)$ *is invertible for any positive definite operator* $\mathsf{L} \in \mathrm{End}^+(\mathcal{T})$.

Lemma 3.18. $\mathsf{M} \in \mathrm{End}(\mathcal{T})$ *has the universal invertibility property if and only if* $\mathcal{R}(\mathsf{M}) = \mathcal{R}(\mathsf{M}^T)$, *where* $\mathcal{R}(\mathsf{M})$ *is the range of* M, *and*

$$(\mathsf{M}^{-1}\boldsymbol{u}, \boldsymbol{u}) \geqslant (\mathsf{L}_0\boldsymbol{u}, \boldsymbol{u}), \qquad \forall \boldsymbol{u} \in \mathcal{R}(\mathsf{M}). \tag{3.39}$$

Here M^{-1} *is understood as the inverse of the bijective map* $\mathsf{M} : \mathcal{R}(\mathsf{M}) \to \mathcal{R}(\mathsf{M})$.

The proof of this lemma is a bit technical. It can be found in section 3.2.4. We remark that if $\mathsf{M}\boldsymbol{v} = \boldsymbol{u}$ for some $\boldsymbol{v} \in \mathcal{T}$, then $\boldsymbol{v} = \mathsf{M}^{-1}\boldsymbol{u} + \boldsymbol{w}$, where $\boldsymbol{w} \in \ker(\mathsf{M})$, but then, since $\ker(\mathsf{M}) = \mathcal{R}(\mathsf{M}^T)^\perp = \mathcal{R}(\mathsf{M})^\perp$, we have $(\boldsymbol{v}, \boldsymbol{u}) = (\mathsf{M}^{-1}\boldsymbol{u}, \boldsymbol{u})$. Hence, $\mathsf{M}^{-1}\boldsymbol{u}$ in the left-hand side of (3.39) can also be interpreted as *any* pre-image of $\boldsymbol{u}$.

Corollary 3.19. *For any unit vector* $\boldsymbol{n}$ *the operator* $\mathsf{M} = \boldsymbol{\Gamma}_0(\boldsymbol{n})$ *has the universal invertibility property.*

Proof. We have $\ker(\boldsymbol{\Gamma}_0(\boldsymbol{n})) = \mathcal{J}_{\boldsymbol{n}}$ and $\mathcal{R}(\boldsymbol{\Gamma}_0(\boldsymbol{n})) = \mathcal{E}_{\boldsymbol{n}} = \ker(\boldsymbol{\Gamma}_0(\boldsymbol{n}))^\perp = \mathcal{R}(\boldsymbol{\Gamma}_0(\boldsymbol{n})^T)$. Also, from the equality $\boldsymbol{\Gamma}_0(\boldsymbol{n})\mathsf{L}_0\boldsymbol{u} = \boldsymbol{u}$ that holds for all $\boldsymbol{u} \in \mathcal{E}_{\boldsymbol{n}}$, we conclude that $\mathsf{L}_0\boldsymbol{u}$ is a pre-image of $\boldsymbol{u}$ under the map $\boldsymbol{\Gamma}_0(\boldsymbol{n})$. Thus, $(\boldsymbol{\Gamma}_0(\boldsymbol{n})^{-1}\boldsymbol{u}, \boldsymbol{u}) = (\mathsf{L}_0\boldsymbol{u}, \boldsymbol{u})$ for all $\boldsymbol{u} \in \mathcal{E}_{\boldsymbol{n}}$. Therefore $\boldsymbol{\Gamma}_0(\boldsymbol{n})$ satisfies (3.39) with equality sign. □

3.2.3 Lamination formula

Let us apply formula (3.37) to a laminar microstructure, where $\mathsf{L}(z) = \mathsf{L}(z_1)$ and the period cell is chosen to be $Q = [0, 1]^d$. Then $W_{\mathsf{M}}(\mathsf{L})$ is also a function of z_1 only. In that case it is easy to see that

$$\Lambda_{\mathsf{M}} W_{\mathsf{M}}(\mathsf{L}) = \left(\mathsf{M} - \boldsymbol{\Gamma}_0(\boldsymbol{e}_1)\right)\left(W_{\mathsf{M}}(\mathsf{L}) - \langle W_{\mathsf{M}}(\mathsf{L})\rangle\right).$$

This suggests the choice $\mathsf{M} = \boldsymbol{\Gamma}_0(\boldsymbol{e}_1)$, (which by corollary 3.19 has the universal invertibility property) reducing (3.37) to a surprisingly simple formula: $W_{\mathsf{M}}(\mathsf{L}_*) = \langle W_{\mathsf{M}}(\mathsf{L})\rangle$. In general, if the lamination normal is a unit vector $\boldsymbol{n}$ (see figure 3.1), then we choose $\mathsf{M} = \boldsymbol{\Gamma}_0(\boldsymbol{n})$. The result is summarized as a theorem.

Theorem 3.20. *Let* $\mathsf{L}(z) = \tilde{\mathsf{L}}(z \cdot \boldsymbol{n})$ *and let* L_0 *be an arbitrary positive definite reference medium. Then*

$$\langle W_{\boldsymbol{n}}(\mathsf{L})\rangle = W_{\boldsymbol{n}}(\mathsf{L}_*), \tag{3.40}$$

where

$$W_{\boldsymbol{n}}(\mathsf{L}) = W_{\boldsymbol{\Gamma}_0(\boldsymbol{n})}(\mathsf{L}). \tag{3.41}$$

Corollary 3.21. *A G-closed set is an image of a convex subset of* $\mathrm{End}(\mathcal{T})$ *under the analytic diffeomorphism* $W_{\boldsymbol{n}}^{-1}$.

Proof. The statement follows from the fact that if the set G is G-closed and $\{\mathsf{L}_1, \mathsf{L}_2\} \subset G$ then $\mathsf{L}_* \in G$, where L_* is the effective tensor of the laminate made with materials L_1 and L_2 taken in volume fractions θ_1 and $\theta_2 = 1 - \theta_1$ with lamination normal $\boldsymbol{n}$. By theorem 3.20 we obtain

$$W_{\boldsymbol{n}}(\mathsf{L}_*) = \theta_1 W_{\boldsymbol{n}}(\mathsf{L}_1) + \theta_2 W_{\boldsymbol{n}}(\mathsf{L}_2). \tag{3.42}$$

Hence, $W_{\boldsymbol{n}}(G)$ is convex for all $\boldsymbol{n} \in \mathbb{S}^{d-1}$. The map $W_{\boldsymbol{n}}$ is an analytic diffeomorphism by lemma 3.24 in section 3.2.4, since, by corollary 3.19, $\mathsf{M} = \boldsymbol{\Gamma}_0(\boldsymbol{n})$ has the universal invertibility property. □

Formula (3.42) was first derived by Milton [14] and independently by Zhikov [25]. Other linear lamination formulas were derived by Backus [3] and Tartar [22] for elasticity.

Corollary 3.21 implies substantial regularity of the set $G(U)$ *regardless of the regularity of* U, which itself can be any measurable subset of $\mathcal{M}(\alpha, \beta)$, defined in (2.15). Specifically, the set $G(U)$ must be an analytic submanifold of $\mathrm{End}^+(\mathcal{T})$ with a Lipschitz boundary. We remark that if $W_{\boldsymbol{n}}(G)$ is a convex set for all $\boldsymbol{n} \in \mathbb{S}^2$ then the set G is L-closed (or stable under lamination). In almost all cases the G-closure and L-closure coincide. However, there is an example by Milton of a set that is L-closed but not G-closed in section 31.9 of [15]. Another example is given in this book in appendix B. Francfort and Milton [7] gave an algorithm for constructing general polycrystalline G-closures in the context of two-dimensional conductivity using corollary 3.21.

3.2.4 Properties of transformation $W_{\mathsf{M}}(\mathsf{L})$†

In this section we study properties of the transformation $W_{\mathsf{M}}(\mathsf{L})$, where M has the universal invertibility property. We first prove the universal invertibility criterion lemma 3.18.

Proof of lemma 3.18. We begin the proof with a technical lemma.

Lemma 3.22. *Let* $\{\boldsymbol{a}, \boldsymbol{b}\} \subset \mathcal{T}\backslash\{0\}$. *Then there exists* $\mathsf{L} \in \mathrm{Sym}^+(\mathcal{T})$, *such that* $\mathsf{L}\boldsymbol{a} = \boldsymbol{b}$ if and only if $(\boldsymbol{a}, \boldsymbol{b}) > 0$.

Proof. The necessity of the inequality $(\boldsymbol{a}, \boldsymbol{b}) > 0$ is obvious. Now suppose that it is satisfied. If $\boldsymbol{b} = \lambda\boldsymbol{a}$, then $\lambda > 0$ and $\mathsf{L} = \lambda\mathsf{I}$ does the job. Now assume that $\boldsymbol{a}$ and $\boldsymbol{b}$ are linearly independent. We then define $\mathsf{L} = \mathsf{I}$ on the orthogonal complement of $\mathcal{S} = \mathrm{Span}\{\boldsymbol{a}, \boldsymbol{b}\}$ and define $\mathsf{L}\boldsymbol{b} = \boldsymbol{p} \in \mathcal{S}$, where $\boldsymbol{p}$ will be chosen in such a way that L is symmetric and positive definite on $\mathcal{S}$ (and therefore on $\mathcal{T}$). We note that symmetry of L implies that $(\boldsymbol{p}, \boldsymbol{a}) = |\boldsymbol{b}|^2$. In that case the number $\beta = (\boldsymbol{p}, \boldsymbol{b})$ determines vector $\boldsymbol{p} \in \mathcal{S}$ and, hence, $\mathsf{L} \in \mathrm{Sym}(\mathcal{T})$ uniquely. We require that $(\mathsf{L}(x\boldsymbol{a} + y\boldsymbol{b}), (x\boldsymbol{a} + y\boldsymbol{b})) > 0$ for all $(x, y) \neq (0, 0)$. We compute

$$\left(\mathsf{L}(x\boldsymbol{a} + y\boldsymbol{b}), (x\boldsymbol{a} + y\boldsymbol{b})\right) = x^2(\boldsymbol{a}, \boldsymbol{b}) + 2xy\left|\boldsymbol{b}\right|^2 + y^2\beta.$$

Hence, the positivity of L on $\mathcal{S} = \mathrm{Span}\{\boldsymbol{a}, \boldsymbol{b}\}$ is equivalent to $|\boldsymbol{b}|^4 < \beta(\boldsymbol{a}, \boldsymbol{b})$. Hence, any choice of $\beta > |\boldsymbol{b}|^4/(\boldsymbol{a}, \boldsymbol{b})$ will result in a positive definite $\mathsf{L} \in \mathrm{Sym}(\mathcal{T})$, satisfying $\mathsf{L}\boldsymbol{a} = \boldsymbol{b}$. □

To prove lemma 3.18 we observe that $\mathcal{R}(\mathsf{M})$ is an invariant subspace of $\mathsf{F} = \mathsf{I} + \mathsf{M}(\mathsf{L} - \mathsf{L}_0)$ and that F is singular on $\mathcal{T}$ if and only if it is singular on $\mathcal{R}(\mathsf{M})$. Indeed, if $\mathsf{F}\boldsymbol{t} = 0$ for some $\boldsymbol{t} \in \mathcal{T}\backslash\{0\}$, then $\boldsymbol{t} = -\mathsf{M}(\mathsf{L} - \mathsf{L}_0)\boldsymbol{t} \in \mathcal{R}(\mathsf{M})$. Conversely, if $\boldsymbol{u} \in \mathcal{R}(\mathsf{M})\backslash\{0\}$ is such that $\mathsf{F}\boldsymbol{u} = 0$, then F is certainly singular on $\mathcal{T}$. In turn, $\mathsf{F}\colon \mathcal{R}(\mathsf{M}) \to \mathcal{R}(\mathsf{M})$ is singular if and only if we can choose $\boldsymbol{u} \in \mathcal{R}(\mathsf{M})\backslash\{0\}$ such that $(\mathsf{L} - \mathsf{L}_0)\boldsymbol{u} = -\boldsymbol{v}$, where $\mathsf{M}\boldsymbol{v} = \boldsymbol{u}$. By lemma 3.22 we can always choose a positive definite operator L, such that $\mathsf{L}\boldsymbol{u} = \mathsf{L}_0\boldsymbol{u} - \boldsymbol{v}$, provided $(\mathsf{L}_0\boldsymbol{u} - \boldsymbol{v}, \boldsymbol{u}) > 0$. Hence, in order to ensure that F is non-singular we must require that $(\mathsf{L}_0\boldsymbol{u} - \boldsymbol{v}, \boldsymbol{u}) \leqslant 0$ for every $\boldsymbol{u} \in \mathcal{R}(\mathsf{M})$ and every $\boldsymbol{v} \in \mathcal{T}$, satisfying $\boldsymbol{u} = \mathsf{M}\boldsymbol{v}$. Recall that $\ker(\mathsf{M})^{\perp} = \mathcal{R}(\mathsf{M}^T)$ and therefore $\mathsf{M} : \mathcal{R}(\mathsf{M}^T) \to \mathcal{R}(\mathsf{M})$ is an invertible transformation. Hence, there exists a unique $\boldsymbol{v}_0 \in \mathcal{R}(\mathsf{M}^T)$, such that $\mathsf{M}\boldsymbol{v}_0 = \boldsymbol{u}$. We denote $\boldsymbol{v}_0 = \mathsf{M}^{-1}\boldsymbol{u}$. Thus, if $\mathsf{M}\boldsymbol{v} = \boldsymbol{u}$, then $\boldsymbol{v} = \boldsymbol{v}_0 + \boldsymbol{w}$, where $\boldsymbol{w} \in \ker(\mathsf{M})$ can be arbitrary. Our conclusion is that F is non-singular for every positive definite L if and only if

$$(\mathsf{L}_0\boldsymbol{u} - \mathsf{M}^{-1}\boldsymbol{u} - \boldsymbol{w}, \boldsymbol{u}) \leqslant 0, \qquad \forall\boldsymbol{u} \in \mathcal{R}(\mathsf{M}), \forall\boldsymbol{w} \in \ker(\mathsf{M}). \tag{3.43}$$

In particular, we must have $(\boldsymbol{u}, \boldsymbol{w}) = 0$ for all $\boldsymbol{u} \in \mathcal{R}(\mathsf{M})$ and all $\boldsymbol{w} \in \ker(\mathsf{M})$. Thus, $\mathcal{R}(\mathsf{M}) = (\ker(\mathsf{M}))^{\perp} = \mathcal{R}(\mathsf{M}^T)$, and inequality (3.43) reduces to (3.39). □

Theorem 3.27 (Compensated compactness). *Assume that the symbol* $\mathbf{\Gamma}(\boldsymbol{\eta})$ *is* $\mathcal{E}$ *and* $\mathcal{J}$*-regular. Let* $\boldsymbol{f}_\epsilon \rightharpoonup \boldsymbol{f}_0$ *and* $\boldsymbol{g}_\epsilon \rightharpoonup \boldsymbol{g}_0$ *as* $\epsilon \to 0$ *weakly in* $L^2(\Omega; \mathcal{T})$. *Let us assume that* $\Gamma_\Omega \boldsymbol{f}_\epsilon \to \Gamma_\Omega \boldsymbol{f}_0$ *and* $\Pi_\Omega \boldsymbol{g}_\epsilon \to \Pi_\Omega \boldsymbol{g}_0$ *as* $\epsilon \to 0$ *in* $L^2(\Omega; \mathcal{T})$ *norm. Then* $(\boldsymbol{f}_\epsilon, \boldsymbol{g}_\epsilon) \overset{*}{\rightharpoonup} (\boldsymbol{f}_0, \boldsymbol{g}_0)$ *in the sense of measures on* Ω.

Proof. Observe that a finite Borel measure is completely determined by its values on the balls. It follows that it is sufficient to prove the theorem when Ω is a ball. The translation and scaling invariance means that without loss of generality we may assume that $\Omega = B$–the unit ball. Next we observe that replacing $\boldsymbol{f}_\epsilon$ with $\boldsymbol{f}_\epsilon - \boldsymbol{f}_0$ and $\boldsymbol{g}_\epsilon$ with $\boldsymbol{g}_\epsilon - \boldsymbol{g}_0$ we may assume without loss of generality, that $\boldsymbol{f}_0 = \boldsymbol{g}_0 = \mathbf{0}$. Observe that $\tilde{\boldsymbol{f}}_\epsilon = \boldsymbol{f}_\epsilon - \Gamma_B \boldsymbol{f}_\epsilon \in \overline{\mathcal{J}(B)}$ and $\tilde{\boldsymbol{g}}_\epsilon = \boldsymbol{g}_\epsilon - \Pi_B \boldsymbol{g}_\epsilon \in \overline{\mathcal{E}(B)}$. By assumption of the theorem it is sufficient to prove that

$$\lim_{\epsilon \to 0} \int_B \phi(\boldsymbol{x})\left(\tilde{\boldsymbol{f}}_\epsilon, \tilde{\boldsymbol{g}}_\epsilon\right) \mathrm{d}\boldsymbol{x} = 0$$

for every $\phi \in C_0(B)$. By assumption, $\overline{\mathcal{J}(B)} = \mathcal{J}(B)$ and $\overline{\mathcal{E}(B)} = \mathcal{E}(B)$. By lemma 2.6 there exist bounded extension operators $X_\mathcal{E}(B) : \mathcal{E}(B) \to \mathcal{E}$ and $X_\mathcal{J}(B) : \mathcal{J}(B) \to \mathcal{J}$. Let $\boldsymbol{E}_\epsilon = X_\mathcal{E}(B)\tilde{\boldsymbol{g}}_\epsilon \in \mathcal{E}$ and $\boldsymbol{J}_\epsilon = X_\mathcal{J}(B)\tilde{\boldsymbol{f}}_\epsilon \in \mathcal{J}$. Then, $\boldsymbol{E}_\epsilon \rightharpoonup \mathbf{0}$ and $\boldsymbol{J}_\epsilon \rightharpoonup \mathbf{0}$. Extending ϕ by zero to all of $\mathbb{R}^d$, we can write

$$\phi(\boldsymbol{x})\left(\tilde{\boldsymbol{f}}_\epsilon(\boldsymbol{x}), \tilde{\boldsymbol{g}}_\epsilon(\boldsymbol{x})\right) = \phi(\boldsymbol{x})\left(\boldsymbol{E}_\epsilon(\boldsymbol{x}), \boldsymbol{J}_\epsilon(\boldsymbol{x})\right), \qquad \boldsymbol{x} \in \mathbb{R}^d.$$

Recall from (2.9) that Γ is a Fourier multiplier operator with continuous and homogeneous of degree zero symbol, while $\phi \in C_0(\mathbb{R}^d)$. Therefore, by the first commutation lemma [23]

$$\mathsf{K}\boldsymbol{f} = \phi\Gamma\boldsymbol{f} - \Gamma(\phi\boldsymbol{f})$$

is a compact operator on $L^2(\mathbb{R}^d; \mathcal{T})$. In particular $\mathsf{K}\boldsymbol{E}_\epsilon \to \mathbf{0}$ in the L^2 norm. Hence,

$$\begin{aligned}
\lim_{\epsilon \to 0} \int_B \phi(\boldsymbol{x})(\boldsymbol{f}_\epsilon, \boldsymbol{g}_\epsilon)\, \mathrm{d}\boldsymbol{x} &= \lim_{\epsilon \to 0} \int_B \phi(\boldsymbol{x})\left(\tilde{\boldsymbol{f}}_\epsilon, \tilde{\boldsymbol{g}}_\epsilon\right) \mathrm{d}\boldsymbol{x} = \lim_{\epsilon \to 0} \int_{\mathbb{R}^d} \phi(\boldsymbol{E}_\epsilon, \boldsymbol{J}_\epsilon)\, \mathrm{d}\boldsymbol{x} \\
&= \lim_{\epsilon \to 0} \int_{\mathbb{R}^d} \phi(\Gamma\boldsymbol{E}_\epsilon, \boldsymbol{J}_\epsilon)\, \mathrm{d}\boldsymbol{x} \\
&= \lim_{\epsilon \to 0} \left\{ \int_{\mathbb{R}^d} (\Gamma(\phi\boldsymbol{E}_\epsilon), \boldsymbol{J}_\epsilon)\, \mathrm{d}\boldsymbol{x} + \int_{\mathbb{R}^d} (\mathsf{K}\boldsymbol{E}_\epsilon, \boldsymbol{J}_\epsilon)\, \mathrm{d}\boldsymbol{x} \right\} = 0.
\end{aligned}$$

The theorem is proved now. □

The compensated compactness theorem is remarkable, since the product of two weakly convergent sequences should not converge weakly to the product of the weak limits. The imposed constraints are too weak to force the sequences themselves to converge strongly, yet the orthogonal nature of the spaces $\mathcal{E}$ and $\mathcal{J}$ has the effect that in the product the convergences of the two sequences have ‘complimentary strengths' that compensate for each one's 'weakness'. In fact, for the homogenization theory in this book, it is only the compensated compactness property that is needed[1] to establish the classical results of homogenization. Thus, we make the following definition.

Definition 3.28. *We will say that the symbol* $\boldsymbol{\Gamma}(\boldsymbol{\eta})$ *has the compensated compactness property, if* $(\boldsymbol{E}_\epsilon, \boldsymbol{J}_\epsilon) \overset{*}{\rightharpoonup} 0$ *in the sense of measures on* B_1, *for every* $\boldsymbol{E}_\epsilon \in \mathcal{E}(B_1)$, $\boldsymbol{J}_\epsilon \in \mathcal{J}(B_1)$ *such that* $\boldsymbol{E}_\epsilon \rightharpoonup \mathbf{0}$, $\boldsymbol{J}_\epsilon \rightharpoonup \mathbf{0}$ *in* $L^2(B_1; \mathcal{T})$.

It is easy to see that the conclusion of theorem 3.27 holds if we replace the $\mathcal{E}$- and $\mathcal{J}$-regularity of the symbol $\boldsymbol{\Gamma}(\boldsymbol{\eta})$ with the compensated compactness property in the sense of definition 3.28. Indeed, in order to conclude that $(\boldsymbol{f}_\epsilon, \boldsymbol{g}_\epsilon) \overset{*}{\rightharpoonup} (\boldsymbol{f}_0, \boldsymbol{g}_0)$ in Ω it is enough to prove that the convergence holds in every ball in Ω, which follows from the convergence in the unit ball and scaling and translation invariance of the spaces $\mathcal{E}$ and $\mathcal{J}$. One only needs to verify that for any ball $B \subset \Omega$ one has $\Gamma_B \boldsymbol{f}_\epsilon \to \mathbf{0}$ and $\Pi_B \boldsymbol{g}_\epsilon \to \mathbf{0}$. This follows from the properties of the orthogonal projections

$$\Gamma_B \Gamma_\Omega = \Gamma_\Omega \Gamma_B = \Gamma_B, \qquad \Pi_B \Pi_\Omega = \Pi_\Omega \Pi_B = \Pi_B.$$

Theorem 3.29 (Compensated compactness criterion). *For* $\phi \in C_0(B_1)$ *we define the operator* $K_\phi\colon \mathcal{E}(B_1) \to \overline{\mathcal{J}(B_1)}$ *by the formula*

$$K_\phi \boldsymbol{f} = \phi \boldsymbol{f} - \Gamma_{B_1}(\phi \boldsymbol{f}).$$

Then, the symbol $\boldsymbol{\Gamma}(\boldsymbol{\eta})$ *has the compensated compactness property if and only if* K_ϕ *is a compact operator.*

Proof. Let us first assume that operators K_ϕ are compact. Then for $\boldsymbol{E}_\epsilon$, $\boldsymbol{J}_\epsilon$ as in definition 3.28 we have $K_\phi \boldsymbol{E}_\epsilon \to 0$, as $\epsilon \to 0$, in the L^2-norm and, hence, $(K_\phi \boldsymbol{E}_\epsilon, \boldsymbol{J}_\epsilon)_{L^2(B_1)} \to 0$. It remains to observe that

$$\left(K_\phi \boldsymbol{E}_\epsilon, \boldsymbol{J}_\epsilon\right)_{L^2(B_1)} = \left(\phi \boldsymbol{E}_\epsilon, \boldsymbol{J}_\epsilon\right)_{L^2(B_1)}.$$

Now, let us assume that the compensated compactness property holds. Let $\boldsymbol{E}_\epsilon \in \mathcal{E}(B_1)$, $\boldsymbol{E}_\epsilon \rightharpoonup 0$ in L^2, as $\epsilon \to 0$. We prove that K_ϕ is compact by showing that $K_\phi \boldsymbol{E}_\epsilon \to 0$ in the L^2-norm. To this end we observe that $K_\phi \boldsymbol{E}_\epsilon \in \overline{J(B_1)}$ and $K_\phi \boldsymbol{E}_\epsilon \rightharpoonup 0$ in L^2, as $\epsilon \to 0$. Therefore,

$$\|K_\phi \boldsymbol{E}_\epsilon\|^2_{L^2(B_1)} = \left(\phi \boldsymbol{E}_\epsilon, K_\phi \boldsymbol{E}_\epsilon\right)_{L^2(B_1)} \to 0,$$

as $\epsilon \to 0$, by the compensated compactness property. □

[1] We do not know if the compensated compactness property is genuinely weaker than $\mathcal{E}$ and $\mathcal{J}$-regularity of $\boldsymbol{\Gamma}(\boldsymbol{\eta})$.

Corollary 3.30. *The symbol* $\boldsymbol{\Gamma}(\boldsymbol{\eta})$ *has the compensated compactness property if and only if there exists a linear operator* $A : L^2(B_1; \mathcal{T}) \to \mathcal{E}_0(B_1)$, *such that the operators* $T_\phi : \mathcal{E}(B_1) \to L^2(B_1; \mathcal{T})$, *given by*

$$T_\phi \boldsymbol{f} = \phi \boldsymbol{f} - A(\phi \boldsymbol{f}),$$

are compact for every $\phi \in C_0(B_1)$.

Proof. If $\boldsymbol{\Gamma}(\boldsymbol{\eta})$ has the compensated compactness property, then $A = \Gamma_{B_1}$ has the required property by theorem 3.29.

To prove the converse, we observe that $\|K_\phi \boldsymbol{f}\|_{L^2(B_1)} \leqslant \|T_\phi \boldsymbol{f}\|_{L^2(B_1)}$ for every $\boldsymbol{f} \in \mathcal{E}(B_1)$. Thus, if $\boldsymbol{f}_\epsilon \rightharpoonup 0$, in $\mathcal{E}(B_1)$ as $\epsilon \to 0$, then, $\|K_\phi \boldsymbol{f}_\epsilon\|_{L^2(B_1)} \to 0$ by compactness of T_ϕ. Hence, K_ϕ is compact and the compensated compactness property follows from theorem 3.29. □

We will conclude the discussion of compensated compactness by showing that compensated compactness guarantees the density of compactly supported functions in $\mathcal{E}$. In other words, theorem 2.15 is valid with $\mathcal{J}$-regularity replaced by the compensated compactness property.

Theorem 3.31. *Suppose that the symbol* $\boldsymbol{\Gamma}(\boldsymbol{\eta})$ *has the compensated compactness property in the sense of definition 3.28. Then the set of compactly supported functions* $C(\mathcal{E})$, *given by (2.29) is dense in* $\mathcal{E}$. *The statement with* $\mathcal{E}$ *and* $\mathcal{J}$ *interchanged is also true.*

Proof. We begin by proving several simple lemmas about weak convergence in Lebesgue spaces.

Lemma 3.32. *Let* $\phi \in C(\overline{B_1})$. *We define*

$$\psi_n(\boldsymbol{x}) = \begin{cases} \phi\left(\dfrac{\boldsymbol{x}}{n}\right), & \boldsymbol{x} \in B_n, \\ 0, & \boldsymbol{x} \in \mathbb{R}^d \backslash B_n. \end{cases}$$

Then $\psi_n \stackrel{*}{\rightharpoonup} \phi(0)$, *as* $n \to \infty$, *in* $L^\infty(\mathbb{R}^d)$.

Proof. It is obvious that

$$\lim_{n\to\infty} \psi_n(\boldsymbol{x}) = \phi(0)$$

for every $\boldsymbol{x} \in \mathbb{R}^d$. Let $\eta \in L^1(\mathbb{R}^d)$ be arbitrary. Then $|\psi_n(\boldsymbol{x})\eta(\boldsymbol{x})| \leqslant \|\phi\|_{C(\overline{B_1})}|\eta(\boldsymbol{x})|$. Thus, by the Lebesgue dominated convergence theorem

$$\lim_{n\to\infty} \int_{\mathbb{R}^d} \psi_n(\boldsymbol{x})\eta(\boldsymbol{x})\,\mathrm{d}\boldsymbol{x} = \phi(0) \int_{\mathbb{R}^d} \eta(\boldsymbol{x})\,\mathrm{d}\boldsymbol{x}.$$

The lemma is proved. □

Lemma 3.33. *Let $\phi \in C(\overline{B_1})$. We define*

$$\psi_n(\boldsymbol{x}) = \begin{cases} n^{-d/2}\phi\left(\dfrac{\boldsymbol{x}}{n}\right), & \boldsymbol{x} \in B_n, \\ 0, & \boldsymbol{x} \in \mathbb{R}^d \backslash B_n. \end{cases}$$

Then $\psi_n \rightharpoonup 0$, as $n \to \infty$, in $L^2(\mathbb{R}^d)$.

Proof. It is easy to verify that ψ_n is a bounded sequence in $L^2(\mathbb{R}^d)$. It is obvious that

$$\lim_{n\to\infty} \psi_n(\boldsymbol{x}) = 0$$

for every $\boldsymbol{x} \in \mathbb{R}^d$. Let $\eta \in L^1(\mathbb{R}^d) \cap L^2(\mathbb{R}^d)$ be arbitrary. Then $|\psi_n(\boldsymbol{x})\eta(\boldsymbol{x})| \leqslant \|\phi\|_{C(\overline{B_1})} |\eta(\boldsymbol{x})|$. Thus, by the Lebesgue dominated convergence theorem

$$\lim_{n\to\infty} \int_{\mathbb{R}^d} \psi_n(\boldsymbol{x})\eta(\boldsymbol{x})\,\mathrm{d}\boldsymbol{x} = 0.$$

The statement of the lemma follows from the density of $L^1(\mathbb{R}^d) \cap L^2(\mathbb{R}^d)$ in $L^2(\mathbb{R}^d)$. □

Lemma 3.34. *Let $f \in L^2(\mathbb{R}^d)$. We define $g_n(z) = n^{d/2} f(nz)$, $z \in B_1$. Then $g_n \rightharpoonup 0$, as $n \to \infty$, in $L^2(B_1)$.*

Proof. It is easy to check that $\|g_n\|_{L^2(B_1)} \leqslant \|f\|_{L^2(\mathbb{R}^d)}$. For every $\phi \in C(\overline{B_1})$ we have

$$(g_n, \phi)_{L^2(B_1)} = n^{-d/2} \int_{B_n} f(\boldsymbol{x})\phi\left(\frac{\boldsymbol{x}}{n}\right) \mathrm{d}\boldsymbol{x} \to 0,$$

as $n \to \infty$, by lemma 3.33. The statement of the lemma follows from the density of $C(\overline{B_1})$ in $L^2(B_1)$. □

We are now ready to prove the theorem. Let $\boldsymbol{E}_0 \in \mathcal{E}$ be orthogonal to all functions in $C(\mathcal{E})$. We define a sequence $\boldsymbol{E}_n(z) = n^{d/2}\boldsymbol{E}_0(nz)$, $z \in B_1$. By assumption $\boldsymbol{E}_0|B_n$ is orthogonal to $\mathcal{E}_0(B_n)$. Therefore, $\boldsymbol{E}_0|B_n \in \overline{\mathcal{J}(B_n)}$. However, $\boldsymbol{E}_0|B_n \in \mathcal{E}(B_n)$, since $\boldsymbol{E}_0 \in \mathcal{E}$. Thus, $\boldsymbol{E}_n \in \overline{\mathcal{J}(B_1)} \cap \mathcal{E}(B_1)$ for every $n \geqslant 1$. By lemma 3.34, $\boldsymbol{E}_n \rightharpoonup 0$ in $L^2(B_1; \mathcal{T})$. But then, by the compensated compactness property of the symbol, we must have

$$\lim_{n\to\infty} \int_{B_1} \phi(z)(\boldsymbol{E}_n(z), \boldsymbol{E}_n(z))\,\mathrm{d}z = 0$$

for every $\phi \in C_0(B_1)$. However,

$$\left(\phi \boldsymbol{E}_n, \boldsymbol{E}_n\right)_{L^2(B_1)} = \int_{Bn} \phi\left(\frac{\boldsymbol{x}}{n}\right) |\boldsymbol{E}_0(\boldsymbol{x})|^2\,\mathrm{d}\boldsymbol{x},$$

and by lemma 3.32

$$\lim_{n\to\infty} \int_{Bn} \phi\left(\frac{\boldsymbol{x}}{n}\right) |\boldsymbol{E}_0(\boldsymbol{x})|^2\,\mathrm{d}\boldsymbol{x} = \phi(0)\|\boldsymbol{E}_0\|_{L^2(\mathbb{R}^d)}.$$

We conclude that $\boldsymbol{E}_0 = 0$ is the only choice that avoids a contradiction. □

Corollary 3.35. *Combining theorem 3.31 with part (d) of theorem 2.16 we deduce that the compensated compactness property of the symbol implies* $V_{\mathcal{E}} = \mathcal{T}_{\mathcal{E}}$.

3.3.2 Proof of the periodic homogenization theorem

We start by establishing the relation between the periodic and local versions of the spaces $\mathcal{E}$ and $\mathcal{J}$.

Lemma 3.36. *Suppose* $\boldsymbol{E} \in \mathcal{E}_{\text{per}}$ *and* $\boldsymbol{J} \in \mathcal{J}_{\text{per}}$. *Then for any bounded measurable subset* $\Omega \subset \mathbb{R}^d$

$$\lim_{\epsilon\to 0} \left\| \Pi_\Omega \boldsymbol{E}\left(\frac{\boldsymbol{x}}{\epsilon}\right) \right\|_{L^2(\Omega)} = \lim_{\epsilon\to 0} \left\| \Gamma_\Omega \boldsymbol{J}\left(\frac{\boldsymbol{x}}{\epsilon}\right) \right\|_{L^2(\Omega)} = 0,$$

where Γ_Ω *and* Π_Ω *are the orthogonal projections onto* $\mathcal{E}_0(\Omega)$ *and* $\mathcal{J}_0(\Omega)$, *respectively.*

Proof. The proof follows [23], Example 2.1. It is enough to prove only one of the statements in the lemma. The other is obtained by renaming $\mathcal{E}$ into $\mathcal{J}$ and vice versa. By a property of orthogonal projections

$$\left\| \Gamma_\Omega \boldsymbol{J}\left(\frac{\boldsymbol{x}}{\epsilon}\right) \right\|_{L^2(\Omega)} = \inf_{\boldsymbol{j}\in\mathcal{J}(\Omega)} \left\| \boldsymbol{J}\left(\frac{\boldsymbol{x}}{\epsilon}\right) - \boldsymbol{j} \right\|_{L^2(\Omega)}. \tag{3.48}$$

The idea is to construct a test function $\boldsymbol{j} = \boldsymbol{j}_\epsilon \in \mathcal{J}(\Omega)$ in (3.48), so that $\boldsymbol{J}(\boldsymbol{x}/\epsilon) - \boldsymbol{j}_\epsilon \to 0$, as $\epsilon \to 0$, in $L^2(\Omega; \mathcal{T})$. Since Ω is bounded, there exists $a > 0$, such that $\Omega \subset [-a/4, a/4]^d$. Let $\phi(\boldsymbol{x})$ be given by its Fourier transform:

$$\widehat{\phi}(\xi) = a^d \chi_{[-\frac{1}{a},\frac{1}{a}]^d}(\xi), \qquad \phi(x) = \prod_{j=1}^{d} \frac{a\sin(2\pi x_j/a)}{\pi x_j}.$$

It is easy to see that $(4/\pi)^d \leqslant \phi(x) \leqslant 2^d$ on $[-a/4, a/4]^d \supset \Omega$. Let $\psi \in C_0(\mathbb{R}^d)$ be such that $\psi(x) = 1/\phi(x)$ for all $x \in \Omega$. We now define

$$j_\epsilon(x) = J\left(\frac{x}{\epsilon}\right)\phi(x)\psi(x) - \Gamma\left(J\left(\frac{x}{\epsilon}\right)\phi(x)\psi(x)\right) \in \mathcal{J},$$

and therefore,

$$j_\epsilon(x)\Big|\Omega = J\left(\frac{x}{\epsilon}\right) - \Gamma\left(J\left(\frac{x}{\epsilon}\right)\phi(x)\psi(x)\right)\Big|\,\Omega \in \mathcal{J}(\Omega).$$

Using this as a test function in (3.48) we obtain

$$\left\| \Gamma_\Omega J\left(\frac{x}{\epsilon}\right) \right\|_{L^2(\Omega)} \leqslant \left\| \Gamma\left(J\left(\frac{x}{\epsilon}\right)\phi(x)\psi(x)\right) \right\|_{L^2(\Omega)} \leqslant \left\| \Gamma\left(J\left(\frac{x}{\epsilon}\right)\phi(x)\psi(x)\right) \right\|_{L^2(\mathbb{R}^d)}.$$

Let $\mathsf{K} = \Gamma\psi - \psi\Gamma$. By the first commutation lemma, [23], K is a compact operator on $L^2(\mathbb{R}^d; \mathcal{T})$. This would allow us to conclude that $\mathsf{K}(\phi(x)J(x/\epsilon)) \to 0$ in the norm of $L^2(\mathbb{R}^d; \mathcal{T})$, if we can show that $\phi(x)J(x/\epsilon) \rightharpoonup 0$, as $\epsilon \to 0$, in the weak topology of $L^2(\mathbb{R}^d; \mathcal{T})$. This is a consequence of the following lemma.

Lemma 3.37. *Suppose that $\phi \in L^2(\mathbb{R}^d)$, such that $\widehat{\phi}$ has a compact support. Let $f \in L^2_{\text{per}}(Q)$. Let $g_\epsilon(x) = \phi(x)f(x/\epsilon)$. Then $g_\epsilon \rightharpoonup \phi(x)\langle f\rangle$, as $\epsilon \to 0$, in the weak topology of $L^2(\mathbb{R}^d)$.*

Proof. The first observation is that it is sufficient to prove the lemma only for the case $\langle f\rangle = 0$. Let us first show that g_ϵ is bounded in $L^2(\mathbb{R}^d)$. Using the expansion

$$f(z) = \sum_{k\in\mathbb{Z}^d} \hat{f}(k)e^{2\pi i Qz\cdot k}$$

we compute

$$\hat{g}_\epsilon(\xi) = \sum_{k\in\mathbb{Z}^d} \hat{f}(k)\widehat{\phi}\left(\xi - \frac{Q^T k}{\epsilon}\right). \tag{3.49}$$

Then,

$$\| g_\epsilon \|^2_{L^2(\mathbb{R}^d)} = \| \hat{g}_\epsilon \|^2_{L^2(\mathbb{R}^d)} = \sum_{k\in\mathbb{Z}^d} \int_{\mathbb{R}^d} \left| \hat{f}(k)\widehat{\phi}\left(\xi - \frac{Q^T k}{\epsilon}\right)\right|^2 d\xi$$

because for all sufficiently small $\epsilon > 0$ functions $\widehat{\phi}(\boldsymbol{\xi} - \boldsymbol{Q}^T\boldsymbol{k}/\epsilon)$ and $\widehat{\phi}(\boldsymbol{\xi} - \boldsymbol{Q}^T\boldsymbol{m}/\epsilon)$ have disjoint support, whenever $\boldsymbol{m} \neq \boldsymbol{k}$. But then

$$\| g_\epsilon \|^2_{L^2(\mathbb{R}^d)} = \sum_{\boldsymbol{k}\in\mathbb{Z}^d} |\hat{f}(\boldsymbol{k})|^2 \| \widehat{\phi} \|^2_{L^2(\mathbb{R}^d)} = \det\boldsymbol{Q} \, \| f \|^2_{L^2(Q)} \| \phi \|^2_{L^2(\mathbb{R}^d)}$$

for all sufficiently small $\epsilon > 0$. In particular, g_ϵ is bounded in $L^2(\mathbb{R}^d)$.

It is well-known that the set of functions with compact support is dense in $L^2(\mathbb{R}^d)$. Therefore, the set of functions, whose Fourier transform has compact support, is also dense in $L^2(\mathbb{R}^d)$. Let ψ be such a function. Then, by Plancherel's formula and (3.49), we obtain

$$(g_\epsilon, \psi)_{L^2(\mathbb{R}^d)} = \sum_{\boldsymbol{k}\in\mathbb{Z}^d\setminus\{\boldsymbol{0}\}} \hat{f}(\boldsymbol{k}) \int_{\mathbb{R}^d} \widehat{\phi}\left(\boldsymbol{\xi} - \frac{\boldsymbol{Q}^T\boldsymbol{k}}{\epsilon}\right)\overline{\widehat{\psi}(\boldsymbol{\xi})}d\boldsymbol{\xi} = 0,$$

for all sufficiently small $\epsilon > 0$, because functions $\widehat{\phi}(\boldsymbol{\xi} - \boldsymbol{Q}^T\boldsymbol{k}/\epsilon)$ and $\widehat{\psi}(\boldsymbol{\xi})$ have disjoint support for any $\boldsymbol{k} \in \mathbb{Z}^d\setminus\{\boldsymbol{0}\}$. We conclude that $g_\epsilon \rightharpoonup 0$, as $\epsilon \to 0$, in the weak topology of $L^2(\mathbb{R}^d)$. □

Writing

$$\Gamma\left(\boldsymbol{J}\left(\frac{\boldsymbol{x}}{\epsilon}\right)\phi(\boldsymbol{x})\psi(\boldsymbol{x})\right) = \mathsf{K}\left(\boldsymbol{J}\left(\frac{\boldsymbol{x}}{\epsilon}\right)\phi(\boldsymbol{x})\right) + \psi(\boldsymbol{x})\Gamma\left(\boldsymbol{J}\left(\frac{\boldsymbol{x}}{\epsilon}\right)\phi(\boldsymbol{x})\right),$$

we see that by compactness of K and Lemma 3.37, the first term goes to 0 in the norm of $L^2(\mathbb{R}^d; \mathcal{T})$. Hence, in order to prove the lemma it remains to show that $\Gamma(\phi(\boldsymbol{x})\boldsymbol{J}(\boldsymbol{x}/\epsilon)) \to 0$, as $\epsilon \to 0$ in the norm of $L^2(\mathbb{R}^d; \mathcal{T})$. Replacing $\boldsymbol{J}(\boldsymbol{z})$ by its Fourier expansion (3.7) and applying the Plancherel identity we obtain

$$\left\| \Gamma\left(\boldsymbol{J}\left(\frac{\boldsymbol{x}}{c}\right)\phi(\boldsymbol{x})\right) \right\|^2_{L^2(\mathbb{R}^d)} = \int_{\mathbb{R}^d} \left| \Gamma\left(\frac{\boldsymbol{\xi}}{|\boldsymbol{\xi}|}\right) \sum_{\boldsymbol{k}\in\mathbb{Z}^d\setminus\{\boldsymbol{0}\}} \hat{\boldsymbol{J}}(\boldsymbol{k})\widehat{\phi}\left(\boldsymbol{\xi} - \frac{\boldsymbol{Q}^T\boldsymbol{k}}{\epsilon}\right) \right|^2 d\boldsymbol{\xi}.$$

Recall that for all sufficiently small $\epsilon > 0$ functions $\widehat{\phi}(\boldsymbol{\xi} - \boldsymbol{Q}^T\boldsymbol{k}/\epsilon)$ and $\widehat{\phi}(\boldsymbol{\xi} - \boldsymbol{Q}^T\boldsymbol{m}/\epsilon)$ have disjoint support, whenever $\boldsymbol{m} \neq \boldsymbol{k}$. Therefore,

$$\int_{\mathbb{R}^d} \left| \Gamma\left(\frac{\boldsymbol{\xi}}{|\boldsymbol{\xi}|}\right) \sum_{\boldsymbol{k}\in\mathbb{Z}^d\setminus\{\boldsymbol{0}\}} \hat{\boldsymbol{J}}(\boldsymbol{k})\widehat{\phi}\left(\boldsymbol{\xi} - \frac{\boldsymbol{Q}^T\boldsymbol{k}}{\epsilon}\right) \right|^2 d\boldsymbol{\xi}$$
$$= \sum_{\boldsymbol{k}\in\mathbb{Z}^d\setminus\{\boldsymbol{0}\}} \int_{\mathbb{R}^d} \left| \Gamma\left(\frac{\boldsymbol{\xi}}{|\boldsymbol{\xi}|}\right) \hat{\boldsymbol{J}}(\boldsymbol{k})\widehat{\phi}\left(\boldsymbol{\xi} - \frac{\boldsymbol{Q}^T\boldsymbol{k}}{\epsilon}\right) \right|^2 d\boldsymbol{\xi}.$$

Changing variables $\boldsymbol{\xi}' = \boldsymbol{\xi} - \boldsymbol{Q}^T\boldsymbol{k}/\epsilon$ in the integral we obtain

$$\left\| \Gamma\left(J\left(\frac{x}{\epsilon}\right)\phi(x)\right) \right\|^2_{L^2(\mathbb{R}^d)} = \sum_{k\in\mathbb{Z}^d\setminus\{0\}} \int_{\mathbb{R}^d} \left| \Gamma\left(\frac{Q^T k + \epsilon\xi'}{|Q^T k + \epsilon\xi'|}\right) \hat{J}(k)\hat{\phi}(\xi') \right|^2 d\xi'.$$

Since $\Gamma(\eta)$ is uniformly bounded on $\mathbb{S}^{d-1}$ we have

$$\left| \Gamma\left(\frac{Q^T k + \epsilon\xi'}{|Q^T k + \epsilon\xi'|}\right) \hat{J}(k)\hat{\phi}(\xi') \right| \leqslant C|\hat{J}(k)\hat{\phi}(\xi')|.$$

Observing that

$$\sum_{k\in\mathbb{Z}^d\setminus\{0\}} \int_{\mathbb{R}^d} |\hat{J}(k)\hat{\phi}(\xi')|^2 d\xi' = \det Q \|J\|^2_{L^2_{\text{per}}(Q)} \|\phi\|^2_{L^2(\mathbb{R}^d)} < +\infty,$$

we conclude that the Lebesgue dominated convergence theorem is applicable and hence

$$\lim_{\epsilon\to 0} \sum_{k\in\mathbb{Z}^d\setminus\{0\}} \int_{\mathbb{R}^d} \left| \Gamma\left(\frac{Q^T k + \epsilon\xi'}{|Q^T k + \epsilon\xi'|}\right) \hat{J}(k)\hat{\phi}(\xi') \right|^2 d\xi'$$
$$= \sum_{k\in\mathbb{Z}^d\setminus\{0\}} \int_{\mathbb{R}^d} \left| \Gamma\left(\frac{Q^T k}{|Q^T k|}\right) \hat{J}(k)\hat{\phi}(\xi') \right|^2 d\xi' = 0.$$

□

For a given $E_0 \in \mathcal{U}$, let $E \in \mathcal{E}_{\text{per}}$ be the unique solution of the cell problem

$$E \in \mathcal{E}_{\text{per}}, \qquad J(z) = \mathsf{L}^T(z)(E(z) + E_0), \qquad J \in \mathcal{J}_{\text{per}} \oplus \mathcal{U}. \tag{3.50}$$

We note that the cell problem (3.50) is slightly different from the cell problem (3.9). At the end of this section we will establish their equivalence. At the moment, however, the periodic cell problem in the form (3.50) will be more convenient.

Theorem 3.38 (Periodic homogenization). *Assume that the symbol $\Gamma(\eta)$ has the compensated compactness property in the sense of definition 3.28. Let $\mathsf{L}(z) \in L^\infty(\mathbb{R}^d; \text{End}(\mathcal{T}))$ be Q-periodic, satisfying $\mathsf{L}(z) \in \mathcal{M}(\alpha, \beta)$ for all $z \in Q$. Let Ω be an open and bounded domain in $\mathbb{R}^d$, and let $\mathsf{L}_\epsilon(x) = \mathsf{L}(x/\epsilon)$, $x \in \Omega$. Then L_ϵ H-converges to the constant tensor field $\mathsf{L}_* \in \text{End}(\mathcal{T})$ on Ω, defined by its action on the arbitrary vector $e_0 \in \mathcal{T}$ by*

$$\mathsf{L}_*^T e_0 = \frac{1}{|Q|} \int_Q \mathsf{L}^T(z)(E(z) + e_0)\, \mathrm{d}z, \tag{3.51}$$

where $E(z)$ solves the cell problem (3.50).

Proof. Here we present the elegant proof of Murat and Tartar [17] based on the compensated compactness property. We fix $f \in \mathcal{E}_0(\Omega)$ and let $E_\epsilon = \mathsf{T}^{-1}_{\mathsf{L}(x/\epsilon)} f$. The sequences E_ϵ and $J_\epsilon = \mathsf{L}(x/\epsilon)E_\epsilon$ are uniformly bounded in $L^2(\Omega; \mathcal{T})$ because of (2.18). By the Banach–Alaoglu theorem there is a subsequence ϵ_k such that $E_{\epsilon_k} \rightharpoonup E_*$ and $J_{\epsilon_k} \rightharpoonup J_*$ in $L^2(\Omega; \mathcal{T})$. We observe that $\Gamma_\Omega J_{\epsilon_k} = f$ and $\Pi_\Omega E_{\epsilon_k} = \mathbf{0}$. Observe that

$$\left(E\left(\frac{x}{\epsilon_k}\right) + e_0, J_{\epsilon_k}\right) = \left(\mathsf{L}^T\left(\frac{x}{\epsilon_k}\right)\left(E\left(\frac{x}{\epsilon_k}\right) + e_0\right), E_{\epsilon_k}\right).$$

By lemma 3.36 and the compensated compactness property we obtain

$$(e_0, J_*) = \left(\mathsf{L}_*^T e_0, E_*\right) = (e_0, \mathsf{L}_* E_*)$$

for any $e_0 \in \mathcal{T}$. Thus, $J_* = \mathsf{L}_* E_*$, and therefore, $E_* = \mathsf{T}_{\mathsf{L}_*}^{-1} f$. Observe that the weak limits E_* and J_* are independent of the choice of the subsequence ϵ_k. We conclude that the original families E_ϵ and J_ϵ have weak limits E_* and J_*, respectively. Hence, $\mathsf{L}(x/\epsilon)$ H-converges to L_*, and the theorem is proved. □

Corollary 3.39. *Let* L_* *be the effective tensor of the periodic composite with period cell* Q *and local tensor* $\mathsf{L}(x)$. *Then* L_* *is given by (3.10).*

Proof. By theorem 3.38 $\mathsf{L}^T(x/\epsilon) \overset{H}{\rightharpoonup} \mathsf{K}_*$ and $\mathsf{K}_*^T e_0 = \langle \mathsf{L}(z)(E(z) + e_0)\rangle$, for all $e_0 \in \mathcal{T}$, where $E(z)$ solves the cell problem (3.9). Let $E' \in \mathcal{E}_{\text{per}}$ be the solution of the cell problem (3.50) with e_0 replaced by e_0'. We use the observation that for any $\tilde{E} \in \mathcal{E}_{\text{per}} \oplus \mathcal{U}$ and any $\tilde{J} \in \mathcal{J}_{\text{per}} \oplus \mathcal{U}$ we have $\langle(\tilde{E}, \tilde{J})\rangle = (\langle\tilde{E}\rangle, \langle\tilde{J}\rangle)$. Therefore,

$$(e_0, \mathsf{L}_*^T e_0') = \left\langle (E + e_0, \mathsf{L}^T(E' + e_0'))\right\rangle = \left\langle (\mathsf{L}(E + e_0), E' + e_0')\right\rangle = (\mathsf{K}_*^T e_0, e_0').$$

We conclude that $\mathsf{L}_* = \mathsf{K}_*^T$. □

3.3.3 Basic properties of H-convergence

In this section we prove a number of fundamental properties of H-convergence, given by definition 3.1.

Theorem 3.40. *Assume that:*

(i) $\mathbf{\Gamma}(\eta)$ *is* $\mathcal{E}$*-nondegenerate.*
(ii) $\mathbf{\Gamma}(\eta)$ *satisfies the compensated compactness property.*
(iii) $\mathsf{L}_\epsilon \in \mathcal{M}(\alpha, \beta)^\Omega$.
(iv) $\mathsf{L}_\epsilon \overset{H}{\rightharpoonup} \mathsf{L}_* \in \mathcal{M}(\alpha, \beta)^\Omega$, as $\epsilon \to 0$.

Then the following statements are true.

(a) $\mathsf{L}_\epsilon^T \overset{H}{\rightharpoonup} \mathsf{L}_*^T$, *as* $\epsilon \to 0$ *on* Ω.

(b) *Boundary conditions do not affect H-limits. Suppose* $\omega \subset \Omega$, $\boldsymbol{E}_\epsilon \in \mathcal{E}(\omega)$, $\boldsymbol{E}_\epsilon \rightharpoonup \boldsymbol{E}_*$ *in* $L^2(\omega; \mathcal{T})$, $\boldsymbol{g} \in L^2(\omega; \mathcal{T})$ *and* $\Gamma_\omega(\mathsf{L}_\epsilon(\boldsymbol{g} + \boldsymbol{E}_\epsilon)) \to \boldsymbol{f}$ *in* $L^2(\omega; \mathcal{T})$. *Then* $\mathsf{L}_\epsilon(\boldsymbol{g} + \boldsymbol{E}_\epsilon) \rightharpoonup \mathsf{L}_*(\boldsymbol{g} + \boldsymbol{E}_*)$ *in* $L^2(\omega; \mathcal{T})$.

(c) *The H-limit is local. If* $\omega \subset \Omega$, *then* $\mathsf{L}_\epsilon \overset{H}{\rightharpoonup} \mathsf{L}_*$ *in* ω.

(d) *Suppose* Ω *is a disjoint union of a finite number of sets* ω_i, $i = 1, \ldots, N$ *in the sense that* $\omega_i \cap \omega_j = \varnothing$, *if* $i \neq j$ *and*

$$\left|\Omega \setminus \bigcup_{i=1}^{N} \omega_i\right| = 0.$$

Suppose that $\mathsf{L}_\epsilon \in \mathcal{M}(\alpha, \beta)^\Omega$ *and* $\mathsf{L}_\epsilon|\omega_i \overset{H}{\rightharpoonup} \mathsf{L}_i$ *in* ω_i, $i = 1, \ldots, N$. *Then* $\mathsf{L}_\epsilon \overset{H}{\rightharpoonup} \mathsf{L}_*$ *in* Ω, *where* $\mathsf{L}_*(\boldsymbol{x}) = \mathsf{L}_i(\boldsymbol{x})$, *if* $\boldsymbol{x} \in \omega_i$.

Proof. Part (a). Let $\{\boldsymbol{f}, \boldsymbol{g}\} \subset \mathcal{E}_0(\Omega)$ be arbitrary. Let $\boldsymbol{E}_\epsilon^{(T)} = \mathsf{T}^{-1}_{\mathsf{L}_\epsilon^T}\boldsymbol{f}$ and $\boldsymbol{E}_\epsilon = \mathsf{T}^{-1}_{\mathsf{L}_\epsilon}\boldsymbol{g}$. Then

$$\left(\boldsymbol{E}_\epsilon^{(T)}, \boldsymbol{g}\right)_{L^2(\Omega)} = \left(\boldsymbol{E}_\epsilon^{(T)}, \mathsf{T}_{\mathsf{L}_\epsilon}\boldsymbol{E}_\epsilon\right)_{L^2(\Omega)} = \left(\mathsf{T}_{\mathsf{L}_\epsilon^T}\boldsymbol{E}_\epsilon^{(T)}, \boldsymbol{E}_\epsilon\right)_{L^2(\Omega)} = (\boldsymbol{f}, \boldsymbol{E}_\epsilon)_{L^2(\Omega)}.$$

By definition of H-convergence

$$\lim_{\epsilon \to 0}\left(\boldsymbol{E}_\epsilon^{(T)}, \boldsymbol{g}\right)_{L^2(\Omega)} = \lim_{\epsilon \to 0}(\boldsymbol{f}, \boldsymbol{E}_\epsilon)_{L^2(\Omega)} = \left(\boldsymbol{f}, \mathsf{T}^{-1}_{\mathsf{L}_*}\boldsymbol{g}\right)_{L^2(\Omega)}.$$

Hence, $\boldsymbol{E}_\epsilon^{(T)} \rightharpoonup \boldsymbol{E}_*$, such that

$$(\boldsymbol{E}_*, \boldsymbol{g})_{L^2(\Omega)} = \left(\boldsymbol{f}, \mathsf{T}^{-1}_{\mathsf{L}_*}\boldsymbol{g}\right)_{L^2(\Omega)}.$$

Now we extract a weakly convergent subsequence $\boldsymbol{J}_{\epsilon_k}$ from $\boldsymbol{J}_\epsilon = \mathsf{L}_\epsilon^T\boldsymbol{E}_\epsilon^{(T)}$ and apply the compensated compactness property. On the one hand we have

$$\left(\mathsf{L}_{\epsilon_k}^T\boldsymbol{E}_{\epsilon_k}^{(T)}, \boldsymbol{E}_{\epsilon_k}\right) \overset{*}{\rightharpoonup} \left(\boldsymbol{J}_*, \mathsf{T}^{-1}_{\mathsf{L}_*}\boldsymbol{g}\right).$$

On the other hand

$$\left(\mathsf{L}_{\epsilon_k}^T\boldsymbol{E}_{\epsilon_k}^{(T)}, \boldsymbol{E}_{\epsilon_k}\right) = \left(\boldsymbol{E}_{\epsilon_k}^{(T)}, \mathsf{L}_{\epsilon_k}\boldsymbol{E}_{\epsilon_k}\right) \overset{*}{\rightharpoonup} \left(\boldsymbol{E}_*, \mathsf{L}_*\mathsf{T}^{-1}_{\mathsf{L}_*}\boldsymbol{g}\right).$$

We conclude that for a.e. $\boldsymbol{x} \in \Omega$

$$\left(\boldsymbol{J}_* - \mathsf{L}_*^T\boldsymbol{E}_*, \boldsymbol{h}\right) = 0 \text{ for any } \boldsymbol{h} \in \mathcal{E}_0(\Omega).$$

In particular,

$$f = \Gamma_\Omega(J_*) = \Gamma_\Omega\left(\mathsf{L}_*^T E_*\right) = \mathsf{T}_{\mathsf{L}_*^T} E_*.$$

We also have $J_* - \mathsf{L}_*^T E_* \in \mathcal{N}(\Omega)$, where $\mathcal{N}(\Omega)$ is defined in (2.30). By corollaries 2.17 and 3.35 and the $\mathcal{E}$-nondegeneracy we conclude that $\mathcal{N}(\Omega) = \{0\}$. Thus, the weak limit $J_* = \mathsf{L}_*^T E_*$ is independent of the weakly convergent subsequence of J_ϵ. Thus, the entire family $J_\epsilon \rightharpoonup J_*$ and therefore, $\mathsf{L}_\epsilon^T \overset{H}{\rightharpoonup} \mathsf{L}_*^T$.

Part (b). Fix any $E_0 \in \mathcal{E}_0(\Omega)$ and let $\tilde{E}_\epsilon \in \mathcal{E}_0(\Omega)$ be the solution of

$$\Gamma_\Omega\left(\mathsf{L}_\epsilon^T \tilde{E}_\epsilon\right) = \Gamma_\Omega\left(\mathsf{L}_*^T E_0\right).$$

By part (a) $\mathsf{L}_\epsilon^T \overset{H}{\rightharpoonup} \mathsf{L}_*^T$ and therefore, $\tilde{E}_\epsilon \rightharpoonup E_0$ and $\mathsf{L}_\epsilon^T \tilde{E}_\epsilon \rightharpoonup \mathsf{L}_*^T E_0$, as $\epsilon \to 0$. We now want to apply the compensated compactness theorem on ω. We have $X_0(\Omega)\tilde{E}_\epsilon \in \mathcal{E}$, and hence $\tilde{E}_\epsilon|\omega \in \mathcal{E}(\omega)$. Observe that $\mathsf{L}_\epsilon^T \tilde{E}_\epsilon - \mathsf{L}_*^T E_0 \in \overline{\mathcal{J}(\Omega)}$. Therefore, $(\mathsf{L}_\epsilon^T \tilde{E}_\epsilon - \mathsf{L}_*^T E_0)|\omega \in \overline{\mathcal{J}(\omega)}$, and thus

$$\Gamma_\omega(\mathsf{L}_\epsilon(g + E_\epsilon)) \to f, \quad \Pi_\omega\left(\tilde{E}_\epsilon\middle|\omega\right) = 0,$$

$$\Gamma_\omega\left(\mathsf{L}_\epsilon^T \tilde{E}_\epsilon\middle|\omega\right) = \Gamma_\omega\left(\mathsf{L}_*^T E_0\middle|\omega\right), \quad \Pi_\omega(g + E_\epsilon) = \Pi_\omega g.$$

The family $\mathsf{L}_\epsilon(g + E_\epsilon)$ is bounded in $L^2(\omega; \mathcal{T})$ therefore, there exists a sequence $\epsilon_k \to 0$ such that $\mathsf{L}_{\epsilon_k}(g + E_{\epsilon_k}) \rightharpoonup q$ in $L^2(\omega; \mathcal{T})$. By the compensated compactness theorem we have

$$\left(\mathsf{L}_{\epsilon_k}\left(g + E_{\epsilon_k}\right), \tilde{E}_{\epsilon_k}\middle|\omega\right) \overset{*}{\rightharpoonup} \left(q, E_0\middle|\omega\right).$$

But

$$\left(\mathsf{L}_{\epsilon_k}\left(g + E_{\epsilon_k}\right), \tilde{E}_{\epsilon_k}\middle|\omega\right) = \left(g + E_{\epsilon_k}, \mathsf{L}_{\epsilon_k}^T\left(\tilde{E}_{\epsilon_k}\middle|\omega\right)\right),$$

and by the compensated compactness theorem

$$\left(g + E_{\epsilon_k}, \mathsf{L}_{\epsilon_k}^T\left(\tilde{E}_{\epsilon_k}\middle|\omega\right)\right) \overset{*}{\rightharpoonup} \left(g + E_*, \mathsf{L}_*^T E_0\middle|\omega\right).$$

Thus, we have proved that

$$\left(q - \mathsf{L}_*(g + E_*), E_0\middle|\omega\right) = 0 \text{ for a.e. } x \in \omega.$$

Equivalently,

$$\left(X_0(\omega)\left(\boldsymbol{q} - \mathsf{L}_*(\boldsymbol{g} + \boldsymbol{E}_*)\right), \boldsymbol{E}_0\right) = 0 \text{ for a.e. } \boldsymbol{x} \in \Omega.$$

We conclude that $X_0(\omega)(\boldsymbol{q} - \mathsf{L}_*(\boldsymbol{g} + \boldsymbol{E}_*)) \in \mathcal{N}(\Omega)$. In part (a) we have proved that $\mathcal{N}(\Omega) = \{0\}$. Thus, the weak limit $\boldsymbol{q} = \mathsf{L}_*(\boldsymbol{g} + \boldsymbol{E}_*)$ is independent of the weakly convergent subsequence of $\mathsf{L}_\epsilon(\boldsymbol{g} + \boldsymbol{E}_\epsilon)$ on ω. Thus, the entire family $\mathsf{L}_\epsilon(\boldsymbol{g} + \boldsymbol{E}_\epsilon) \rightharpoonup \mathsf{L}_*(\boldsymbol{g} + \boldsymbol{E}_*)$, as $\epsilon \to 0$ in $L^2(\omega; \mathcal{T})$.

Part (c). Let $\boldsymbol{f} \in \mathcal{E}_0(\omega)$ and $\boldsymbol{E}_\epsilon \in \mathcal{E}_0(\omega)$ be the unique solution of $\Gamma_\omega(\mathsf{L}_\epsilon \boldsymbol{E}_\epsilon) = \boldsymbol{f}$. Then, the family $\boldsymbol{E}_\epsilon = \mathsf{T}_{\mathsf{L}\epsilon}^{-1}\boldsymbol{f}$ is bounded, since due to (2.18) we have

$$\|\boldsymbol{E}_\epsilon\| \leqslant \left\|\mathsf{T}_{\mathsf{L}\epsilon}^{-1}\right\| \|\boldsymbol{f}\| \leqslant \frac{1}{\alpha}\|\boldsymbol{f}\|.$$

It follows that we can extract a weakly convergent subsequence $\boldsymbol{E}_{\epsilon_k} \rightharpoonup \boldsymbol{E}_* \in \mathcal{E}_0(\omega)$, as $k \to \infty$ in $L^2(\omega; \mathcal{T})$. By part (b) we conclude that $\mathsf{L}_{\epsilon_k}\boldsymbol{E}_{\epsilon_k} \rightharpoonup \mathsf{L}_*\boldsymbol{E}_*$, as $k \to \infty$ in $L^2(\omega; \mathcal{T})$. Therefore, $\boldsymbol{E}_* \in \mathcal{E}_0(\omega)$ satisfies $\Gamma_\omega(\mathsf{L}_*\boldsymbol{E}_*) = \boldsymbol{f}$. Hence, the weak limit $\boldsymbol{E}_*$ of a subsequence $\boldsymbol{E}_{\epsilon_k}$ is uniquely determined by $\boldsymbol{f}$ and L_* and hence, the entire family $\boldsymbol{E}_\epsilon$ converges weakly to $\boldsymbol{E}_*$ in $L^2(\omega; \mathcal{T})$. Thus, $\mathsf{L}_\epsilon \overset{H}{\rightharpoonup} \mathsf{L}_*$ in ω.

Part (d). Let $\boldsymbol{f} \in \mathcal{E}_0(\Omega)$ and $\boldsymbol{E}_\epsilon \in \mathcal{E}_0(\Omega)$ be the unique solution of $\Gamma_\Omega(\mathsf{L}_\epsilon \boldsymbol{E}_\epsilon) = \boldsymbol{f}$. The family $\boldsymbol{E}_\epsilon$ is bounded in $L^2(\Omega; \mathcal{T})$. Therefore, we can extract a weakly convergent subsequence $\boldsymbol{E}_{\epsilon_k} \rightharpoonup \boldsymbol{E}_0$ as $k \to \infty$ in $L^2(\Omega; \mathcal{T})$. Observe that $\Gamma_{\omega_i}(\mathsf{L}_{\epsilon_k}\boldsymbol{E}_{\epsilon_k}|\omega_i) = \Gamma_{\omega_i}(\boldsymbol{f}|\omega_i)$. By part (b) $\mathsf{L}_{\epsilon_k}\boldsymbol{E}_{\epsilon_k}|\omega_i \rightharpoonup \mathsf{L}_i\boldsymbol{E}_0$ in $L^2(\omega_i; \mathcal{T})$. Therefore, $\mathsf{L}_{\epsilon_k}\boldsymbol{E}_{\epsilon_k} \rightharpoonup \mathsf{L}_*\boldsymbol{E}_0$ in $L^2(\Omega; \mathcal{T})$. But then $\boldsymbol{E}_0$ satisfies $\Gamma_\Omega(\mathsf{L}_*\boldsymbol{E}_0) = \boldsymbol{f}$. This determines $\boldsymbol{E}_0$ uniquely, since $\mathsf{L}_* \in \mathcal{M}(\alpha, \beta)^\Omega$. It follows that the entire family $\boldsymbol{E}_\epsilon$ weakly converges to $\boldsymbol{E}_0$ in $L^2(\Omega; \mathcal{T})$, since the limit of every weakly convergent subsequence must be $\boldsymbol{E}_0$. Hence, $\mathsf{L}_\epsilon \overset{H}{\rightharpoonup} \mathsf{L}_*$. □

3.3.4 H-compactness

Now that the basic properties of H-convergence are established we are ready to prove theorem 3.3 describing the remarkable compactness property of H-convergence. To be precise, we will prove sequential H-compactness of $\mathcal{M}(\alpha, \beta)^\Omega$. We note that by definition, H-topology on $\mathcal{M}(\alpha, \beta)^\Omega$ is a pull-back topology of the weak topology on $\mathcal{B}(\mathcal{E}_0(\Omega), \mathcal{E}_0(\Omega) \times L^2(\Omega; \mathcal{T}))$ via

$$\Phi : \mathcal{M}(\alpha, \beta)^\Omega \to \mathcal{B}\left(\mathcal{E}_0(\Omega), \mathcal{E}_0(\Omega) \times L^2(\Omega; \mathcal{T})\right), \quad \Phi(L)\boldsymbol{f} = \begin{bmatrix} \mathsf{T}_\mathsf{L}^{-1}\boldsymbol{f} \\ \mathsf{L}\mathsf{T}_\mathsf{L}^{-1}\boldsymbol{f} \end{bmatrix}$$

that maps $\mathcal{M}(\alpha, \beta)^\Omega$ into a norm-bounded subset of $\mathcal{B}(\mathcal{E}_0(\Omega), \mathcal{E}_0(\Omega) \times L^2(\Omega; \mathcal{T}))$, on which the weak topology is metrizable, since both $\mathcal{E}_0(\Omega)$ and $\mathcal{E}_0(\Omega) \times L^2(\Omega; \mathcal{T})$ are separable. Hence, the H-topology on $\mathcal{M}(\alpha, \beta)^\Omega$ is also metrizable. In this case, there is no difference between topological compactness and sequential compactness. With that said, we care only about the sequential compactness property. Here we restate theorem 3.3 with slightly weaker assumptions, showing that the weaker compensated compactness property of the Fourier multiplier symbol $\boldsymbol{\Gamma}(\boldsymbol{\eta})$ can replace $\mathcal{E}$-and $\mathcal{J}$-regularity in the theory of homogenization.

Theorem 3.41. *Assume that Ω is open and bounded, and*

(i) $\boldsymbol{\Gamma}(\boldsymbol{\eta})$ *is $\mathcal{E}$-nondegenerate.*

(ii) $\boldsymbol{\Gamma}(\boldsymbol{\eta})$ *satisfies the compensated compactness property.*

(iii) $\mathsf{L}_\epsilon \in \mathcal{M}(\alpha, \beta)^\Omega$.

Then there exists a subsequence $\epsilon_k \to 0$ and $\mathsf{L}_* \in \mathcal{M}(\alpha, \beta)^\Omega$, such that $\mathsf{L}_{\epsilon_k} \overset{H}{\rightharpoonup} \mathsf{L}_*$.

Proof.

Lemma 3.42. *Suppose $\mathsf{L}_\epsilon \in \mathcal{M}(\alpha, \beta)^\Omega$ and $\mathsf{A} \in \mathcal{B}(\mathcal{E}_0(\Omega))$ is such that $\mathsf{T}_{\mathsf{L}_\epsilon}^{-1} \rightharpoonup \mathsf{A}$ in the weak topology of $\mathcal{B}(\mathcal{E}_0(\Omega))$. Then for any $\boldsymbol{f} \in \mathcal{E}_0(\Omega)$ we have*

$$\|\mathsf{A}\| \leqslant \frac{1}{\alpha}, \qquad (\mathsf{A}\boldsymbol{f}, \boldsymbol{f})_{L^2(\Omega)} \geqslant \frac{\alpha}{\beta^2} \|\boldsymbol{f}\|^2.$$

In particular, $\mathsf{A} \in \mathcal{B}(\mathcal{E}_0(\Omega))$ is invertible.

Proof. By (2.18) we have $\|\mathsf{T}_{\mathsf{L}_\epsilon}^{-1}\| \leqslant 1/\alpha$. Compactness of the unit ball in $\mathcal{B}(\mathcal{E}_0(\Omega))$ in the weak topology implies that $\|\mathsf{A}\| \leqslant 1/\alpha$.

Observe that for any $\mathsf{L} \in \mathcal{M}(\alpha, \beta)$ and any $\boldsymbol{e} \in \mathcal{T}$ we have

$$|\mathsf{L}\boldsymbol{e}| \leqslant \beta|\boldsymbol{e}|, \qquad (\mathsf{L}\boldsymbol{e}, \boldsymbol{e}) \geqslant \alpha|\boldsymbol{e}|^2.$$

Therefore, for any $\mathsf{L} \in \mathcal{M}(\alpha, \beta)^\Omega$ and any $\boldsymbol{E} \in \mathcal{E}_0(\Omega)$

$$\|\mathsf{T}_\mathsf{L}\boldsymbol{E}\| = \|\Gamma_\Omega(\mathsf{L}\boldsymbol{E})\| \leqslant \|\mathsf{L}\boldsymbol{E}\| \leqslant \beta\|\boldsymbol{E}\|$$

and

$$(\mathsf{T}_\mathsf{L}\boldsymbol{E}, \boldsymbol{E})_{L^2(\Omega)} = (\mathsf{L}\boldsymbol{E}, \boldsymbol{E})_{L^2(\Omega)} \geqslant \alpha\, \|\boldsymbol{E}\|^2.$$

Then, for any $\boldsymbol{f} \in \mathcal{E}_0(\Omega)$ we set $\boldsymbol{E} = \mathsf{T}_\mathsf{L}^{-1}\boldsymbol{f}$ and obtain

$$\|\mathsf{T}_\mathsf{L}^{-1}\boldsymbol{f}\| \geqslant \frac{1}{\beta}\|\boldsymbol{f}\|.$$

Hence, combining the two inequalities, we obtain

$$\left(\mathsf{T}_{\mathsf{L}}^{-1}\boldsymbol{f},\boldsymbol{f}\right)_{L^2(\Omega)} = \left(\mathsf{T}_{\mathsf{L}}^{-1}\boldsymbol{f},\ \mathsf{T}_{\mathsf{L}}\mathsf{T}_{\mathsf{L}}^{-1}\boldsymbol{f}\right)_{L^2(\Omega)} \geqslant \alpha\ \|\mathsf{T}_{\mathsf{L}}^{-1}\boldsymbol{f}\|^2 \geqslant \frac{\alpha}{\beta^2}\ \|\boldsymbol{f}\|^2,$$

which holds for any $\mathsf{L} \in \mathcal{M}(\alpha, \beta)^{\Omega}$ and any $\boldsymbol{f} \in \mathcal{E}_0(\Omega)$. This implies that

$$(\mathsf{A}\boldsymbol{f},\boldsymbol{f})_{L^2(\Omega)} = \lim_{\epsilon\to 0}\left(\mathsf{T}_{\mathsf{L}\epsilon}^{-1}\boldsymbol{f},\boldsymbol{f}\right) \geqslant \frac{\alpha}{\beta^2}\ \|\boldsymbol{f}\|^2.$$

□

We now take a domain Ω' such that $\Omega \subset \Omega'$. We define

$$\mathsf{K}_\epsilon(\boldsymbol{x}) = \begin{cases} \mathsf{L}_\epsilon(\boldsymbol{x}), & \boldsymbol{x} \in \Omega, \\ \alpha\boldsymbol{I}_{\mathcal{T}}, & \boldsymbol{x} \in \Omega'\backslash\Omega. \end{cases}$$

Clearly, $\mathsf{K}_\epsilon \in \mathcal{M}(\alpha, \beta)^{\Omega'}$. We construct the sequence ϵ_k as a subsequence for which the sequences of operators $\mathsf{T}_{\mathsf{K}_{\epsilon_k}}^{-1}$, $\mathsf{T}_{\mathsf{K}_{\epsilon_k}^T}^{-1}$, $\mathsf{K}_{\epsilon_k}\mathsf{T}_{\mathsf{K}_{\epsilon_k}}^{-1}$ and $\mathsf{K}_{\epsilon_k}^T\mathsf{T}_{\mathsf{K}_{\epsilon_k}^T}^{-1}$ converge weakly in $\mathcal{B}(\mathcal{E}_0(\Omega'))$ and $\mathcal{B}(\mathcal{E}_0(\Omega'), L^2(\Omega'; \mathcal{T}))$, respectively. Such a subsequence exists, since the Hilbert spaces $L^2(\Omega'; \mathcal{T})$ and $\mathcal{E}_0(\Omega')$ are separable. Thus,

$$\mathsf{T}_{\mathsf{K}_{\epsilon_k}}^{-1} \rightharpoonup \mathsf{A}_1, \quad \mathsf{T}_{\mathsf{K}_{\epsilon_k}^T}^{-1} \rightharpoonup \mathsf{A}_2, \quad \mathsf{K}_{\epsilon_k}\mathsf{T}_{\mathsf{K}_{\epsilon_k}}^{-1} \rightharpoonup \mathsf{A}_3, \quad \mathsf{K}_{\epsilon_k}^T\mathsf{T}_{\mathsf{K}_{\epsilon_k}^T}^{-1} \rightharpoonup \mathsf{A}_4, \quad k \to \infty$$

in $\mathcal{B}(\mathcal{E}_0(\Omega'))$ and $\mathcal{B}(\mathcal{E}_0(\Omega'), L^2(\Omega'; \mathcal{T}))$, respectively. We now fix arbitrary fields $\{\boldsymbol{f}_1, \boldsymbol{f}_2\} \subset \mathcal{E}_0(\Omega')$ and let $\{\boldsymbol{E}_k^{(1)}, \boldsymbol{E}_k^{(1)}\} \subset \mathcal{E}_0(\Omega')$ be the unique solutions of

$$\Gamma_{\Omega'}\left(\mathsf{K}_{\epsilon_k}\boldsymbol{E}_k^{(1)}\right) = \mathsf{A}_1^{-1}\boldsymbol{f}_1, \qquad \Gamma_{\Omega'}\left(\mathsf{K}_{\epsilon_k}^T\boldsymbol{E}_k^{(2)}\right) = \mathsf{A}_2^{-1}\boldsymbol{f}_2.$$

Here, the existence of bounded operators A_1^{-1} and A_2^{-1} is guaranteed by lemma 3.42. We observe that

$$\boldsymbol{E}_k^{(1)} = \mathsf{T}_{\mathsf{K}_{\epsilon_k}}^{-1}\mathsf{A}_1^{-1}\boldsymbol{f}_1 \rightharpoonup \boldsymbol{f}_1, \qquad \boldsymbol{E}_k^{(2)} = \mathsf{T}_{\mathsf{K}_{\epsilon_k}^T}^{-1}\mathsf{A}_2^{-1}\boldsymbol{f}_2 \rightharpoonup \boldsymbol{f}_2.$$

$$\mathsf{K}_{\epsilon_k}\boldsymbol{E}_k^{(1)} = \mathsf{K}_{\epsilon_k}\mathsf{T}_{\mathsf{K}_{\epsilon_k}}^{-1}\mathsf{A}_1^{-1}\boldsymbol{f}_1 \rightharpoonup \mathsf{A}_3\mathsf{A}_1^{-1}\boldsymbol{f}_1, \quad \mathsf{K}_{\epsilon_k}^T\boldsymbol{E}_k^{(2)} = \mathsf{K}_{\epsilon_k}^T\mathsf{T}_{\mathsf{K}_{\epsilon_k}^T}^{-1}\mathsf{A}_2^{-1}\boldsymbol{f}_2 \rightharpoonup \mathsf{A}_4\mathsf{A}_2^{-1}\boldsymbol{f}_2$$

in $L^2(\Omega'; \mathcal{T})$.

We now apply the compensated compactness theorem on Ω. We have

$$\Gamma_\Omega\left(\mathsf{K}_{\epsilon_k}\boldsymbol{E}_k^{(1)}\right) = \Gamma_\Omega\left(\mathsf{A}_1^{-1}\boldsymbol{f}_1\right), \qquad \Pi_\Omega\boldsymbol{E}_k^{(1)} = 0,$$

$$\Gamma_\Omega\left(\mathsf{K}_{\epsilon_k}^T\boldsymbol{E}_k^{(2)}\right) = \Gamma_\Omega\left(\mathsf{A}_2^{-1}\boldsymbol{f}_2\right), \qquad \Pi_\Omega\boldsymbol{E}_k^{(2)} = 0.$$

Thus, by the compensated compactness theorem

$$\left(\mathsf{K}_{\epsilon_k}\boldsymbol{E}_k^{(1)}, \boldsymbol{E}_k^{(2)}\right) \stackrel{*}{\rightharpoonup} \left(\mathsf{A}_3\mathsf{A}_1^{-1}f_1, f_2\right)$$

in the sense of measures on Ω. However,

$$\left(\mathsf{K}_{\epsilon_k}\boldsymbol{E}_k^{(1)}, \boldsymbol{E}_k^{(2)}\right) = \left(\boldsymbol{E}_k^{(1)}, \mathsf{K}_{\epsilon_k}^T\boldsymbol{E}_k^{(2)}\right).$$

Applying the compensated compactness theorem again we obtain

$$\left(\boldsymbol{E}_k^{(1)}, \mathsf{K}_{\epsilon_k}^T\boldsymbol{E}_k^{(2)}\right) \stackrel{*}{\rightharpoonup} \left(f_1, \mathsf{A}_4\mathsf{A}_2^{-1}f_2\right).$$

Thus,

$$\left(\mathsf{A}_3\mathsf{A}_1^{-1}f_1, f_2\right) = \left(f_1, \mathsf{A}_4\mathsf{A}_2^{-1}f_2\right)$$

for a.e. $\boldsymbol{x} \in \Omega$. Hence, for a.e. $\boldsymbol{x} \in \Omega$ this is a bilinear form on $V_{\mathcal{E}}$. Under our assumptions $V_{\mathcal{E}} = \mathcal{T}$, by corollary 3.35. Thus, there exists a measurable function $\mathsf{L}_* : \Omega \to \mathrm{End}(\mathcal{T})$ such that for a.e. $\boldsymbol{x} \in \Omega$

$$\left(\mathsf{A}_3\mathsf{A}_1^{-1}f_1, f_2\right) = \left(f_1, \mathsf{A}_4\mathsf{A}_2^{-1}f_2\right) = \left(\mathsf{L}_*f_1, f_2\right).$$

In other words, we have proved that

$$\mathsf{L}_{\epsilon_k}\boldsymbol{E}_k^{(1)} \rightharpoonup \mathsf{L}_*f_1, \qquad \mathsf{L}_{\epsilon_k}^T\boldsymbol{E}_k^{(2)} \rightharpoonup \mathsf{L}_*^Tf_2,$$

in $L^2(\Omega; \mathcal{T})$.

Let us show that $\mathsf{L}_* \in \mathcal{M}(\alpha, \beta)^\Omega$. We have

$$\left(\mathsf{L}_{\epsilon_k}\boldsymbol{E}_k^{(1)}, \boldsymbol{E}_k^{(1)}\right) \geqslant \alpha\left|\boldsymbol{E}_k^{(1)}\right|^2 \geqslant \alpha\left(2\left(\boldsymbol{E}_k^{(1)}, f_1\right) - |f_1|^2\right).$$

Passing to the limit as $k \to \infty$ and using the compensated compactness theorem we obtain

$$\left(\mathsf{L}_*f_1, f_1\right) \geqslant \alpha|f_1|^2 \quad \text{for all } f_1 \in \mathcal{E}_0(\Omega').$$

Since $V_{\mathcal{E}} = \mathcal{T}$, for any $\boldsymbol{e} \in \mathcal{T}$ and $\boldsymbol{x}_0 \in \Omega$ there is $f_1 \in \mathcal{E}_0(\Omega')$ such that $f_1(\boldsymbol{x}_0) = \boldsymbol{e}$ and $\boldsymbol{x}_0$ is a Lebesgue point of f_1. Thus for a.e. $\boldsymbol{x} \in \Omega$ and every $\boldsymbol{e} \in \mathcal{T}$ we have

$$(\mathsf{L}_*(\boldsymbol{x})\boldsymbol{e}, \boldsymbol{e}) \geqslant \alpha|\boldsymbol{e}|^2.$$

To prove the other inequality in the definition of the set $\mathcal{M}(\alpha, \beta)$ we let

$$\boldsymbol{J}_k = \mathsf{L}_{\epsilon_k}\boldsymbol{E}_k^{(1)}.$$

Recall that $J_k \rightharpoonup \mathsf{L}_* f_1$ in $L^2(\Omega; \mathcal{T})$. We have

$$\left(\mathsf{L}_{\epsilon_k} E_k^{(1)}, E_k^{(1)}\right) = \left(J_k, \mathsf{L}_{\epsilon_k}^{-1} J_k\right) \geqslant \frac{1}{\beta}|J_k|^2 \geqslant \frac{1}{\beta}\left(2(J_k, \mathsf{L}_* f_1) - |\mathsf{L}_* f_1|^2\right).$$

Passing to the limit in the above inequality we obtain

$$\left(\mathsf{L}_* f_1, f_1\right) \geqslant \frac{1}{\beta}|\mathsf{L}_* f_1|^2 \text{ for all } f_1 \in \mathcal{E}_0(\Omega').$$

As before, we conclude that for a.e. $x \in \Omega$ and every $e \in \mathcal{T}$ we have

$$\left(\mathsf{L}_*(x)e, e\right) \geqslant \frac{1}{\beta}|\mathsf{L}_* e|^2,$$

or equivalently,

$$\left(\mathsf{L}_*^{-1}(x)j, j\right) \geqslant \frac{1}{\beta}|j|^2, \text{ for all } j \in \mathcal{T}.$$

Hence, $\mathsf{L}_* \in \mathcal{M}(\alpha, \beta)^{\Omega}$.

To complete the proof of the compactness theorem we need to show that $\mathsf{L}_{\epsilon_k} \overset{H}{\rightharpoonup} \mathsf{L}_*$, as $k \to \infty$. For that purpose we fix an arbitrary $f \in \mathcal{E}_0(\Omega)$ and define $E_k \in \mathcal{E}_0(\Omega)$ as the unique solution of

$$\Gamma_\Omega\left(\mathsf{L}_{\epsilon_k} E_k\right) = f.$$

The sequences $\{E_k : k \geqslant 1\}$ and $\{J_k = \mathsf{L}_{\epsilon_k} E_k : k \geqslant 1\}$ are bounded in $L^2(\Omega; \mathcal{T})$. Therefore, there exists a subsequence k' such that $E_{k'} \rightharpoonup E_*$ and $J_{k'} \rightharpoonup J_*$ in $L^2(\Omega; \mathcal{T})$. Observe that

$$\left(E_{k'}^{(2)}, \mathsf{L}_{\epsilon_{k'}} E_{k'}\right) = \left(\mathsf{L}_{\epsilon_{k'}}^T E_{k'}^{(2)}, E_{k'}\right).$$

By the compensated compactness theorem we obtain

$$\left(f_2, J_*\right) = \left(\mathsf{L}_*^T f_2, E_*\right) = \left(f_2, \mathsf{L}_* E_*\right).$$

We conclude, taking into account that $V_{\mathcal{E}} = \mathcal{T}$, that $J_* = \mathsf{L}_* E_*$. But then $E_* \in \mathcal{E}_0(\Omega)$ is determined uniquely by the equation

$$f = \Gamma_\Omega J_* = \Gamma_\Omega(\mathsf{L}_* E_*), \tag{3.52}$$

since $\mathsf{L}_* \in \mathcal{M}(\alpha, \beta)^{\Omega}$. Since the limits E_* and J_* do not depend on the choice of the weakly convergent subsequence of E_k and J_k, we conclude that $E_k \rightharpoonup E_*$ and $\mathsf{L}_{\epsilon_k} E_k \rightharpoonup \mathsf{L}_* E_*$, where E_* solves (3.52). Thus, $\mathsf{L}_{\epsilon_k} \overset{H}{\rightharpoonup} \mathsf{L}_*$, as $k \to \infty$ by definition of H-convergence. □

Corollary 3.43. *The sets $\mathcal{M}(\alpha, \beta)$ are G-closed for all $0 < \alpha < \beta$.*

The corollary is an immediate consequence of theorems 3.38 and 3.41.

3.3.5 H-convergence and G-closure

So far we have regarded periodic homogenization as one specific example of the application of a more general H-convergence theory (see section 3.3.2). However, the similarity of results in these theories suggested that there should be a deeper connection between the two. This connection was first conjectured by Kohn and DalMaso and proved by Tartar [24] and generalized by Raitums [18] for nonlinear monotone constitutive relations. See also [1]. In this book we restate this theorem in the language of the abstract Hilbert space formalism. Under some technical assumptions on the symbol $\boldsymbol{\Gamma}(\boldsymbol{\eta})$ the set of all H-limits of composites made with materials in U consists of all measurable functions with values in $G(U)$. We begin the proof of this result with a lemma that underscores the intimate relation between the periodic and general composites.

Lemma 3.44. *Let Q be a parallelepiped of periods. Let $\boldsymbol{J} \in \mathcal{J}_{\text{per}}$ and $\boldsymbol{E} \in \mathcal{E}_{\text{per}}$, where $\mathcal{E}_{\text{per}}$ and $\mathcal{J}_{\text{per}}$ are defined in (3.5) and (3.6). Then $\boldsymbol{J}|Q \in \overline{\mathcal{J}(Q)}$ and $\boldsymbol{E}|Q \in \overline{\mathcal{E}(Q)}$. In other words, $\Gamma_Q \boldsymbol{J} = \Pi_Q \boldsymbol{E} = 0$, where Γ_Q and Π_Q are the orthogonal projections onto $\mathcal{E}_0(Q)$ and $\mathcal{J}_0(Q)$, respectively.*

Proof. $\boldsymbol{J} \in \mathcal{J}_{\text{per}}$ means that

$$\boldsymbol{J}(\boldsymbol{x}) = \sum_{\boldsymbol{k} \in \mathbb{Z}^d \setminus \{\boldsymbol{0}\}} \widehat{\boldsymbol{J}}(\boldsymbol{k}) \mathrm{e}^{2\pi i Q z \cdot \boldsymbol{k}}, \quad \boldsymbol{\Gamma}\left(\frac{Q^T \boldsymbol{k}}{\left|Q^T \boldsymbol{k}\right|}\right) \widehat{\boldsymbol{J}}(\boldsymbol{k}) = \boldsymbol{0}, \quad \boldsymbol{k} \in \mathbb{Z} \setminus \{\boldsymbol{0}\}. \tag{3.53}$$

Let $\boldsymbol{E} \in \mathcal{E}_0(Q)$ be arbitrary. By the Parceval identity

$$\int_Q (\boldsymbol{J}(\boldsymbol{x}), \boldsymbol{E}(\boldsymbol{x}))\, \mathrm{d}\boldsymbol{x} = \sum_{\boldsymbol{k} \in \mathbb{Z}^d \setminus \{\boldsymbol{0}\}} \left(\widehat{\boldsymbol{J}}(\boldsymbol{k}), \boldsymbol{E}_{\boldsymbol{k}}\right), \tag{3.54}$$

where

$$\boldsymbol{E}_k = \frac{1}{|Q|} \int_Q \boldsymbol{E}(\boldsymbol{x}) \mathrm{e}^{-2\pi i Q x \cdot \boldsymbol{k}}\, \mathrm{d}\boldsymbol{x}, \quad \boldsymbol{k} \in \mathbb{Z}^d.$$

Now we observe that

$$|Q| \boldsymbol{E}_k = \int_{\mathbb{R}^d} (X_0(Q) \boldsymbol{E})(\boldsymbol{x}) \mathrm{e}^{-2\pi i Q x \cdot \boldsymbol{k}}\, \mathrm{d}\boldsymbol{x} = \widehat{X_0(Q)\boldsymbol{E}}\left(Q^T \boldsymbol{k}\right).$$

By assumption $X_0(Q)\boldsymbol{E} \in \mathcal{E}$ and therefore,

$$|Q| \boldsymbol{\Gamma}\left(\frac{Q^T \boldsymbol{k}}{\left|Q^T \boldsymbol{k}\right|}\right) \boldsymbol{E}_k = \boldsymbol{\Gamma}\left(\frac{Q^T \boldsymbol{k}}{\left|Q^T \boldsymbol{k}\right|}\right) \widehat{X_0(Q)\boldsymbol{E}}\left(Q^T \boldsymbol{k}\right) = \widehat{X_0(Q)\boldsymbol{E}}\left(Q^T \boldsymbol{k}\right) = |Q| \boldsymbol{E}_k. \tag{3.55}$$

Hence, from (3.54) and (3.55) we obtain

$$(\boldsymbol{J}, \boldsymbol{E})_{L^2_{\rm per}(Q)} = \sum_{\boldsymbol{k}\in\mathbb{Z}^d\setminus\{\boldsymbol{0}\}} \left(\hat{\boldsymbol{J}}(\boldsymbol{k}), \boldsymbol{E}_k\right) = \sum_{\boldsymbol{k}\in\mathbb{Z}^d\setminus\{\boldsymbol{0}\}} \left(\hat{\boldsymbol{J}}(\boldsymbol{k}), \boldsymbol{\Gamma}\left(\frac{Q^T\boldsymbol{k}}{\left|Q^T\boldsymbol{k}\right|}\right)\boldsymbol{E}_k\right)$$

$$= \sum_{\boldsymbol{k}\in\mathbb{Z}^d\setminus\{\boldsymbol{0}\}} \left(\boldsymbol{\Gamma}\left(\frac{Q^T\boldsymbol{k}}{\left|Q^T\boldsymbol{k}\right|}\right)\hat{\boldsymbol{J}}(\boldsymbol{k}), \boldsymbol{E}_k\right) = 0,$$

where the last equality is due to (3.53). Thus, we have shown that $\boldsymbol{J} \in \mathcal{E}_0(Q)^\perp = \overline{\mathcal{J}(Q)}$. The statement about $\mathcal{E}_{\rm per}$ is proved in the same way. □

It is natural to use composite materials as constituents of new periodic composites. These 'second generation' composites should still be regarded as composites made with the same set of materials with which the original composites were made. The following iterated homogenization lemma shows that the notion of H-convergence supports this intuition mathematically.

Lemma 3.45 (Iterated homogenization). *Assume that the symbol* $\boldsymbol{\Gamma}(\boldsymbol{\eta})$ *and the family of local tensors* L_ϵ *satisfy all conditions of theorem* 3.40 *with* Ω *replaced by the parallelepiped of periods* Q*. For every* $\epsilon > 0$ *let* $\mathsf{K}_*(\epsilon)$ *be the effective tensor of a periodic composite with period cell* Q *and the local tensor* $\mathsf{L}_\epsilon(\boldsymbol{x})$*. Let* K_* *be the effective tensor of a periodic composite with period cell* Q *and the local tensor* $\mathsf{L}_*(\boldsymbol{x})$*. Then*

$$\lim_{\epsilon\to 0} \mathsf{K}_*(\epsilon) = \mathsf{K}_*.$$

In other words

$$\lim_{\epsilon\to 0}\ \mathop{\rm Hlim}_{\delta\to 0} \mathsf{L}_\epsilon\left(\frac{\boldsymbol{x}}{\delta}\right) = \mathop{\rm Hlim}_{\delta\to 0}\ \mathop{\rm Hlim}_{\epsilon\to 0} \mathsf{L}_\epsilon\left(\frac{\boldsymbol{x}}{\delta}\right) = \mathsf{K}_*.$$

Proof. According to theorem 3.38

$$\mathsf{K}_*^T(\epsilon)\boldsymbol{e}_0 = \frac{1}{|Q|}\int_Q \mathsf{L}_\epsilon^T(\boldsymbol{z})(\boldsymbol{E}_\epsilon(\boldsymbol{z}) + \boldsymbol{e}_0)\,\mathrm{d}\boldsymbol{z}, \tag{3.56}$$

where $\boldsymbol{E}_\epsilon$ is the unique solution of the periodic cell problem

$$\boldsymbol{E}_\epsilon \in \mathcal{E}_{\rm per}, \qquad \boldsymbol{J}_\epsilon(\boldsymbol{z}) = \mathsf{L}_\epsilon^T(\boldsymbol{z})(\boldsymbol{E}_\epsilon(\boldsymbol{z}) + \boldsymbol{e}_0), \qquad \boldsymbol{J}_\epsilon \in \mathcal{J}_{\rm per} \oplus \mathcal{U}.$$

Let ϵ_k be a subsequence such that $\boldsymbol{E}_{\epsilon_k} \rightharpoonup \boldsymbol{E}_*$ in $L^2_{\rm per}(Q, \mathcal{T})$ and $\mathsf{K}_*^T(\epsilon_k)\boldsymbol{e}_0 \to \boldsymbol{j}_*$ in $\mathcal{T}$. By lemma 3.44 we have $\boldsymbol{J}_{\epsilon_k} \in \overline{\mathcal{J}(Q)}$ and

$$\Gamma_Q\left(\mathsf{L}_{\epsilon_k}^T(\boldsymbol{z})\left(\boldsymbol{E}_{\epsilon_k}(\boldsymbol{z}) + \boldsymbol{e}_0\right)\right) = \Gamma_Q\left(\mathsf{K}_*^T(\epsilon_k)\boldsymbol{e}_0\right) \to \Gamma_Q(\boldsymbol{j}_*).$$

By part (a) of theorem 3.40 we have $\mathsf{L}_\epsilon^T \overset{H}{\rightharpoonup} \mathsf{L}_*^T$ and then part (b) of theorem 3.40 applies: $\mathsf{L}_{\epsilon_k}^T(\boldsymbol{E}_{\epsilon_k} + \boldsymbol{e}_0) \rightharpoonup \mathsf{L}_*^T(\boldsymbol{E}_* + \boldsymbol{e}_0)$. Hence, the weak limit $\boldsymbol{E}_*$ is the solution of the cell problem

$$\boldsymbol{E}_* \in \mathcal{E}_{\rm per}, \quad \boldsymbol{J}_* = \mathsf{L}_*^T(\boldsymbol{E}_* + \boldsymbol{e}_0), \quad \boldsymbol{J}_* \in \mathcal{J}_{\rm per} \oplus \mathcal{U}. \tag{3.57}$$

The solution $\boldsymbol{E}_*$ of (3.57) is unique, since $\mathsf{L}_* \in \mathcal{M}(\alpha, \beta)^Q$ by theorem 3.3. We conclude that the entire family $\boldsymbol{E}_\epsilon$ converges weakly in $L^2_{\rm per}(Q; \mathcal{T})$ to $\boldsymbol{E}_*$. Thus, according to (3.56)

$$\lim_{\epsilon \to 0} \mathsf{K}_*^T(\epsilon)\boldsymbol{e}_0 = \frac{1}{|Q|}\int_Q \mathsf{L}_*^T(\boldsymbol{z})(\boldsymbol{E}_*(\boldsymbol{z}) + \boldsymbol{e}_0)\,d\boldsymbol{z} = \mathsf{K}_*^T\boldsymbol{e}_0.$$

□

We are now ready to prove theorem 3.7 under slightly weaker technical assumptions.

Theorem 3.46. *Suppose that the symbol $\boldsymbol{\Gamma}(\boldsymbol{\eta})$ is $\mathcal{E}$-nondegenerate and satisfies the compensated compactness property. Let U be any closed subset of $\mathcal{M}(\alpha, \beta)$.*

(i) *Assume that $\mathsf{L}_\epsilon \in U^\Omega$. Suppose that $\mathsf{L}_\epsilon \overset{H}{\rightharpoonup} \mathsf{L}_*$. Then $\mathsf{L}_* \in G(U)^\Omega$.*

(ii) *Assume that $\mathsf{L}_* \in G(U)^\Omega$. Then there exists $\mathsf{L}_\epsilon \in U^\Omega$, such that $\mathsf{L}_\epsilon \overset{H}{\rightharpoonup} \mathsf{L}_*$.*

Proof. Part (i). Let us fix $\boldsymbol{x}_0 \in \Omega$—a Lebesgue point of $\mathsf{L}_*(\boldsymbol{x})$, $\epsilon > 0$, and a cube Q_r of side r centered at the point $\boldsymbol{x}_0$. Next, let us make a periodic composite with period cell Q_r and local tensor $\mathsf{L}_\epsilon|Q_r$. Let $\mathsf{K}_*(r, \epsilon)$ be its effective tensor. Clearly, $\mathsf{K}_*(r, \epsilon) \in G(U)$. By part (c) of theorem 3.40 $\mathsf{L}_\epsilon|Q_r \overset{H}{\rightharpoonup} \mathsf{L}_*|Q_r$. Therefore, by the iterated homogenization lemma 3.45

$$\lim_{\epsilon \to 0} \mathsf{K}_*(r, \epsilon) = \mathsf{K}_*(r),$$

where $\mathsf{K}_*(r)$ is the effective tensor of a periodic composite with period cell Q_r and the local tensor $\mathsf{L}_*(\boldsymbol{x})$, $\boldsymbol{x} \in Q_r$. Rescaling the cube Q_r to the unit cube Q we see that $\mathsf{K}_*(r)$ is also the effective tensor of a Q-periodic composite with the local tensor $\mathsf{L}_*(\boldsymbol{x}_0 + r\boldsymbol{z})$, $\boldsymbol{z} \in Q$. We observe that $\mathsf{L}_*(\boldsymbol{x}_0 + r\boldsymbol{z}) \to \mathsf{L}_0(\boldsymbol{x}_0)$ in $L^1(Q)$, as $r \to 0$, since $\boldsymbol{x}_0$ is a Lebesgue point of $\mathsf{L}_*(\boldsymbol{x})$. Hence, by lemma 3.2 $\mathsf{L}_*(\boldsymbol{x}_0 + r\boldsymbol{z}) \overset{H}{\rightharpoonup} \mathsf{L}_*(\boldsymbol{x}_0)$. Applying the iterated homogenization lemma 3.45 again we conclude that

$$\lim_{r \to 0} \mathsf{K}_*(r) = \mathsf{L}_*(\boldsymbol{x}_0).$$

Thus, $\mathsf{L}_*(\boldsymbol{x}_0) \in G(U)$.

Part (ii). We can use the geometric structure of G-closed sets established in corollary 3.21 to represent each function $\mathsf{L} \in G(U)^\Omega$ as $\mathsf{L} = \mathsf{A}(\mathsf{F})$, where $\mathsf{F} \in \mathcal{C}^\Omega$, A is a fixed analytic diffeomorphism and $\mathcal{C}$ is a closed convex subset of $\mathrm{End}(\mathcal{T})$. Let $\mathsf{L}_0 \in G(U)$ and define

$$\mathsf{F}_0(\boldsymbol{x}) = \begin{cases} \mathsf{F}(\boldsymbol{x}), & \boldsymbol{x} \in \Omega, \\ \mathsf{A}^{-1}(\mathsf{L}_0), & \boldsymbol{x} \in \Omega^c. \end{cases}$$

Thus, $\mathsf{F} : \mathbb{R}^d \to \mathcal{C}$. Let $\rho : \mathbb{R}^d \to [0, +\infty)$ be a standard convolution kernel, i.e. $\rho(\boldsymbol{x})$ is supported on the unit ball B_1 and

$$\int_{B_1} \rho(\boldsymbol{x})\, \mathrm{d}\boldsymbol{x} = 1.$$

Then the function

$$\mathsf{F}_\epsilon(\boldsymbol{x}) = \int_{B_1} \mathsf{F}_0(\boldsymbol{x} + \epsilon \boldsymbol{z})\rho(\boldsymbol{z})\, \mathrm{d}\boldsymbol{z}$$

is continuous and takes values in $\mathcal{C}$, since $\mathcal{C}$ is a closed convex set. Additionally, $\mathsf{F}_\epsilon \to \mathsf{F}$, as $\epsilon \to 0$ in measure in Ω. Therefore, $\mathsf{L}_\epsilon = \mathsf{A}(\mathsf{F}_\epsilon)$ converges to L, as $\epsilon \to 0$ in measure in Ω. By lemma 3.2 it follows that $\mathsf{L}_\epsilon \overset{H}{\rightharpoonup} \mathsf{L}$. We have constructed the family of continuous functions $\mathsf{L}_\epsilon \in G(U)^\Omega$ such that L is the H-limit of L_ϵ.

Now let us fix a continuous function $\mathsf{L} \in G(U)^\Omega$. Let us cover Ω by a lattice of cubes $Q_{\boldsymbol{k}}(2r) = (-r, r)^d + 2r\boldsymbol{k}$ of side $2r > 0$, $\boldsymbol{k} \in \mathbb{Z}^d$:

$$\Omega \subset \bigcup_{\boldsymbol{k}\in I_r} \overline{Q_{\boldsymbol{k}}(2r)}, \qquad I_r = \left\{\boldsymbol{k} \in \mathbb{Z}^d : Q_{\boldsymbol{k}}(2r) \bigcap \Omega \neq \varnothing\right\}.$$

For each $\boldsymbol{k} \in I_r$ we choose a point $\boldsymbol{\xi}_{\boldsymbol{k}}^r \in Q_{\boldsymbol{k}}(2r) \cap \Omega$. The function

$$\mathsf{L}_r(\boldsymbol{x}) = \mathsf{L}\left(\boldsymbol{\xi}_{\boldsymbol{k}}^r\right), \quad \boldsymbol{x} \in Q_{\boldsymbol{k}}(2r)$$

is piecewise constant on the cubes of the lattice and converges to $\mathsf{L}(\boldsymbol{x})$, as $r \to 0$ pointwise (regardless of the specific choice of points $\boldsymbol{\xi}_{\boldsymbol{k}}^r$), since $\mathsf{L}(\boldsymbol{x})$ is continuous. Therefore, we have constructed a family of functions $\mathsf{L}_r \in G(U)^\Omega$ that are piecewise constant on the cubes of the lattice and H-converge to L.

Now let us fix an arbitrary lattice of cubes and a function $\mathsf{L} \in G(U)^\Omega$, constant on each cube of the lattice. Then, in each cube we place a periodic composite, whose constituents are taken from U, H-convergent to the constant value of $\mathsf{L}(\boldsymbol{x})$ in that cube. By part (d) of theorem 3.40 we have constructed a sequence of functions in U^Ω that H-converges to $\mathsf{L}(\boldsymbol{x})$. It remains to recall that the set $\mathcal{M}(\alpha, \beta)^\Omega$ is metrizable in the H-topology. Thus, U^Ω is H-dense in $G(U)^\Omega$. □

Corollary 3.47. *The set $G(U)$ is G-closed.*

Proof. Let L_* be the effective tensor of a periodic composite with local tensor $\mathsf{L} \in G(U)^{\mathcal{Q}}$. By part (ii) of theorem 3.7 there is $\mathsf{L}_\epsilon \in U^{\mathcal{Q}}$ such that $\mathsf{L}_\epsilon \overset{H}{\rightharpoonup} \mathsf{L}$. Let $\mathsf{K}_*(\epsilon)$ be the effective tensor of the periodic composite with local tensor $\mathsf{L}_\epsilon(\boldsymbol{x})$. Then $\mathsf{K}_*(\epsilon) \in G(U)$. By the iterated homogenization lemma 3.45 $\mathsf{K}_*(\epsilon) \to \mathsf{L}_*$, as $\epsilon \to 0$. Therefore, $\mathsf{L}_* \in G(U)$. □

3.3.6 On the convergence of fluxes

In this section we explain the role of convergence of fluxes in the definition of H-convergence. Whether or not this requirement is essential depends on the presence of non-zero bilinear null-Lagrangians. In the early work of Spagnolo [20] the convergence of fluxes was not necessary because in the context of conductivity there are no non-zero *quadratic* null-Lagrangians. In general, the assumption of symmetry of material tensors eliminates the need for requiring convergence of fluxes only if the space of quadratic null-Lagrangians is $\{0\}$. In particular, it is easy to check that for conductivity, elasticity[2] and piezoelectricity there are no non-zero quadratic null-Lagrangians. However, the full definition 3.1 of H-convergence must be used in the context of thermoelectricity (both in two dimensions and three dimensions) as well as in the two-dimensional and three-dimensional conductivity in the presence of the magnetic field (Hall effect).

Definition 3.48. *A bilinear form $B(\boldsymbol{u}, \boldsymbol{v})$ on $\mathcal{T}$ is called a null-Lagrangian if*

$$\int_{\mathbb{R}^d} B(\boldsymbol{E}_1(\boldsymbol{x}), \boldsymbol{E}_2(\boldsymbol{x}))\, \mathrm{d}\boldsymbol{x} = 0 \tag{3.58}$$

for all $\{\boldsymbol{E}_1, \boldsymbol{E}_2\} \subset \mathcal{E}$.

The null-Lagrangians can be characterized geometrically by means of the Plancherel identity. It is easy to see that B is a null-Lagrangian if and only if

$$B(\boldsymbol{e}_1, \boldsymbol{e}_2) = 0 \text{ for all } \{\boldsymbol{e}_1, \boldsymbol{e}_2\} \in \mathcal{E}_{\boldsymbol{\eta}}, \text{ and all } |\boldsymbol{\eta}| = 1. \tag{3.59}$$

In our theory null-Lagrangians appear in a somewhat hidden form. The following lemma is our link to null-Lagrangians.

Lemma 3.49. *Assume that the symbol $\boldsymbol{\Gamma}(\boldsymbol{\eta})$ is such that $C_0^\infty(\mathbb{R}^d; \mathcal{T}) \cap \mathcal{E}$ is dense in $\mathcal{E}$. Let Ω be an open and bounded subset of $\mathbb{R}^d$ and $\mathsf{N} \in L^\infty(\Omega; \mathrm{End}(\mathcal{T}))$ is such that*

$$\int_\Omega (\mathsf{N}(\boldsymbol{x})\boldsymbol{E}_1(\boldsymbol{x}), \boldsymbol{E}_2(\boldsymbol{x}))\, \mathrm{d}\boldsymbol{x} = 0 \tag{3.60}$$

for all $\{\boldsymbol{E}_1, \boldsymbol{E}_2\} \subset \mathcal{E}_0(\Omega)$. Then the bilinear forms $B_{\boldsymbol{x}}(\boldsymbol{u}, \boldsymbol{v}) = (\mathsf{N}(\boldsymbol{x})\boldsymbol{u}, \boldsymbol{v})$ are null-Lagrangians for a.e. $\boldsymbol{x} \in \Omega$.

Proof. We split the proof into two steps. In the first step we show that

$$\int_\Omega (\mathsf{N}(\boldsymbol{x}_0)\boldsymbol{E}_1(\boldsymbol{x}), \boldsymbol{E}_2(\boldsymbol{x}))\, \mathrm{d}\boldsymbol{x} = 0 \tag{3.61}$$

for a.e. $\boldsymbol{x}_0 \in \Omega$ and all $\{\boldsymbol{E}_1, \boldsymbol{E}_2\} \subset \mathcal{E}_0(\Omega)$. In the second step we show that if $\mathsf{N} \in \mathrm{End}(\mathcal{T})$ satisfies (3.61) then $B(\boldsymbol{u}, \boldsymbol{v}) = (\mathsf{N}\boldsymbol{u}, \boldsymbol{v})$ must be a null-Lagrangian.

---

[2] Curiously, if the dimension of space is $d \geqslant 4$, then there are non-zero quadratic null-Lagrangians in elasticity.

Step 1. We start with the observation that by lemma 2.3, relation (3.60) is equivalent to $\mathsf{N}\boldsymbol{E} \in \overline{\mathcal{J}(\Omega)}$ for any $\boldsymbol{E} \in \mathcal{E}_0(\Omega)$. Let us fix a Lebesgue point $\boldsymbol{x}_0 \in \Omega$ of $\mathsf{N}(\boldsymbol{x})$ and choose $\delta_0 > 0$ so small that $\boldsymbol{E}_\delta(\boldsymbol{x}) = \boldsymbol{E}((\boldsymbol{x} - \boldsymbol{x}_0)/\delta) \in \mathcal{E}_0(\Omega)$ for all $\delta \in (0, \delta_0)$. But then $\mathsf{N}\boldsymbol{E}_\delta \in \overline{\mathcal{J}(\Omega)}$. Hence,

$$\mathsf{N}(\boldsymbol{x}_0 + \delta \boldsymbol{z})\boldsymbol{E}(\boldsymbol{z}) \in \overline{\mathcal{J}(\Omega)}.$$

Observe that

$$\mathsf{N}(\boldsymbol{x}_0 + \delta \boldsymbol{z})\boldsymbol{E}(\boldsymbol{z}) \to \mathsf{N}(\boldsymbol{x}_0)\boldsymbol{E}(\boldsymbol{z}), \qquad \delta \to 0$$

in $L^2(\Omega; \mathcal{T})$, since $\boldsymbol{x}_0$ is a Lebesgue point of $\mathsf{N}(\boldsymbol{x})$. Thus, $\mathsf{N}(\boldsymbol{x}_0)\boldsymbol{E} \in \overline{\mathcal{J}(\Omega)}$ for a.e. $\boldsymbol{x}_0 \in \Omega$ and all $\boldsymbol{E} \in \mathcal{E}_0(\Omega)$. Hence, the constant tensor $\mathsf{N}(\boldsymbol{x}_0)$ satisfies (3.61).

Step 2. In this step $\mathsf{N} \in \mathrm{End}(\mathcal{T})$ is a constant tensor satisfying (3.61) and $B(\boldsymbol{u}, \boldsymbol{v}) = (\mathsf{N}\boldsymbol{u}, \boldsymbol{v})$. Let D be an arbitrary open and bounded subset of $\mathbb{R}^d$, and let $\boldsymbol{x}_0 \in D$ and $\boldsymbol{x}_1 \in \Omega$ be arbitrary but fixed. Then there exists $\delta > 0$, so that $\tilde{\boldsymbol{E}}(\boldsymbol{x}) = \boldsymbol{E}(\boldsymbol{x}_0 + (\boldsymbol{x} - \boldsymbol{x}_1)/\delta) \in \mathcal{E}_0(\Omega)$ for any $\boldsymbol{E} \in \mathcal{E}_0(D)$. Hence, changing variables $\boldsymbol{z} = \boldsymbol{x}_0 + (\boldsymbol{x} - \boldsymbol{x}_1)/\delta$, we have for any $\{\boldsymbol{E}_1, \boldsymbol{E}_2\} \subset \mathcal{E}_0(D)$

$$0 = \int_\Omega B\left(\tilde{\boldsymbol{E}}_1, \tilde{\boldsymbol{E}}_2\right) \mathrm{d}\boldsymbol{x} = \delta^d \int_{\mathbb{R}^d} B(\boldsymbol{E}_1, \boldsymbol{E}_2)\, \mathrm{d}\boldsymbol{z} = \delta^d \int_D B(\boldsymbol{E}_1, \boldsymbol{E}_2)\, \mathrm{d}\boldsymbol{z}.$$

Thus, B satisfies (3.61) with Ω replaced by D. In particular, B satisfies (3.61) with Ω replaced by B_n for all $n \geqslant 1$. Now, let us fix $\{\boldsymbol{E}_1, \boldsymbol{E}_2\} \subset \mathcal{E}$. By assumption there exists a sequence $\boldsymbol{E}_n^{(j)} \in \mathcal{E}_0(B_n)$, $j = 1, 2$, such that take $X_0(B_n)\boldsymbol{E}_n^{(j)} \to \boldsymbol{E}_j$ in $L^2(\mathbb{R}^d; \mathcal{T})$, as $n \to \infty$. Thus,

$$\begin{aligned}\int_{\mathbb{R}^d} B(\boldsymbol{E}_1, \boldsymbol{E}_2)\, \mathrm{d}\boldsymbol{x} &= \lim_{n\to\infty} \int_{\mathbb{R}^d} B\left(X_0(B_n)\boldsymbol{E}_n^{(1)}, X_0(B_n)\boldsymbol{E}_n^{(2)}\right) \mathrm{d}\boldsymbol{x} \\ &= \lim_{n\to\infty} \int_{Bn} B\left(\boldsymbol{E}_n^{(1)}, \boldsymbol{E}_n^{(2)}\right) \mathrm{d}\boldsymbol{x} = 0.\end{aligned}$$

Thus, $B(\boldsymbol{u}, \boldsymbol{v}) = (\mathsf{N}\boldsymbol{u}, \boldsymbol{v})$ is a null-Lagrangian and the lemma is proved. □

We remark that property (3.60) is equivalent to

$$\Gamma_\Omega(\mathsf{N}(\boldsymbol{x})\boldsymbol{E}) = 0 \text{ for all } \boldsymbol{E} \in \mathcal{E}_0(\Omega). \tag{3.62}$$

The set $\mathfrak{N}$ of null-Lagrangians is a subspace of $\mathrm{End}(\mathcal{T})$, which is closed with respect to transposition: $\mathsf{N}^T \in \mathfrak{N}$, provided $\mathsf{N} \in \mathfrak{N}$. Therefore, both the symmetric and the antisymmetric parts of a null-Lagrangian are null-Lagrangians. Thus,

$$\mathfrak{N} = \mathfrak{N}_s \oplus \mathfrak{N}_a, \qquad \mathfrak{N}_s = \mathfrak{N} \bigcap \mathrm{Sym}(\mathcal{T}), \qquad \mathfrak{N}_a = \mathfrak{N} \bigcap \mathrm{Skew}(\mathcal{T}),$$

where $\mathrm{Skew}(\mathcal{T})$ is the set of all antisymmetric operators on $\mathcal{T}$.

Now, we are ready to examine the necessity of convergence of fluxes for H-convergence. For this purpose we assume that

(A1) $\{\mathsf{L}_\epsilon, \mathsf{L}_*\} \subset \mathcal{M}(\alpha, \beta)^\Omega$.

(A2) For every $\boldsymbol{f} \in \mathcal{E}_0(\Omega)$ $\boldsymbol{E}_\epsilon \rightharpoonup \boldsymbol{E}_*$, as $\epsilon \to 0$ in $L^2(\Omega; \mathcal{T})$, where $\boldsymbol{E}_\epsilon \in \mathcal{E}_0(\Omega)$ is the unique solution of $\Gamma_\Omega(\mathsf{L}_\epsilon \boldsymbol{E}_\epsilon) = \boldsymbol{f}$ and $\boldsymbol{E}_* \in \mathcal{E}_0(\Omega)$ is the unique solution of $\Gamma_\Omega(\mathsf{L}_* \boldsymbol{E}_*) = \boldsymbol{f}$.

Our goal is either to prove that $\mathsf{L}_\epsilon \boldsymbol{E}_\epsilon \rightharpoonup \mathsf{L}_* \boldsymbol{E}_*$, establishing H-convergence $\mathsf{L}_\epsilon \overset{H}{\rightharpoonup} \mathsf{L}_*$, or find an example, where the weak limit of $\mathsf{L}_\epsilon \boldsymbol{E}_\epsilon$ differs from $\mathsf{L}_* \boldsymbol{E}_*$. The latter is easy to construct, provided $\mathfrak{N} \neq \{0\}$. Let $\mathsf{N} \in \mathfrak{N} \backslash \{0\}$. Let $\delta \neq 0$ be so small in absolute value that $\mathsf{K}_* = \mathsf{L}_* + \delta \mathsf{N} \in \mathcal{M}(\alpha, \beta)^\Omega$. Then K_* satisfies (A1) and (A2).

Let $\boldsymbol{E}_* \in \mathcal{E}_0(\Omega)$ be such that $\mathsf{N}\boldsymbol{E}_* \neq \mathbf{0}$. This can always be arranged if $V_\mathcal{E} = \mathcal{T}$. Let $\boldsymbol{E}_\epsilon$ be the unique solution of $\Gamma_\Omega(\mathsf{L}_\epsilon \boldsymbol{E}_\epsilon) = \Gamma_\Omega(\mathsf{L}_* \boldsymbol{E}_*) = \Gamma_\Omega(\mathsf{K}_* \boldsymbol{E}_*)$. Then the weak limit of $\mathsf{L}_\epsilon \boldsymbol{E}_\epsilon$ (or a weakly convergent subsequence thereof) cannot be equal to both $\mathsf{L}_* \boldsymbol{E}_*$ and $(\mathsf{L}_* + \delta \mathsf{N})\boldsymbol{E}_*$.

We now prove the converse, namely that if $\mathfrak{N} = \{0\}$ then conditions (A1) and (A2) imply that $\mathsf{L}_\epsilon \overset{H}{\rightharpoonup} \mathsf{L}_*$. The following simple lemma gives the reason.

Lemma 3.50. *Suppose L_ϵ and L_* satisfy assumptions (A1) and (A2) above. Suppose $\mathsf{L}_\epsilon \overset{H}{\rightharpoonup} \mathsf{K}_*$, as $\epsilon \to 0$. Then $\mathsf{N}(\boldsymbol{x}) = \mathsf{K}_*(\boldsymbol{x}) - \mathsf{L}_*(\boldsymbol{x})$ satisfies (3.60).*

Proof. Let $\boldsymbol{E} \in \mathcal{E}_0(\Omega)$ be arbitrary. Let $\boldsymbol{E}_\epsilon$ be the unique solution of $\Gamma_\Omega(\mathsf{L}_\epsilon \boldsymbol{E}_\epsilon) = \Gamma_\Omega(\mathsf{L}_* \boldsymbol{E})$. Then $\boldsymbol{E}_\epsilon \rightharpoonup \boldsymbol{E}$, as $\epsilon \to 0$ in $L^2(\Omega; \mathcal{T})$ by assumption (A1). But, by definition of H-convergence we conclude that $\boldsymbol{E}$ also solves $\Gamma_\Omega(\mathsf{K}_* \boldsymbol{E}) = \Gamma_\Omega(\mathsf{L}_* \boldsymbol{E})$. Thus, $\mathsf{N}(\boldsymbol{x})$ satisfies (3.62), which is equivalent to (3.60). □

Combining this lemma with compactness theorem (theorem 3.41) and lemma 3.49 gives the desired result, which we state formally as a theorem.

Theorem 3.51. *Assume that the symbol $\boldsymbol{\Gamma}(\boldsymbol{\eta})$ is $\mathcal{E}$-nondegenerate and satisfies the compensated compactness property. Suppose $\mathsf{L}_\epsilon(\boldsymbol{x})$ and $\mathsf{L}_*(\boldsymbol{x})$ have the properties* (A1), (A2) *and that $\mathfrak{N} = \{0\}$. Then $\mathsf{L}_\epsilon \overset{H}{\rightharpoonup} \mathsf{L}_*$.*

Proof. By the compactness theorem, there exists an H-convergent subsequence $\mathsf{L}_{\epsilon_k} \overset{H}{\rightharpoonup} \mathsf{K}_*$. Then, by lemma 3.50 the difference $\mathsf{N}(\boldsymbol{x}) = \mathsf{L}_*(\boldsymbol{x}) - \mathsf{K}_*(\boldsymbol{x})$ satisfies (3.60). Applying lemma 3.49 we conclude that $\mathsf{N}(\boldsymbol{x}) \in \mathfrak{N} = \{0\}$ for a.e. $\boldsymbol{x} \in \Omega$. Thus, all H-limits of H-convergent subsequences of L_ϵ are equal to L_*. The theorem follows. □

In most physical contexts the tensors of material properties are symmetric. In that case we have the following theorem.

Theorem 3.52. *Assume that the symbol $\boldsymbol{\Gamma}(\boldsymbol{\eta})$ is $\mathcal{E}$-nondegenerate and satisfies the compensated compactness property. Suppose $\mathsf{L}_\epsilon(\boldsymbol{x})$ and $\mathsf{L}_*(\boldsymbol{x})$ have the properties (A1), (A2) and that $\mathsf{L}_\epsilon^T = \mathsf{L}_\epsilon$:*

(i) *If $\mathfrak{N}_a = \{0\}$, then $\mathsf{L}_*^T = \mathsf{L}_*$.*

(ii) *If $\mathsf{L}_*^T = \mathsf{L}_*$ and $\mathfrak{N}_s = \{0\}$, then $\mathsf{L}_\epsilon \overset{H}{\rightharpoonup} \mathsf{L}_*$.*

Proof. Part (i). Let $\{\boldsymbol{E}_1, \boldsymbol{E}_2\} \subset \mathcal{E}_0(\Omega)$ be arbitrary. Let $\boldsymbol{E}_\epsilon^{(j)} \in \mathcal{E}_0(\Omega)$ solve

$$\Gamma_\Omega\left(\mathsf{L}_\epsilon \boldsymbol{E}_\epsilon^{(j)}\right) = \Gamma_\Omega\left(\mathsf{L}_* \boldsymbol{E}_j\right), \quad j = 1, 2.$$

Then $\boldsymbol{E}_\epsilon^{(j)} \rightharpoonup \boldsymbol{E}_j$, $j = 1, 2$ in $L^2(\Omega; \mathcal{T})$ by assumption. Then

$$\left(\boldsymbol{E}_\epsilon^{(1)}, \mathsf{L}_* \boldsymbol{E}_2\right)_{L^2(\Omega)} = \left(\boldsymbol{E}_\epsilon^{(1)}, \Gamma_\Omega(\mathsf{L}_* \boldsymbol{E}_2)\right)_{L^2(\Omega)} = \left(\boldsymbol{E}_\epsilon^{(1)}, \Gamma_\Omega\left(\mathsf{L}_\epsilon \boldsymbol{E}_\epsilon^{(2)}\right)\right)_{L^2(\Omega)}$$

$$= \left(\Gamma_\Omega\left(\mathsf{L}_\epsilon \boldsymbol{E}_\epsilon^{(1)}\right), \boldsymbol{E}_\epsilon^{(2)}\right)_{L^2(\Omega)} = \left(\mathsf{L}_* \boldsymbol{E}_1, \boldsymbol{E}_\epsilon^{(2)}\right)_{L^2(\Omega)}.$$

Passing to the limit as $\epsilon \to 0$ we obtain that

$$\left(\left(\mathsf{L}_* - \mathsf{L}_*^T\right)\boldsymbol{E}_1, \boldsymbol{E}_2\right)_{L^2(\Omega)} = 0$$

for all $\{\boldsymbol{E}_1, \boldsymbol{E}_2\} \subset \mathcal{E}_0(\Omega)$. Therefore, according to lemma 3.49, $\mathsf{N}(\boldsymbol{x}) = \mathsf{L}_*(\boldsymbol{x}) - \mathsf{L}_*^T(\boldsymbol{x}) \in \mathfrak{N}_s = \{0\}$ for a.e. $\boldsymbol{x} \in \Omega$.

Part (ii). The proof repeats the proof of theorem 3.51. We extract an H-convergent subsequence $\mathsf{L}_{\epsilon_k} \overset{H}{\rightharpoonup} \mathsf{K}_*$. By part (a) of theorem 3.40 we conclude that $\mathsf{K}_*^T = \mathsf{K}_*$. Therefore, the difference $\mathsf{N}(\boldsymbol{x}) = \mathsf{L}_*(\boldsymbol{x}) - \mathsf{K}_*(\boldsymbol{x}) \in \mathfrak{N}_s = \{0\}$ for a.e. $\boldsymbol{x} \in \Omega$. Thus, all H-limits of H-convergent subsequences of L_ϵ are equal to L_*. Part (ii) is proved. □

Bibliography

[1] Allaire G 2002 *Shape Optimization by the Homogenization Method* (*Applied Mathematical Sciences* vol 146) (New York: Springer)

[2] Babadjian J-F and Barchiesi M 2009 A variational approach to the local character of g-closure: the convex case *Ann. Inst. Henri Poincaré* C **26** 351–73

[3] Backus G E 1962 Long-wave elastic anisotropy produced by horizontal layering *J. Geophys. Res.* **67** 4427–40

[4] Cherkaev A and Zhang Y 2011 Optimal anisotropic three-phase conducting composites: plane problem *Int. J. Solids Struct.* **48** 2800–13

[5] De Giorgi E and Spagnolo S 1973 Sulla convergenza degli integrali dell'energia per operatori ellittici del secondo ordine *Boll. Un. Mater. Ital.* **8** 391–411

[6] Francfort G A and Milton G W 1987 Optimal bounds for conduction in two-dimensional, multiphase, polycrystalline media *J. Stat. Phys.* **46** 161–77

[7] Francfort G A and Milton G W 1994 Sets of conductivity and elasticity tensors stable under lamination *Commun. Pure Appl. Math.* **47** 257–79

[8] Grabovsky Y 1993 The G-closure of two well-ordered anisotropic conductors *Proc. R. Soc. Edin.* A **123** 423–32

[9] Grabovsky Y and Milton G W 1998 Rank one plus a null-Lagrangian is an inherited property of two-dimensional compliance tensors under homogenization *Proc. R. Soc. Edin.* A **128** 283–99

[10] Hollister S J and Kikuchi N 1992 A comparison of homogenization and standard mechanics analyses for periodic porous composites *Comput. Mech.* **10** 73–95
[11] Lurie K A and Cherkaev A V 1981 *G*-closure of a set of anisotropic conducting media in the case of two dimensions *Dokl. Akad. Nauk SSSR* **259** 328–31 (in Russian)
[12] Lurie K A and Cherkaev A V 1984 Exact estimates of conductivity of composites formed by two isotropically conducting media taken in prescribed proportion *Proc. R. Soc. Edin.* A **99** 71–87
[13] Lurie K A and Cherkaev A V 1986 Exact estimates of a binary mixture of isotropic components *Proc. R. Soc. Edin.* A **104** 21–38
[14] Milton G W 1990 On characterizing the set of possible effective tensors of composites: the variational method and the translation method *Commun. Pure Appl. Math.* **43** 63–125
[15] Milton G W 2002 *The Theory of Composites* (*Cambridge Monographs on Applied and Computational Mathematics* vol 6) (Cambridge: Cambridge University Press)
[16] Murat F 1981 Compacité par compensation: condition nécessaire et suffisante de continuité faible sous une hypothèse de rang constant *Ann. Scuola Norm. Sup. Pisa Cl. Sci.* **8** 69–102
[17] Murat F and Tartar L 1997 H-convergence *Topics in the Mathematical Modelling of Composite Materials* ed A Cherkaev and R V Kohn (*Progress in Non-linear Differential equations and Their Applications* vol 31) (Boston, MA: Birkhäuser) ch 3, pp 21–43
[18] Raitums U 2001 On the local representation of G-closure *Arch. Ration. Mech. Anal.* **158** 213–34
[19] Sánchez-Palencia E 1980 *Nonhomogeneous Media and Vibration Theory* (Berlin: Springer)
[20] Spagnolo S 1968 Sulla convergenza di soluzioni di equazioni paraboliche ed ellittiche *Ann. Scuola Norm. Sup. Pisa.* **22** 571–97
Spagnolo S 1968 Sulla convergenza di soluzioni di equazioni paraboliche ed ellittiche *Ann. Scuola Norm. Sup. Pisa.* **22** 673 (erratum)
[21] Tartar L 1979 Compensated compactness and applications to partial differential equations *Nonlinear Analysis and Mechanics: Heriot-Watt Symposium* vol 4 (*Research Notes in Mathematics* vol 38) (Boston, MA: Pitman) pp 136–212
[22] Tartar L 1979 Estimation de coefficients homogénéisé *Proc. Third Int. Symp. Computing Methods in Applied Sciences and Engineering (Versailles, 1977)* (*Lecture Notes in Mathematics* vol 704) (Berlin: Springer) pp 364–73
[23] Tartar L 1990 *H*-measures, a new approach for studying homogenisation, oscillations and concentration effects in partial differential equations *Proc. R. Soc. Edin.* A **115** 193–230
[24] Tartar L 2000 An introduction to the homogenization method in optimal design *Optimal Shape Design* (*Lecture Notes in Mathematics* vol 1740) (Berlin: Springer) pp 47–156
[25] Zhikov V V 1991 Estimates for the homogenized matrix and the homogenized tensor *Russ. Math. Surv.* **46** 65–136

Part II

General theory of exact relations and links

Chapter 4

Exact relations

4.1 Introduction

Formulas for effective behavior of composite materials have been sought out throughout the entire history of mathematical analysis of composite media. Clausius–Mossotti formulas [4, 33] (also known as Maxwell–Garnett formulas [11, 27]) for effective conductivity of composites[1] and Einstein's formula for effective viscosity of dilute suspensions [10] are probably the most well-known examples in physics. It is understood that these 'formulas' are actually approximations that are valid asymptotically in the small volume fraction limit. The actual effective behavior of a composite depends on its microstructure. Nevertheless, exact algebraic formulas relating bulk properties of composites and their constituents do exist, and have been attracting steady attention of physicists and engineers. Not only are such formulas surprising, they possess a peculiar aesthetic beauty. This part of the book will explain how *all* such formulas can be obtained systematically for every material property of interest.

As an illustration, let us describe the two earliest and most well-known examples of such formulas. The first, due to Hill [19, 20], says that any composite made with isotropic elastic materials with the same shear modulus μ_0 must always be isotropic (even if geometrically its microstructure is highly anisotropic!) and have the shear modulus μ_0. Moreover, it bulk modulus κ^* can be expressed in terms of the volume average of the local elastic tensor:

$$\left(3\kappa^* + 4\mu_0\right)^{-1} = \left\langle\left(3\kappa(z) + 4\mu_0\right)^{-1}\right\rangle. \tag{4.1}$$

The second example refers to two-dimensional conductivity and was first observed by Keller [23] in a periodic array of circular cylindrical inclusions (see [8, 29, 34] for its subsequent development). If the local conductivity tensor $\boldsymbol{\sigma}(z)$ has the property

[1] For a history of this discovery see, e.g. [24].

doi:10.1088/978-0-7503-1048-2ch4

$\det \boldsymbol{\sigma}(z) = d_0$ for all z, then the effective conductivity tensor $\boldsymbol{\sigma}^*$ must have the same property, $\det \boldsymbol{\sigma}^* = d_0$, *regardless of the microstructure.*

Examples of such exact algebraic formulas for coupled field material properties, such as thermoelasticity and piezoelectricity, abound. In this part of the book we are going to describe a theory that can generate *complete lists* of all such formulas. Some of these lists are presented in part 3 of the book, which is intended to be used as a reference. To begin, we need to give a formal definition of an *exact relation.* We will use Hilbert space formalism developed in the first part of this book in order to obtain a unified theory of exact relations equally applicable to all material properties to which the homogenization theory of part 1 applies.

Definition 4.1. *An* exact relation *is a G-closed smooth submanifold* $\mathbb{M} \subset \mathrm{End}^+(\mathcal{T})$ of positive codimension.

According to this definition, algebraic formulas relating effective behavior of composites to that of their constituents are now viewed geometrically, as describing a smooth submanifold $\mathbb{M}$ in the space $\mathrm{End}^+(\mathcal{T})$ of tensors of material properties. For a submanifold $\mathbb{M}$ to be an exact relation it has to be G-closed, or *stable under homogenization.* In other words, the effective tensor of every composite made of any number of materials taken from $\mathbb{M}$ must again belong to $\mathbb{M}$.

Our first example of Hill's exact relation describes a whole family of such manifolds $\mathbb{M}_{\mu_0}$ parametrized by the constant shear modulus μ_0. Geometrically, each manifold $\mathbb{M}_{\mu_0}$ looks like a (part of a) one-dimensional affine line in the 21-dimensional space of elasticity tensors given by

$$\mathbb{M}_{\mu_0} = \left\{\mathsf{C} \in \mathbb{E}^3_+ : \mathsf{C}\boldsymbol{\varepsilon} = \lambda(\mathrm{Tr}\,\boldsymbol{\varepsilon})\boldsymbol{I} + 2\mu_0\boldsymbol{\varepsilon},\ \boldsymbol{\varepsilon} \in \mathrm{Sym}\left(\mathbb{R}^3\right),\ \lambda \in \mathbb{R}\right\}, \tag{4.2}$$

where $\mathbb{E}^3_+ = \mathrm{Sym}^+(\mathrm{Sym}(\mathbb{R}^3))$ denotes the set of all positive definite three-dimensional elasticity tensors.

Our second example of a Keller–Dykhne exact relation also describes a whole family of such manifolds $\mathbb{M}_{d_0}$, parametrized by the constant determinant d_0 of two-dimensional conductivity tensors. Geometrically, it is a (part of a) sheet of a two-sheeted hyperboloid in the three-dimensional space of two-dimensional conductivity tensors (pictured on the cover of this book), given by

$$\mathbb{M}_{d_0} = \left\{\boldsymbol{\sigma} \in \mathrm{Sym}^+\left(\mathbb{R}^2\right) : \det \boldsymbol{\sigma} = d_0\right\}. \tag{4.3}$$

Observe that formula (4.1) does not seem to fit into the concept of an exact relation that we have just defined, at first glance at least. Nevertheless, the theory of exact relations developed in this chapter is perfectly capable of capturing all such formulas as well. In chapter 5 the abstract nature of our approach will become an advantage, permitting us to use a single theory to obtain volume fraction relations, such a (4.1), as well as links between the effective behavior of a pair of composites sharing the same microstructure, but differing in the properties of their constituents.

There is one class of exact relations that is simple and physically natural. Let us consider composites made with a given set of r materials $\mathsf{L}_1, \ldots, \mathsf{L}_r$. Suppose that there are field values $\{e_1, \ldots, e_k\} \subset \mathcal{T}$, such that

$$\mathsf{L}_1 e_j = \mathsf{L}_2 e_j = \ldots = \mathsf{L}_r e_j = \tau_j, \quad j = 1, \ldots, k.$$

Then, the effective tensor L^* of any composite made with r such materials will also satisfy $\mathsf{L}^* e_j = \tau_j$, $j = 1, \ldots, k$. Physically, this is clear, since all r of the constituents behave as a single material in response to the applied fields $e_1, \ldots, e_k$. Mathematically, this is explained by observing that the cell problem (3.9) admits uniform field solutions $\boldsymbol{E}(z) = e_j$, $j = 1, \ldots, k$. For this reason we will call such exact relations *uniform field relations* (UFR) [6, 7, 25]. Hill's exact relation is an example of UFR, where the fields $\{e_1, \ldots, e_k\}$ can be chosen arbitrarily from the subspace of symmetric, trace-free matrices, as long as they span the subspace. Curiously, formula (4.1) cannot be derived by means of a uniform field argument.

As we will see, the UFRs do not exhaust all possible exact relations in all but a few of the simplest cases, such as three-dimensional conductivity. Elastic composites, for example, enjoy the beautiful 'rank-one plus a null-Lagrangian' (RPN) exact relation in elasticity [16, 17]. To describe it we recall that the isotropic elastic tensor T_0, given by

$$\mathsf{T}_0 \boldsymbol{\varepsilon} = \boldsymbol{\varepsilon} - (\mathrm{Tr}\, \boldsymbol{\varepsilon}) \boldsymbol{I},$$

is a null-Lagrangian in the sense of definition 3.48 for the space $\mathcal{E}$ of curl-free 3×3 matrix fields. The RPN exact relation consists of those elastic tensors C for which there exists a constant $\mu_0 > 0$, so that the tensor $\mathsf{C} - 2\mu_0 \mathsf{T}_0$ is rank-one. Curiously, there does not seem to be a natural physical explanation for the validity of such an exact relation. The mathematical theory of exact relations, on the other hand, places *all* exact relations in every physical context into a unified theoretical framework.

The theory emerges from the simple observation that if a submanifold $\mathbb{M}$ is an exact relation, then it must contain the effective tensor of any simple laminate whose constituents are in $\mathbb{M}$. The general lamination formula (3.40) implies that the submanifold $\mathbb{K}_n = W_n(\mathbb{M})$ must be convex and, hence, flat. Choosing the reference medium $\mathsf{L}_0 \in \mathbb{M}$, and recalling that $W_n(\mathsf{L}_0) = 0$ we can ensure that the convex sets $\mathbb{K}_n$ belong to subspaces Π_n of the same dimension as $\mathbb{M}$. These simple observations can be expressed as a family of algebraic constraints on $\mathbb{M}$, parametrized by unit vectors $\boldsymbol{n}$. The analysis of these algebraic constraints leads to us to conclude that the subspaces Π_n do not depend on $\boldsymbol{n}$ and satisfy a set of beautiful algebraic conditions, which are necessary for the submanifold $\mathbb{M}$ to describe an exact relation. The physical meaning of these conditions is that the submanifold $\mathbb{M}$ is closed under lamination, i.e. $\mathbb{M}$ must contain the effective tensor of every laminate made of materials taken from $\mathbb{M}$.

The passage from laminar to general microstructures occurs via formula (3.37), valid for all composites. This formula can be expanded into the Neumann series (3.38), when $\mathsf{L}(z)$ is sufficiently close to L_0. Each term of the Neumann series (3.38) represents a sum of products[2] of operators (matrices) in $\Pi = \Pi_n$ interlaced with operators $\Lambda_{\Gamma_0(n)}(\boldsymbol{k})$, defined in (3.27). We can conclude that each term is in Π if Π is a multialgebra (i.e. a subspace of operators on $\mathcal{T}$ closed with respect to a set of interlacing matrix products). This explanation, while a little vague, shows how and why the theory works, which would otherwise seem like a mathematical hocus pocus.

This chapter is organized as follows. In section 4.2 we examine the algebraic constraints on $\mathbb{M}$ (or rather the subspace Π) that come from the requirement that $\mathbb{M}$ is closed under lamination. In section 4.3 we analyze the Neumann series (3.38) and prove that each term belongs to Π, provided Π satisfies some additional, but easily verifiable algebraic conditions. In section 4.4 we define UFRs formally and introduce *polycrystalline exact relations*, which are the actual focus of this book, since they are both the most beautiful members of the exact relation family and feasibly computable in every physical context. Section 4.5 concludes the chapter with rigorous proofs of statements made in section 4.3.

4.2 L-relations

The goal of this section is to characterize all submanifolds $\mathbb{M} \subset \mathrm{End}^+(\mathcal{T})$ of non-zero co-dimension that are stable under lamination [14].

Definition 4.2. *A submanifold $\mathbb{M} \subset \mathrm{End}^+(\mathcal{T})$ is called an L-relation if the effective tensor of any laminate made with any two materials, whose tensors lie in $\mathbb{M}$, belongs to $\mathbb{M}$.*

Being an L-relation places rather strong constraints on $\mathbb{M}$, because for each pair of materials from $\mathbb{M}$ both the volume fraction and the direction of lamination can be varied arbitrarily. Figure 4.1 shows lamination paths parametrized by the volume fraction for four different choices of the lamination direction on the exact relation manifold (4.3).

By corollary 3.21, if $\mathbb{M}$ is an L-relation, then the sets $W_n(\mathbb{M})$ must be convex submanifolds of $\mathrm{End}(\mathcal{T})$ of the same dimension as $\mathbb{M}$ for all unit vectors $\boldsymbol{n}$, since all maps W_n, defined in (3.26) and (3.41), are analytic and injective on $\mathrm{End}^+(\mathcal{T})$ by lemma 3.23. Thus, for each unit vector $\boldsymbol{n}$ the set $W_n(\mathbb{M})$ must be a convex, relatively open subset of an affine subspace $\Pi_n \subset \mathrm{End}(\mathcal{T})$. Observing that $W_n(\mathsf{L}_0) = 0$, which is obvious from formula (3.26), we may assume, without loss of generality, that all Π_n are linear subspaces of $\mathrm{End}(\mathcal{T})$, by choosing $\mathsf{L}_0 \in \mathbb{M}$. Let us fix a unit vector $\boldsymbol{n}_0$ and denote $\Pi = \Pi_{n_0}$. If $\mathbb{M}$ is an L-relation we must have

$$\mathbb{M} = \left\{\mathsf{L} \in \mathrm{End}^+(\mathcal{T}) : W_{n_0}(\mathsf{L}) \in \Pi\right\}. \tag{4.4}$$

Now, let us vary the direction of the unit vector $\boldsymbol{n}$. Then the composition map $\Lambda_n = W_n \circ W_{n_0}^{-1}$ must map a sufficiently small neighborhood of 0 in $\Pi = \Pi_{n_0}$ to a

[2] The fact that some of the operators are Fourier multipliers does not change validity of our reasoning, since the Fourier transform of a $\Pi_{\mathbb{C}}$-valued function is also $\Pi_{\mathbb{C}}$-valued, where $\Pi_{\mathbb{C}} = \Pi \otimes \mathbb{C}$ is the complexification of Π.

Figure 4.1. Lamination paths for four different lamination directions on the Keller–Dykhne exact relation manifold (4.3).

neighborhood of 0 in Π_n. We easily compute, using formulas (3.26), (3.41) for W_n,

$$\Lambda_n(\mathsf{K}) = \left[\mathsf{I} - \mathsf{KA}(\boldsymbol{n})\right]^{-1}\mathsf{K}, \qquad \mathsf{A}(\boldsymbol{n}) = \boldsymbol{\Gamma}_0(\boldsymbol{n}_0) - \boldsymbol{\Gamma}_0(\boldsymbol{n}), \tag{4.5}$$

since for K in a sufficiently small neighborhood of 0 the operator $\mathsf{I} - \mathsf{KA}(\boldsymbol{n})$ is always invertible.[3] Expanding $\Lambda_n(\mathsf{K})$ in power series around $\mathsf{K} = 0$

$$\Lambda_n(\mathsf{K}) = \sum_{n=0}^{\infty} \mathsf{K}\left(\mathsf{A}(\boldsymbol{n})\mathsf{K}\right)^n = \mathsf{K} + \mathsf{KA}(\boldsymbol{n})\mathsf{K} + \cdots + \mathsf{K}\left(\mathsf{A}(\boldsymbol{n})\mathsf{K}\right)^n + \cdots \tag{4.6}$$

we conclude that each term of the expansion (4.6) must belong to Π_n. Looking at the first term we see that $\mathsf{K} \in \Pi_n$ for every K in a small neighborhood of $0 \in \Pi$. Thus $\Pi \subset \Pi_n$. But, since all the subspaces Π_n are diffeomorphic images of $\mathbb{M}$ they must have the same dimension as $\mathbb{M}$, and we conclude that $\Pi_n = \Pi$ for all unit vectors $\boldsymbol{n}$.

Looking at the second term in (4.6) we see that the subspace Π must satisfy

$$\mathsf{KA}(\boldsymbol{n})\mathsf{K} \subset \Pi \tag{4.7}$$

for all $\mathsf{K} \in \Pi$ and all unit vectors $\boldsymbol{n}$. We can reformulate (4.7) as follows. If we set $\mathsf{K} = \mathsf{K}_1 + \mathsf{K}_2$ in (4.7) and expand, then we get that for any $\{\mathsf{K}_1, \mathsf{K}_2\} \subset \Pi$ and any direction $\boldsymbol{n}$

$$\mathsf{K}_1\,\mathsf{A}(\boldsymbol{n})\mathsf{K}_2 + \mathsf{K}_2\,\mathsf{A}(\boldsymbol{n})\mathsf{K}_1 \in \Pi. \tag{4.8}$$

Since Π is a subspace we can make all possible linear combinations of expressions in (4.8) with the same K_1 and K_2 and different $\boldsymbol{n}$. Therefore,

$$\mathsf{K}_1\,\mathsf{AK}_2 + \mathsf{K}_2\,\mathsf{AK}_1 \in \Pi, \qquad \text{for all } \{\mathsf{K}_1, \mathsf{K}_2\} \subset \Pi,\ \mathsf{A} \in \mathcal{A}, \tag{4.9}$$

[3] In fact the operator $\mathsf{I} - \mathsf{KA}(\boldsymbol{n})$ is invertible for *any* $\mathsf{K} = W_{\boldsymbol{n}_0}(\mathsf{L})$ by corollary 3.25.

where

$$\mathcal{A} = \text{Span}\left\{\Gamma_0(\boldsymbol{n}) - \Gamma_0(\boldsymbol{n}_0) : \boldsymbol{n} \in \mathbb{R}^d, |\boldsymbol{n}| = 1\right\}. \tag{4.10}$$

It is easy to see that the subspace $\mathcal{A}$ is independent of $\boldsymbol{n}_0$, because

$$\Gamma_0(\boldsymbol{n}) - \Gamma_0(\boldsymbol{n}_1) = \Gamma_0(\boldsymbol{n}) - \Gamma_0(\boldsymbol{n}_0) - \left(\Gamma_0(\boldsymbol{n}_1) - \Gamma_0(\boldsymbol{n}_0)\right).$$

A more natural way to see it is to define $\mathcal{A}$ without reference to $\boldsymbol{n}_0$. Indeed, $\mathcal{A}$ is the vector space parallel to the affine hull of the set $\{\Gamma_0(\boldsymbol{n}) : |\boldsymbol{n}| = 1\}$. In summary, in order for a submanifold $\mathbb{M}$ given by (4.4) to be an exact relation, it is necessary for $W_{\boldsymbol{n}_0}(\mathbb{M})$ to be a relatively open subset of a vector space $\Pi \subset \text{End}(\mathcal{T})$ satisfying (4.9). We observe that no further information about the subspace Π can be obtained from the expansion (4.6), since all other terms in (4.6) will necessarily be in Π, if Π satisfies (4.9). This is proved by induction in n, since

$$\mathsf{K}(\mathsf{A}(\boldsymbol{n})\mathsf{K})^n = \frac{1}{2}\left\{\mathsf{KA}(\boldsymbol{n})\left[\mathsf{K}(\mathsf{A}(\boldsymbol{n})\mathsf{K})^{n-1}\right] + \left[\mathsf{K}(\mathsf{A}(\boldsymbol{n})\mathsf{K})^{n-1}\right]\mathsf{A}(\boldsymbol{n})\mathsf{K}\right\}.$$

The property (4.9) of a subspace $\Pi \subset \text{End}(\mathcal{T})$ can be interpreted algebraically. It means that Π is closed with respect to a family of commutative, but non-associative multiplications

$$\mathsf{K}_1 *_{\mathsf{A}} \mathsf{K}_2 = \frac{1}{2}\left(\mathsf{K}_1\mathsf{AK}_2 + \mathsf{K}_2\mathsf{AK}_1\right), \qquad \mathsf{A} \in \mathcal{A}. \tag{4.11}$$

Each of the products in the family (4.11) is an example of a Jordan product [22, 21], making the subspace Π into a Jordan subalgebra of $\text{End}(\mathcal{T})$. Hence, we introduce the following terminology.

Definition 4.3. *Let $\mathcal{A}$ be a subspace in* $\text{End}(\mathcal{T})$. *The subspace* $\Pi \subset \text{End}(\mathcal{T})$ *is called a Jordan* $\mathcal{A}$-multialgebra *if it is closed with respect to the family of Jordan multiplications (4.11).*

The next theorem shows that algebraic property (4.9) is also sufficient for stability under lamination.

Theorem 4.4. *Suppose that the subspace $\Pi \subset \text{End}(\mathcal{T})$ is a Jordan $\mathcal{A}$-multialgebra, where $\mathcal{A}$ is given by (4.10). Then for any $\mathsf{L}_0 \in \text{End}^+(\mathcal{T})$ and any unit vector $\boldsymbol{n}_0$ the submanifold $\mathbb{M} \subset \text{End}^+(\mathcal{T})$, given by (4.4), is an L-relation and does not depend on the choice of $\boldsymbol{n}_0$ in its definition.*

Proof. Let us take $\{\mathsf{L}_1, \mathsf{L}_2\} \subset \mathbb{M}$ and make a laminate with normal $\boldsymbol{n}$ and volume fractions θ, $1 - \theta$, respectively. By theorem 3.20, its effective tensor L^* must satisfy $W_{\boldsymbol{n}}(\mathsf{L}^*) = \theta W_{\boldsymbol{n}}(\mathsf{L}_1) + (1 - \theta)W_{\boldsymbol{n}}(\mathsf{L}_2)$. We note that corollary 3.43 guarantees that $\mathsf{L}^* \in \text{End}^+(\mathcal{T})$, since $\{\mathsf{L}_1, \mathsf{L}_2\} \subset \text{End}^+(\mathcal{T})$, by definition of $\mathbb{M}$. Our first step is to show that $\{W_{\boldsymbol{n}}(\mathsf{L}_1), W_{\boldsymbol{n}}(\mathsf{L}_2)\} \subset \Pi$.

Lemma 4.5. *Let Π be a Jordan $\mathcal{A}$-multialgebra. If $\mathsf{K}_0 \in \Pi$, $\mathsf{A}_0 \in \mathcal{A}$, and $\mathsf{I} - \mathsf{K}_0\mathsf{A}_0$ is an invertible operator on $\mathcal{T}$, then $(\mathsf{I} - \mathsf{K}_0\mathsf{A}_0)^{-1}\mathsf{K}_0 \in \Pi$.*

Proof. We first observe that the polynomial $p(t) = \det(\mathsf{I} - t\mathsf{K}_0\mathsf{A}_0)$ is such that $p(0) = 1$ and $p(1) \neq 0$. Therefore, there exists a curve $\gamma \subset \mathbb{C}$, joining 0 and 1, and its open neighborhood $U_\gamma \subset \mathbb{C}$, such that there are no zeros of $p(t)$ in U_γ. Then $1/p(t)$ is analytic on U_γ. Let $\mathsf{N} \in \mathrm{End}(\mathcal{T})$ be in the orthogonal complement of Π. Then the function

$$\phi(t) = \left(\left(\mathsf{I} - t\mathsf{K}_0\mathsf{A}_0\right)^{-1}\mathsf{K}_0, \mathsf{N}\right)_{\mathrm{End}(\mathcal{T})}, \qquad t \in \mathbb{C}$$

is also analytic on U_γ, since $p(t)\phi(t)$ is a polynomial in t. However, expanding $(\mathsf{I} - t\mathsf{K}_0\mathsf{A}_0)^{-1}$ into the Neumann series, we see that $\phi(t) = 0$ on a sufficiently small neighborhood of 0 in $\mathbb{C}$. The analyticity of $\phi(t)$ implies that $\phi(t) = 0$ for all $t \in U_\gamma$. In particular, $\phi(1) = 0$. The lemma follows. □

Corollary 4.6. *Let $\mathbb{M}$ be given by (4.4), where Π satisfies (4.9). Then $W_{\boldsymbol{n}}(\mathbb{M}) \subset \Pi$ for any unit vector $\boldsymbol{n}$.*

Proof. Let $\mathsf{L} \in \mathbb{M}$. Then $\mathsf{K} = W_{\boldsymbol{n}_0}(\mathsf{L}) \in \Pi$. For any unit vector $\boldsymbol{n}$ we compute

$$W_{\boldsymbol{n}}(\mathsf{L}) = W_{\boldsymbol{n}}\left(W_{\boldsymbol{n}_0}^{-1}(\mathsf{K})\right) = \boldsymbol{\Lambda}_{\boldsymbol{n}}(\mathsf{K}),$$

where the map $\boldsymbol{\Lambda}_{\boldsymbol{n}}$ is defined in (4.5). The operator $\mathsf{I} - \mathsf{K}\mathsf{A}(\boldsymbol{n})$ is invertible, by corollary 3.25. Then, by lemma 4.5 we conclude that $W_{\boldsymbol{n}}(\mathsf{L}) = \boldsymbol{\Lambda}_{\boldsymbol{n}}(\mathsf{K}) \in \Pi$. □

Corollary 4.6 implies that $\{W_{\boldsymbol{n}}(\mathsf{L}_1), W_{\boldsymbol{n}}(\mathsf{L}_2)\} \subset \Pi$ and hence $W_{\boldsymbol{n}}(\mathsf{L}^*) = \theta W_{\boldsymbol{n}}(\mathsf{L}_1) + (1 - \theta)W_{\boldsymbol{n}}(\mathsf{L}_2) \in \Pi$. Thus, $\mathsf{L}^* \in \mathbb{M}_{\boldsymbol{n}}$, where we (temporarily) define $\mathbb{M}_{\boldsymbol{n}}$ by (4.4) with $\boldsymbol{n}$ playing the role of $\boldsymbol{n}_0$. Now, corollary 4.6 implies that $\mathbb{M}_{\boldsymbol{n}_0} \subset \mathbb{M}_{\boldsymbol{n}}$ for any unit vector $\boldsymbol{n}$. Hence $\mathbb{M}_{\boldsymbol{n}} = \mathbb{M}_{\boldsymbol{n}_0}$, since the unit vector $\boldsymbol{n}_0$ was chosen arbitrarily. Theorem 4.4 is proved. □

4.3 Sufficient conditions for stability under homogenization

In this section we formulate simple algebraic sufficient conditions for an L-relation submanifold $\mathbb{M}$ to be an exact relation, i.e. to be G-closed.

Definition 4.7. *We say that a subspace $\Pi \subset \mathrm{End}(\mathcal{T})$ has the j-chain property, $j \geqslant 2$, if for every $\{\mathsf{K}_1, \ldots, \mathsf{K}_j\} \subset \Pi$ and every $\{\mathsf{A}_1, \ldots, \mathsf{A}_{j-1}\} \subset \mathcal{A}$ we have*

$$\mathsf{K}_1\mathsf{A}_1\mathsf{K}_2\mathsf{A}_2 \ldots \mathsf{K}_{j-1}\mathsf{A}_{j-1}\mathsf{K}_j + \mathsf{K}_j\mathsf{A}_{j-1}\mathsf{K}_{j-1} \ldots \mathsf{A}_2\mathsf{K}_2\mathsf{A}_1\mathsf{K}_1 \in \Pi. \qquad (4.12)$$

We remark that the two-chain property is the definition of Jordan $\mathcal{A}$-multialgebras. We can now state our sufficient conditions.

Theorem 4.8. *Suppose that the subspace* $\Pi \subset \mathrm{End}(\mathcal{T})$ *has two-, three- and four-chain properties, then the submanifold* $\mathbb{M}$ *given by (4.4) is an exact relation.*

The proof is placed in section 4.5.

When tensors of material properties are symmetric, as in most of our case studies in part 3, we can give a neater algebraic interpretation of the sufficient conditions in theorem 4.8. In particular, if the reference medium tensor L_0 is symmetric, then so is $\mathcal{A} \subset \mathrm{Sym}(\mathcal{T})$ due to lemma 3.11.

Definition 4.9. *A subspace* $\Pi' \subset \mathrm{End}(\mathcal{T})$ *is called an associative* $\mathcal{A}$*-multialgebra, if* $\mathsf{K}_1\mathsf{A}\mathsf{K}_2 \in \Pi'$ *for every* $\mathsf{A} \in \mathcal{A}$ *and every* $\{\mathsf{K}_1, \mathsf{K}_2\} \subset \Pi'$.

We observe that for every associative $\mathcal{A}$-multialgebra Π'

$$\Pi'_{\mathrm{sym}} = \Pi' \cap \mathrm{Sym}(\mathcal{T}) \tag{4.13}$$

is a Jordan $\mathcal{A}$-multialgebra that has three- and four-chain properties, since $\mathcal{A} \subset \mathrm{Sym}(\mathcal{T})$.

Definition 4.10. *Let* Π' *be the smallest associative* $\mathcal{A}$*-multialgebra, containing a Jordan* $\mathcal{A}$*-multialgebra* Π*. If* $\Pi = \Pi'_{\mathrm{sym}}$*, then we say that the Jordan* $\mathcal{A}$*-multialgebra* Π *is reflexive.*[4]

The following theorem characterizes reflexive Jordan $\mathcal{A}$-multialgebras.

Theorem 4.11. *Suppose* $\mathcal{A} \subset \mathrm{Sym}(\mathcal{T})$*. Then a Jordan* $\mathcal{A}$*-multialgebra* $\Pi \subset \mathrm{Sym}(\mathcal{T})$ *is reflexive if and only if it has three- and four-chain properties.*

The proof is placed in section 4.5. It is often easier to exhibit explicitly an associative $\mathcal{A}$-multialgebra, whose symmetric operators comprise the given Jordan $\mathcal{A}$-multialgebra, than to check the three- and four-chain properties. Remarkably, *all* Jordan multialgebras corresponding to the results in part 3 of this book are reflexive. While in general, reflexivity, or equivalently, the three- and four-chain properties are only sufficient for the manifold (4.4) to be an exact relation, in two space dimensions we can say more.

[4] Our notion of reflexivity is very similar, but different from the standard terminology, where reflexivity refers to the same property of the Jordan algebra with respect to its natural embedding into the space of symmetric elements of the associative enveloping algebra with natural involution [21].

Theorem 4.12. *In two space dimensions the three-chain property is necessary for a corresponding L-relation to be exact.*

This theorem was proved in [15]. For the sake of completeness the proof is also placed in section 4.5. Using this theorem we have constructed an example in [12] of an L-relation that is not exact in the context of multifield response materials [30–32] coupling four curl-free fields to four divergence-free fields in two space dimensions. The non-reflexive Jordan $\mathcal{A}$-multialgebra and the corresponding L-relation are given for the sake of completeness in appendix B.

We remark that in the classical theory of special Jordan algebras [21, 28] (i.e. when $\dim \mathcal{A} = 1$) the three-chain property is a consequence of the two-chain property. There are also many examples of special Jordan algebras for which the four-chain property is not satisfied. In this light, the fact that each and every Jordan $\mathcal{A}$-multialgebra in physical examples of part 3 is reflexive looks surprising. In order to explain this 'happy coincidence' we have examined the reflexivity of all Jordan subalgebras in $\mathrm{Sym}(\mathbb{R}^n)$ in [13]. As indicated in [28, section 1.5] only subalgebras containing so called 'spin factors' that live in the space of $2^n \times 2^n$ matrices could be non-reflexive. The smallest non-reflexive subalgebra of $\mathrm{Sym}(\mathbb{R}^n)$ occurs when $n = 8$, which is the dimension of $\mathcal{T}$ in our example of an L-relation that is not exact in appendix B. The explanation of the universal reflexivity in part 3 of the book is that, from the algebraic standpoint, all of the examples are sufficiently low-dimensional to contain non-reflexive Jordan $\mathcal{A}$-multialgebras.

4.4 Special types of exact relations

4.4.1 Polycrystalline exact relations

The definition of an exact relation and more generally, of a G-closed set, treats different orientations of the same anisotropic material as distinct materials. However, from the practical point of view it is not natural to demand that an anisotropic constituent be used only in a particular orientation. Composites in which any orientation of an anisotropic constituent is allowed are called *polycrystalline*. When we deal with polycrystalline composites we assume that the tensors of the material properties are symmetric. This is done for physical reasons, since conditions that break the symmetry of material tensors usually also break their rotational equivariance, as in the Hall effect [1, 2] (see chapter 8).

Definition 4.13. *A G-closed submanifold* $\mathbb{M} \subset \mathrm{Sym}^+(\mathcal{T})$ *is called a* polycrystalline exact relation if $\boldsymbol{R} \cdot \mathsf{L} \in \mathbb{M}$ *for all* $\boldsymbol{R} \in \mathrm{SO}(d)$ (d = 2 *or* 3), *where* $\boldsymbol{R} \cdot \mathsf{L}$ *denotes the tensor* L *in a rotated coordinate system.*

Since any material in $\mathbb{M}$ causes $\mathbb{M}$ to contain its entire $\mathrm{SO}(d)$-orbit, there are far fewer polycrystalline exact relations than general ones. Partly for this reason, it is much easier to characterize all polycrystalline exact relations. The rotational equivariance postulate 2.2, that we have not needed thus far, will become essential in much of part 2. We did not want to make postulate 2.2 into a standing assumption in this book because we also address situations where the isotropy of space is broken by an external magnetic field, for example. In the absence of special symmetry-breaking circumstances the tensor of material properties can usually be represented by symmetric matrices (see, e.g. [3]).

It is intuitively clear that a polycrystal with statistically isotropic texture must be isotropic, i.e. have an effective tensor L, such that

$$\boldsymbol{R} \cdot \mathsf{L} = \mathsf{L}, \qquad \forall \boldsymbol{R} \in \mathrm{SO}(d),\ d = 2, 3. \tag{4.14}$$

It means that any polycrystalline exact relation submanifold $\mathbb{M}$ must contain an isotropic tensor. In the theory of polycrystalline exact relations we will always use an isotropic reference medium $\mathsf{L}_0 \in \mathbb{M}$. Then, the rotational equivariance postulate 2.2 together with formula (3.13) for $\boldsymbol{\Gamma}_0(\boldsymbol{n})$ imply

$$\boldsymbol{R} \cdot \boldsymbol{\Gamma}_0(\boldsymbol{n}) = \boldsymbol{\Gamma}_0(\boldsymbol{R}\boldsymbol{n}) \tag{4.15}$$

for every $\boldsymbol{R} \in \mathrm{SO}(d)$ and every unit vector $\boldsymbol{n}$. Hence, we also have

$$\boldsymbol{R} \cdot W_{\boldsymbol{n}}(\mathsf{L}) = W_{\boldsymbol{R}\boldsymbol{n}}(\boldsymbol{R} \cdot \mathsf{L}), \qquad \forall \boldsymbol{R} \in \mathrm{SO}(d). \tag{4.16}$$

Recall that for every $\mathsf{K} \in \Pi$ that is sufficiently close to zero and any unit vector $\boldsymbol{n}$, there exists $\mathsf{L} \in \mathbb{M}$ (depending on both K and $\boldsymbol{n}$), such that $\mathsf{K} = W_{\boldsymbol{n}}(\mathsf{L})$. Thus, for any rotation $\boldsymbol{R} \in \mathrm{SO}(d)$ we have $\boldsymbol{R} \cdot \mathsf{K} = W_{\boldsymbol{R}\boldsymbol{n}}(\boldsymbol{R} \cdot \mathsf{L}) \in \Pi$, since $\boldsymbol{R} \cdot \mathsf{L} \in \mathbb{M}$. It follows that the Jordan multialgebra Π is a rotationally invariant subspace of $\mathrm{Sym}(\mathcal{T})$. It is easy to see that the subspace $\mathcal{A}$, given by (4.10) must also be rotationally invariant:

$$\begin{aligned}\boldsymbol{R} \cdot (\boldsymbol{\Gamma}_0(\boldsymbol{n}) - \boldsymbol{\Gamma}_0(\boldsymbol{n}_0)) &= \boldsymbol{\Gamma}_0(\boldsymbol{R}\boldsymbol{n}) - \boldsymbol{\Gamma}_0(\boldsymbol{R}\boldsymbol{n}_0) \\ &= \big(\boldsymbol{\Gamma}_0(\boldsymbol{R}\boldsymbol{n}) - \boldsymbol{\Gamma}_0(\boldsymbol{n}_0)\big) - \big(\boldsymbol{\Gamma}_0(\boldsymbol{R}\boldsymbol{n}_0) - \boldsymbol{\Gamma}_0(\boldsymbol{n}_0)\big) \in \mathcal{A}.\end{aligned}$$

In fact, there is a simpler alternative description of $\mathcal{A}$ that will permit us to compute the subspace $\mathcal{A}$ in a fairly straightforward way. Let $\mathrm{Sym}_{\mathrm{SO}(d)}(\mathcal{T})$ denote a subspace of all isotropic tensors in $\mathrm{Sym}(\mathcal{T})$, and let $\overline{\boldsymbol{\Gamma}}$ denote the orthogonal projection of $\boldsymbol{\Gamma}_0(\boldsymbol{n})$ onto $\mathrm{Sym}_{\mathrm{SO}(d)}(\mathcal{T})$. The rotational invariance of the orthogonal projection operator together with (4.15) implies that $\overline{\boldsymbol{\Gamma}}$ is independent of $\boldsymbol{n}$.

Lemma 4.14. *Suppose that the rotational equivariance postulate 2.2 holds and the reference medium L_0 is isotropic. Then there exists a single tensor $\tilde{\boldsymbol{\Gamma}} \in \mathrm{Sym}(\mathcal{T})$, such that*

$$\mathcal{A} = \mathrm{Span}\left\{\boldsymbol{R} \cdot \tilde{\boldsymbol{\Gamma}} : \boldsymbol{R} \in \mathrm{SO}(d)\right\}, \quad d = 2, 3. \tag{4.17}$$

Proof. Let $\tilde{\mathcal{A}}$ temporarily denote the right-hand side of (4.17). Let us fix a unit vector $\boldsymbol{n}_0$ and define $\tilde{\boldsymbol{\Gamma}} = \boldsymbol{\Gamma}_0(\boldsymbol{n}_0) - \overline{\boldsymbol{\Gamma}}$. For any other unit vector $\boldsymbol{n}$ we can find a rotation $\boldsymbol{R} \in \mathrm{SO}(d)$, such that $\boldsymbol{R}\boldsymbol{n}_0 = \boldsymbol{n}$. Then, using (4.15), we have

$$\Gamma_0(\boldsymbol{n}) - \Gamma_0(\boldsymbol{n}_0) = \boldsymbol{R} \cdot \Gamma_0(\boldsymbol{n}_0) - \Gamma_0(\boldsymbol{n}_0) = \boldsymbol{R} \cdot \tilde{\Gamma} - \tilde{\Gamma} \in \tilde{\mathcal{A}}. \tag{4.18}$$

It follows that $\mathcal{A} \subset \tilde{\mathcal{A}}$. To prove the reverse inclusion we apply Haar's theorem about the existence of a right- and left-invariant probability measure [18] to the group SO(d). Then,

$$\bar{\Gamma} = \int_{SO(d)} \boldsymbol{Q} \cdot \Gamma_0(\boldsymbol{n}_0)\, d\mu(\boldsymbol{Q}), \tag{4.19}$$

where μ is the Haar measure on SO(d). It follows that

$$\int_{SO(d)} \boldsymbol{R} \cdot \tilde{\Gamma}\, d\mu(\boldsymbol{R}) = 0. \tag{4.20}$$

Integrating (4.18) and applying (4.20) we conclude that $-\tilde{\Gamma} \in \mathcal{A}$, and the reverse inclusion follows. □

There are many fewer rotationally invariant subspaces than arbitrary ones in Sym($\mathcal{T}$). Moreover, the tools of representation theory of SO(2) and SO(3), developed for the study of angular momentum in quantum mechanics (see e.g. [9]), give a relatively straightforward method for describing them. For example, in the case of conductivity, there are only two rotationally invariant proper subspaces in Sym($\mathbb{R}^d$). They are $\mathbb{R}\boldsymbol{I}_d = \{\alpha \boldsymbol{I}_d : \alpha \in \mathbb{R}\}$ and

$$\mathrm{Sym}_0(\mathbb{R}^d) = \left\{\boldsymbol{X} \in \mathrm{End}(\mathbb{R}^d) \colon \boldsymbol{X}^T = \boldsymbol{X},\ \mathrm{Tr}\,\boldsymbol{X} = 0\right\}. \tag{4.21}$$

It is easy to check that for conductivity $\Gamma_0(\boldsymbol{n}) = \sigma_0^{-1}\boldsymbol{n} \otimes \boldsymbol{n}$, where $\sigma_0 \boldsymbol{I}_d$ is an isotropic reference medium. Hence, $\mathcal{A} = \mathrm{Sym}_0(\mathbb{R}^d)$. It is rather easy to verify which of the two rotationally invariant subspaces, if any, are Jordan $\mathcal{A}$-multialgebras. When $d = 3$ we verify that neither of the two is a Jordan $\mathcal{A}$-multialgebra. However, when $d = 2$ we discover that $\Pi = \mathrm{Sym}_0(\mathbb{R}^2)$ is in fact a Jordan $\mathcal{A}$-multialgebra. This proves that any polycrystalline G-closed set in Sym($\mathbb{R}^3$) must have a non-empty interior if it consists of more than one point. For two dimensional conductivity the hypersurfaces $\mathbb{M}_{d_0}$, given by (4.3), are G-closed.

4.4.2 Uniform field relations

There is a simple observation made by Lurie and Cherkaev [25] (although in the context of thermal expansion it dates back to the work of Cribb [6]) that for any given uniform fields $\{\boldsymbol{u}, \boldsymbol{v}\} \subset \mathcal{T}$ the submanifolds of the form

$$\mathcal{U}(\boldsymbol{u}, \boldsymbol{v}) = \left\{\mathsf{L} \in \mathrm{End}^+(\mathcal{T}) : \mathsf{L}\boldsymbol{u} = \boldsymbol{v}\right\} \tag{4.22}$$

are G-closed. Consequently, any intersection of them is also an exact relation, given by

$$\mathbb{M} = \left\{ \mathsf{L} \in \mathrm{End}^+(\mathcal{T}) : \mathsf{L}\boldsymbol{u}_1 = \boldsymbol{v}_1, \ldots, \mathsf{L}\boldsymbol{u}_k = \boldsymbol{v}_k \right\}. \tag{4.23}$$

The vectors $\boldsymbol{u}_i$, $\boldsymbol{v}_i$, $i = 1, \ldots, k$ are non-uniquely determined by the UFR $\mathbb{M}$. It is more convenient to identify the UFR by a subspace $U \subset \mathcal{T}$ and a reference medium $\mathsf{L}_0 \in \mathbb{M}$:

$$\mathbb{M} = \left(\mathsf{L}_0 + \mathrm{Ann}(U)\right) \bigcap \mathrm{End}^+(\mathcal{T}), \tag{4.24}$$

where

$$\mathrm{Ann}(U) = \left\{ \mathsf{K} \in \mathrm{End}(\mathcal{T}) : \mathsf{K}\boldsymbol{u} = 0 \text{ for all } \boldsymbol{u} \in U \right\}$$

is the 'annihilator' of U. In fact, the Jordan $\mathcal{A}$-multialgebra corresponding to $\mathbb{M}$ is exactly $\Pi = \mathrm{Ann}(U)$.

Theorem 4.15. *Suppose* $\mathbb{M}$ *is given by (4.24). Then* $W_n(\mathbb{M})$ *with reference tensor* $\mathsf{L}_0 \in \mathbb{M}$ *is an open subset in* $\Pi = \mathrm{Ann}(U)$. *Moreover,* Π *is always a reflexive Jordan* $\mathcal{A}$*-multialgebra in the sense of definition 4.10.*

Proof. While the problem of computing $\mathbb{M}$ from Π is technically non-trivial (we devote chapter 6 to this issue), computing the subspace Π from $\mathbb{M}$ is straightforward. From elementary differential geometry it is known that the differential $\mathrm{d}W_n$ at L_0 in an isomorphism between the tangent space of $\mathbb{M}$ at L_0 and Π. It is easy to compute that for $\mathrm{d}W_n(\mathsf{K}) = -\mathsf{K}$ for any $\mathsf{K} \in T\mathbb{M}_{\mathsf{L}_0}$ (tangent space of $\mathbb{M}$ at L_0). Thus, $\Pi = T\mathbb{M}_{\mathsf{L}_0}$. The statement of the theorem becomes obvious now. To prove reflexivity we only need to observe that Π is also an associative $\mathcal{A}$-multialgebra. □

It is a simple observation that a UFR is polycrystalline if and only if $U \subset \mathcal{T}$ in (4.24) is rotationally invariant. Since for polycrystals we have assumed that material tensors must be symmetric:

$$\mathbb{M} = \left(\mathsf{L}_0 + \mathrm{Ann}(U)\right) \bigcap \mathrm{Sym}^+(\mathcal{T}), \qquad \Pi = \mathrm{Ann}(U) \bigcap \mathrm{Sym}(\mathcal{T}). \tag{4.25}$$

Example 4.16. The list of all rotationally invariant proper subspaces of $\mathrm{Sym}(\mathbb{R}^d)$ shows that there are exactly two polycrystalline UFRs in both two- and three-dimensional elasticity. The UFR corresponding to $\mathrm{Sym}_0(\mathbb{R}^d)$ is Hill's exact relation [19, 20], that we have mentioned at the beginning of this chapter. It says that any composite made with any number of isotropic materials with constant shear modulus μ_0 must necessarily be isotropic and have the same shear modulus μ_0. The UFR corresponding to $\mathbb{R}\boldsymbol{I}_d$ says that the set of all of Hooke's laws C such that $\mathsf{C}\boldsymbol{I}_d = d\kappa_0\boldsymbol{I}_d$, where κ_0 is a given constant (bulk modulus) is an exact relation. This exact relation is also due to Hill [20], as well as Lurie, Cherkaev and Fedorov [26].

4.5 Proofs of theorems 4.8, 4.11, 4.12†

4.5.1 Proof of theorem 4.8

Observe that $W_n(\mathsf{L}_0) = 0$. Therefore, W_n maps a sufficiently small neighborhood of L_0 in $\mathbb{M}$ onto a small neighborhood of 0 in Π. Let us show that without loss of generality we can assume that $\mathsf{L}(z) \in \mathbb{M}$ is sufficiently close to L_0, or, equivalently, that $W_n(\mathsf{L}(z))$ is sufficiently small (uniformly in $z \in Q$).

Lemma 4.17. *Suppose* $\mathbb{M}$ *is given by (4.4), where* Π *is a Jordan* $\mathcal{A}$*-multialgebra. Let us assume that there exists* $\epsilon > 0$*, such that* $\mathsf{L}^* \in \mathbb{M}$ *for every periodic composite* $\mathsf{L}(z) \in \mathbb{M}$*, satisfying* $|W_n(\mathsf{L}(z))| < \epsilon$*. Then* $\mathsf{L}^* \in \mathbb{M}$ *for any periodic composite with uniformly bounded local tensor* $\mathsf{L}(z) \in \mathbb{M}$.

Proof. Let $\mathsf{L}(z)$ be uniformly bounded and $\mathsf{L}(z) \in \mathbb{M}$ for all $z \in Q$. Then $\mathsf{K}(z) = W_n(\mathsf{L}(z))$ is a uniformly bounded function, satisfying $\mathsf{K}(z) \in \Pi$. Thus, there exists $\delta > 0$ so small that for every $0 < \theta < \delta$ we have, $\theta|\mathsf{K}(z)| < \epsilon$. Let $\mathsf{L}_\theta(z) = W_n^{-1}(\theta \mathsf{K}(z))$. Then, by the lamination formula (3.42), $\mathsf{L}_\theta(z)$ is the effective tensor of the laminate of $\mathsf{L}(z)$ and L_0 taken in volume fractions θ and $1 - \theta$, respectively, with lamination direction $\boldsymbol{n}$, for each fixed $z \in Q$. Theorem 4.4 then implies that $\mathsf{L}_\theta(z) \in \mathbb{M}$, for every $0 \leqslant \theta \leqslant 1$, and every $z \in Q$. By assumption, $\mathsf{L}_\theta^* \in \mathbb{M}$ for every $0 \leqslant \theta < \delta$. That means that $f(\theta) = (W_n(\mathsf{L}_\theta^*), \mathsf{P})_{\mathrm{End}(\mathcal{T})} = 0$ for every P in the orthogonal complement to Π in $\mathrm{End}(\mathcal{T})$ and every $0 \leqslant \theta < \delta$. By formula (3.37)

$$W_n\left(\mathsf{L}_\theta^*\right) = \left\langle \mathsf{K}(z)\left(\theta^{-1}\mathsf{I} - \Lambda_n \mathsf{K}(z)\right)^{-1} \right\rangle \quad \forall \theta \in (0, 1],$$

where $\Lambda_n = \Lambda_{\mathsf{M}}$, given by (3.27) with $\mathsf{M} = \boldsymbol{\Gamma}_0(\boldsymbol{n})$. By corollary 3.19 $\mathsf{M} = \boldsymbol{\Gamma}_0(\boldsymbol{n})$ has the universal invertibility property and hence, by lemma 3.14 the operators $\theta^{-1}\mathsf{I} - \Lambda_n \mathsf{K}$ must be invertible for all $\theta \in (0, 1]$. Thus, θ^{-1} is not in the spectrum of $\Lambda_n \mathsf{K}$ for all $\theta \in (0, 1]$. But then the function $f(\theta)$ must be analytic on a neighborhood of $\theta \in [0, 1]$ in the complex plane [35]. It follows that $f(\theta) = 0$ for all $\theta \in [0, 1]$, since $f(\theta) = 0$ for all $0 \leqslant \theta < \delta$. This proves that $W_n(\mathsf{L}^*) \in \Pi$, which implies that $\mathsf{L}^* \in \mathbb{M}$. □

Next we combine lemma 4.17 with the Neumann expansion (3.38) for the effective tensor of a periodic composite in order to obtain a purely algebraic criterion for stability of a manifold $\mathbb{M}$ under homogenization.

Lemma 4.18. *Let* $\Pi_{\mathbb{C}} = \{\mathsf{K}_1 + i\mathsf{K}_2 \,|\, \{\mathsf{K}_1, \mathsf{K}_2\} \subset \Pi\}$ *be the complexification of* Π*. Then* $\mathbb{M}$*, given by (4.4), is G-closed if and only if*

$$\left\langle \left(\mathsf{K}\Lambda_n\right)^k \mathsf{K}(z) \right\rangle \in \Pi_{\mathbb{C}} \tag{4.26}$$

for every $\mathsf{K} \in L^\infty(Q; \Pi_{\mathbb{C}})$*, which holds if and only if* $\langle (\mathsf{K}\Lambda_n)^k \mathsf{K}(z) \rangle \in \Pi$ *for every* $\mathsf{K} \in L^\infty(Q; \Pi)$.

Proof. If $\mathbb{M}$ is G-closed and $\mathsf{K} \in L^\infty(Q; \Pi)$, then for sufficiently small (in absolute value) ϵ we have $\mathsf{L}_\epsilon(z) = W_n^{-1}(\epsilon\mathsf{K}(x)) \in \mathbb{M}$. Hence, $\mathsf{L}_\epsilon^* \in \mathbb{M}$, which, in turn, implies that $W_n(\mathsf{L}_\epsilon^*) \in \Pi$. By (3.38)

$$W_n\left(\mathsf{L}_\epsilon^*\right) = \sum_{k=0}^{\infty} \epsilon^{k+1}\left\langle \left(\mathsf{K}\Lambda_n\right)^k \mathsf{K}(z)\right\rangle. \tag{4.27}$$

Hence, (4.26) holds for every $\mathsf{K} \in L^\infty(Q; \Pi)$. Conversely, if (4.26) holds for every $\mathsf{K} \in L^\infty(Q; \Pi)$, then formula (4.27) proves that $W_n(\mathsf{L}^*) \in \Pi$, provided $\mathsf{L}(z) \in \mathbb{M}$ and $\mathsf{K}(z) = W_n(\mathsf{L}(z))$ is sufficiently small. Lemma 4.17 now guarantees that $\mathbb{M}$ is G-closed. It is obvious that if (4.26) holds for every $\mathsf{K} \in L^\infty(Q; \Pi_{\mathbb{C}})$, then it also holds for every $\mathsf{K} \in L^\infty(Q; \Pi)$. To finish the proof of the lemma we only need to show that if (4.26) holds for every $\mathsf{K} \in L^\infty(Q; \Pi)$, then it also holds for every $\mathsf{K} \in L^\infty(Q; \Pi_{\mathbb{C}})$.

For any $\{\mathsf{K}_1, \mathsf{K}_2\} \subset L^\infty(Q; \Pi)$ and $\lambda \in \mathbb{C}$ we define $\mathsf{K}_\lambda(z) = \mathsf{K}_1(z) + \lambda\mathsf{K}_2(z)$. Observe that $\mathsf{K}_\lambda \in L^\infty(Q; \Pi)$ for any $\lambda \in \mathbb{R}$, and therefore, for any $\mathsf{P} \in \Pi^\perp$, and any $k \geqslant 1$ we have

$$p(\lambda) = \left(\left\langle \left(\mathsf{K}_\lambda\Lambda_n\right)^k \mathsf{K}_\lambda(z)\right\rangle, \mathsf{P}\right)_{\mathrm{End}(\mathcal{T})} = 0, \qquad \lambda \in \mathbb{R}.$$

Observe that $p(\lambda)$ is a polynomial in λ. Therefore, if it vanishes on $\mathbb{R}$ it must also vanish on $\mathbb{C}$. Hence, $p(i) = 0$, and (4.26) is proved for any $\mathsf{K} \in L^\infty(Q; \Pi_{\mathbb{C}})$. □

Now we convert (4.26) into a rather messy, but entirely algebraic condition. For this purpose we will need extra notation, which will not be used outside of this section. Let S_k, $k \geqslant 2$, denote the symmetric group, i.e. the group of all permutations of k elements. For $\boldsymbol{v} = (\boldsymbol{l}_1, \ldots, \boldsymbol{l}_k) \in (\mathbb{Z}^d)^k$ and for $\sigma \in S_k$ we define

$$\sigma(\boldsymbol{v}) = \left(\boldsymbol{l}_{\sigma(1)}, \ldots, \boldsymbol{l}_{\sigma(k)}\right) \in \left(\mathbb{Z}^d\right)^k. \tag{4.28}$$

Let $\mathcal{O}(\boldsymbol{v}) = \{\sigma(\boldsymbol{v}) \in (\mathbb{Z}^d)^k | \sigma \in S_k\}$ be the orbit of $\boldsymbol{v}$ under the action of the permutation group S_k. The number of elements in $\mathcal{O}(\boldsymbol{v})$ will depend not only on k, but also on how many vectors in the list $\boldsymbol{v}$ coincide. Let $\mathcal{F} \subset \mathbb{Z}^d$ contain at most k vectors. Let $\mathfrak{T}_k(\mathcal{F})$ denote the set of all k-tuples $\boldsymbol{v} = (\boldsymbol{l}_1, \ldots, \boldsymbol{l}_k) \in (\mathbb{Z}^d)^k$ that satisfy

$$\sum_{i=1}^{k} \boldsymbol{l}_i = \mathbf{0}, \tag{4.29}$$

and have the property that $\{\boldsymbol{l}_1, \ldots, \boldsymbol{l}_k\} \subset \mathcal{F}$. Obviously, if $\boldsymbol{v} \in \mathfrak{T}_k(\mathcal{F})$ then so is $\sigma(\boldsymbol{v})$ for any $\sigma \in S_k$. We denote by $\mathcal{Z}_k(\mathcal{F})$ the set of all S_k orbits in $\mathfrak{T}_k(\mathcal{F})$. The set $\mathfrak{T}_k(\mathcal{F})$ is a disjoint union of all orbits in $\mathcal{Z}_k(\mathcal{F})$. Of course, sets $\mathfrak{T}_k(\mathcal{F})$ are empty for many sets $\mathcal{F}$, since (4.29) may never be satisfied. In that case the sets $\mathcal{Z}_k(\mathcal{F})$ are also empty. Finally we denote by $\mathfrak{S}_k(\mathbb{Z}^d)$ the set of all subset $\mathcal{F} \subset \mathbb{Z}^d$, such as $|\mathcal{F}| \leqslant k$. Using this notation we can now formulate an algebraic criterion for (4.26) to hold.

Lemma 4.19. *Condition (4.26) is satisfied if and only if for any* $k \geqslant 2$, *any* $\mathcal{F} \in \mathfrak{S}_k(\mathbb{Z}^d)$ *and for any function* $\hat{\mathsf{K}} : \mathcal{F} \to \Pi_{\mathrm{C}}$ *we have*

$$\sum_{\mathcal{O} \in \mathcal{Z}_k(\mathcal{F})} \sum_{(l_1, \ldots l_k) \in \mathcal{O}} \left(\prod_{s=1}^{k-1} \hat{\mathsf{K}}(l_s) \mathsf{A}_n\left(\sum_{j=1}^{s} l_j \right) \right) \hat{\mathsf{K}}(l_k) \in \Pi_{\mathrm{C}}, \tag{4.30}$$

where

$$\mathsf{A}_n(\boldsymbol{k}) = \begin{cases} \boldsymbol{\Gamma}_0\left(\dfrac{Q^T \boldsymbol{k}}{\left| Q^T \boldsymbol{k} \right|} \right) - \boldsymbol{\Gamma}_0(\boldsymbol{n}), & \boldsymbol{k} \neq 0, \\ 0, & \boldsymbol{k} = 0. \end{cases} \tag{4.31}$$

Proof. For $\mathsf{K} \in L^\infty(Q; \Pi)$ we denote $\mathsf{T}_k(z) = (\mathsf{K}\Lambda_n)^k \mathsf{K}(z)$. Taking the Fourier transform of the recursion formula $\mathsf{T}_k(z) = \mathsf{K}(z)(\Lambda_n \mathsf{T}_{k-1})(z)$ and using induction in k, we can prove that

$$\hat{\mathsf{T}}_{k-1}(\boldsymbol{m}) = \sum_{l_1 + \cdots + l_k = \boldsymbol{m}} \left(\prod_{s=1}^{k-1} \hat{\mathsf{K}}(l_s) \mathsf{A}_n\left(\boldsymbol{m} - \sum_{j=1}^{s} l_j \right) \right) \hat{\mathsf{K}}(l_k),$$

where $\boldsymbol{m}, \boldsymbol{l}_j \in \mathbb{Z}^d$. Thus, we obtain

$$\left\langle (\mathsf{K}\Lambda_n)^{k-1} \mathsf{K}(z) \right\rangle = \hat{\mathsf{T}}_{k-1}(\boldsymbol{0}) = \sum_{l_1 + \cdots + l_k = \boldsymbol{0}} \left(\prod_{s=1}^{k-1} \hat{\mathsf{K}}(l_s) \mathsf{A}_n\left(\sum_{j=1}^{s} l_j \right) \right) \hat{\mathsf{K}}(l_k). \tag{4.32}$$

Observe that the sum in (4.32) can be split into parts. The summation in each part goes over all the distinct permutations of a list of vectors $\boldsymbol{l}_j$, satisfying (4.29). Then we can write

$$\left\langle (\mathsf{K}\Lambda_n)^{k-1} \mathsf{K}(z) \right\rangle = \sum_{\mathcal{O} \in \mathcal{Z}} \sum_{(l_1, \ldots, l_k) \in \mathcal{O}} \left(\prod_{s=1}^{k-1} \hat{\mathsf{K}}(l_s) \mathsf{A}_n\left(\sum_{j=1}^{s} l_j \right) \right) \hat{\mathsf{K}}(l_k), \tag{4.33}$$

where $\mathcal{Z}$ denote the set of all S_k orbits, whose elements satisfy (4.29). The set $\mathcal{Z}$ can be further split into disjointed parts labeled by the set of $\mathbb{Z}^d$ vectors contained in an orbit in $\mathcal{Z}$:

$$\mathcal{Z} = \bigcup_{\mathcal{F} \in \mathfrak{S}_k(\mathbb{Z}^d)} \mathcal{Z}_k(\mathcal{F}).$$

Thus we obtain the representation

$$\left\langle (\mathsf{K}\Lambda_n)^{k-1} \mathsf{K}(z) \right\rangle = \sum_{\mathcal{F} \in \mathfrak{S}_k(\mathbb{Z}^d)} \sum_{\mathcal{O} \in \mathcal{Z}_k(\mathcal{F})} \sum_{(l_1, \ldots, l_k) \in \mathcal{O}} \left(\prod_{s=1}^{k-1} \hat{\mathsf{K}}(l_s) \mathsf{A}_n\left(\sum_{j=1}^{s} l_j \right) \right) \hat{\mathsf{K}}(l_k). \tag{4.34}$$

If (4.30) holds, then every term of the outer sum in (4.34) is in Π_C, since $\hat{K}(p) \in L^\infty(Q; \Pi_C)$ for any $p \in \mathbb{Z}^d$ and any $K \in L^\infty(Q; \Pi)$. Thus, property (4.26) is a consequence of (4.30).

Now suppose that (4.26) holds. Let us fix an arbitrary subset $\mathcal{F} = \{p_1, \ldots, p_s\} \subset \mathbb{Z}^d$, where $s \leqslant k$. Let $\tilde{K} : \mathcal{F} \to \Pi_C$ be an arbitrary function. Let

$$K(z) = \sum_{j=1}^{s} \tilde{K}(p_j) e^{2\pi i p_j \cdot Qz}.$$

Then $K(z) \in L^\infty(Q; \Pi_C)$ and its Fourier coefficients are

$$\hat{K}(m) = \begin{cases} \tilde{K}(m), & \text{if } m \in \mathcal{F}, \\ 0, & \text{otherwise.} \end{cases}$$

According to (4.34)

$$\left\langle (K\Lambda_n)^{k-1} K(z) \right\rangle = \sum_{\mathcal{O} \in \mathcal{Z}_k(\mathcal{F})} \sum_{(l_1, \ldots, l_k) \in \mathcal{O}} \left(\prod_{s=1}^{k-1} \hat{K}(l_s) A_n \left(\sum_{j=1}^{s} l_j \right) \right) \hat{K}(l_k),$$

and (4.30) is proved. □

We now show that lemma 4.19 implies theorem 4.8. Let $\eta \in S_k$ be the permutation defined by $\eta(j) = k + 1 - j$. In other words it maps $1, 2, \ldots, k$ into $k, k-1, \ldots, 2, 1$. Therefore, η is an element of order 2 in the group S_k. It acts on $\mathcal{O}(v)$ and splits $\mathcal{O}(v)$ into a disjoint union of orbits of the group $\{1, \eta\}$. If an orbit contains two elements then their sum has the form (4.12). If the orbit contains a single element then this element is 1/2 of the sum of two copies of itself, which is again of the form (4.12). This shows that if Π has the j-chain property for all $j \geqslant 2$, then (4.29) is satisfied, and hence, by lemmas 4.19 and 4.18, the submanifold $\mathbb{M}$ is G-closed. The proof of theorem 4.8 is completed by the following lemma.

Lemma 4.20. *Suppose* Π *has the j-chain property described in definition 4.7 for* $j = 2$, 3 *and* 4, *then* Π *satisfies the j-chain property for all* $j \geqslant 2$.

Proof. The proof is a straightforward modification of the proof of Cohn's theorem [5, 21]. For the sake of completeness we give a sketch of the proof. The lemma is proved by induction in j. If $j \leqslant 4$ the statement is true by assumption. Let $\{A_1, \ldots, A_{j-1}\} \subset \mathcal{A}$ and $\{K_1, \ldots, K_j\} \subset \Pi$. Let us rewrite the two-chain property defining Jordan multialgebras as

$$\mathsf{K}_1\,\mathsf{A}\mathsf{K}_2 = -\mathsf{K}_2\,\mathsf{A}\mathsf{K}_1 + \mathsf{K}' \tag{4.35}$$

for some $\mathsf{K}' \in \Pi$ and any $\mathsf{A} \in \mathcal{A}$. This identity allows us to swap successively the positions of adjacent Ks in any j-chain, leaving $(j-1)$-chains as remainder. Using a three-chain property we can write

$$\mathsf{K}_1\,\mathsf{A}_1\mathsf{K}_2\,\mathsf{A}_2\mathsf{K}_3 = \mathsf{K}_1\,\mathsf{A}_2\mathsf{K}_2\,\mathsf{A}_1\mathsf{K}_3 + \mathsf{K}'' + \text{ a sum of two-chains}$$

for some $\mathsf{K}'' \in \Pi$. This allows us to swap successively the positions of adjacent As in any j-chain, leaving $(j-1)$-chains and $(j-2)$-chains as the remainder. By first reversing the order of the As and then reversing the order of the Ks we see that

$$\begin{aligned}\mathsf{K}_1\mathsf{A}_1\mathsf{K}_2\mathsf{A}_2\mathsf{K}_3\mathsf{A}_3 \ldots \mathsf{K}_{j-1}\mathsf{A}_{j-1}\mathsf{K}_j = (-1)^{j(j-1)/2}\mathsf{K}_j\mathsf{A}_{j-1}\mathsf{K}_{j-1} \ldots \mathsf{A}_3\mathsf{K}_3\mathsf{A}_2\mathsf{K}_2\mathsf{A}_1\mathsf{K}_1 \\ + \text{a sum of } (j-1)\text{-chains and } (j-2)\text{-chains},\end{aligned}$$

where $j(j-1)/2$ is the number of swaps of adjacent Ks needed to achieve this reordering. If the sign of $(-1)^{j(j-1)/2}$ is negative, then the inductive hypothesis establishes the j-chain property. However, if the sign is positive (as it is when $j = 4$), then we use the four-chain property once, replacing the chain header

$$\mathsf{K}_1\,\mathsf{A}_1\mathsf{K}_2\,\mathsf{A}_2\mathsf{K}_3\,\mathsf{A}_3\mathsf{K}_4 \quad \text{with} \quad -\mathsf{K}_4\,\mathsf{A}_3\mathsf{K}_3\,\mathsf{A}_2\mathsf{K}_2\,\mathsf{A}_1\mathsf{K}_1 + \mathsf{K}'$$

where $\mathsf{K}' \in \Pi$. Now, if we swap the As with the three-chain property and the Ks with the two-chain property, we obtain minus the reversed j-chain plus shorter chains. By induction, the j-chain property will hold for all $j > 4$. □

4.5.2 Proof of theorem 4.12

Let us show that in two dimensions the necessary and sufficient condition (4.30) from lemma 4.19 for $k = 3$ implies the three-chain property. First we observe that any unit vector $\boldsymbol{n}$ can be approximated with any degree of accuracy by $\boldsymbol{Q}^T\boldsymbol{k}_0/|\boldsymbol{Q}^T\boldsymbol{k}_0|$ for some $\boldsymbol{k}_0 \in \mathbb{Z}^2$. Hence, without loss of generality, we can regard that $\boldsymbol{n} = \boldsymbol{Q}^T\boldsymbol{k}_0/|\boldsymbol{Q}^T\boldsymbol{k}_0|$. In that case $\mathsf{A}_{\boldsymbol{n}}(\boldsymbol{k}_0) = 0$, according to (4.31). Let $\boldsymbol{k}_1 \in \mathbb{Z}^2$ be an arbitrary vector linearly independent with $\boldsymbol{k}_0$ and define $\boldsymbol{l}_1 = m_1\boldsymbol{k}_1$, $\boldsymbol{l}_2 = -m_0\boldsymbol{k}_0 - m_1\boldsymbol{k}_1$, $\{m_0, m_1\} \subset \mathbb{Z}\backslash\{0\}$, so that $\boldsymbol{l}_3 = -\boldsymbol{l}_1 - \boldsymbol{l}_2 = m_0\boldsymbol{k}_0$. Observe that $\boldsymbol{l}_1$, $\boldsymbol{l}_2$ and $\boldsymbol{l}_3$ are non-zero and distinct. Therefore $\mathcal{Z}_k(\{\boldsymbol{l}_1, \boldsymbol{l}_2, \boldsymbol{l}_3\}) = \{\mathcal{O}(\boldsymbol{l}_1, \boldsymbol{l}_2, \boldsymbol{l}_3)\}$. Condition (4.30) for $k = 3$ and $\mathcal{F} = \{\boldsymbol{l}_1, \boldsymbol{l}_2, \boldsymbol{l}_3\}$ becomes

$$\mathsf{K}_2\mathsf{A}_{\boldsymbol{n}}(m_0\boldsymbol{k}_0 + m_1\boldsymbol{k}_1)\mathsf{K}_3\mathsf{A}_{\boldsymbol{n}}(\boldsymbol{k}_1)\mathsf{K}_1 + \mathsf{K}_1\mathsf{A}_{\boldsymbol{n}}(\boldsymbol{k}_1)\mathsf{K}_3\mathsf{A}_{\boldsymbol{n}}(m_0\boldsymbol{k}_0 + m_1\boldsymbol{k}_1)\mathsf{K}_2 \in \Pi$$

for all $\{\mathsf{K}_1, \mathsf{K}_2, \mathsf{K}_3\} \subset \Pi$. Now, if we vary $\boldsymbol{k}_1 \in \mathbb{Z}^2\backslash\mathbb{R}\boldsymbol{k}_0$ and $\{m_0, m_1\} \subset \mathbb{Z}\backslash\{0\}$, we observe that the set of pairs of unit vectors

$$\mathcal{P} = \left\{\left(\frac{\boldsymbol{k}_1}{|\boldsymbol{k}_1|}, \frac{m_0\boldsymbol{k}_0 + m_1\boldsymbol{k}_1}{|m_0\boldsymbol{k}_0 + m_1\boldsymbol{k}_1|}\right) : \boldsymbol{k}_1 \in \mathbb{Z}^2\backslash\mathbb{R}\boldsymbol{k}_0, \ \ \{m_0, m_1\} \subset \mathbb{Z}\backslash\{0\}\right\}$$

is a dense subset in $\mathbb{S}^1 \times \mathbb{S}^1$, where $\mathbb{S}^1$ denotes the unit circle. Indeed, for any unit vector $\boldsymbol{u}$ we can choose $\boldsymbol{k}_1 \in \mathbb{Z}^2$, such that $\boldsymbol{k}_1/|\boldsymbol{k}_1|$ approximates $\boldsymbol{u}$ with any degree of accuracy. Then

$$\frac{m_0\boldsymbol{k}_0 + m_1\boldsymbol{k}_1}{|m_0\boldsymbol{k}_0 + m_1\boldsymbol{k}_1|} = \frac{s\boldsymbol{N} + r\boldsymbol{u}}{|s\boldsymbol{N} + r\boldsymbol{u}|}, \quad \boldsymbol{N} = \frac{\boldsymbol{k}_0}{|\boldsymbol{k}_1|}, \quad r = \frac{m_1}{|m_0|}, \quad s = \text{sign}(m_0).$$

By our construction the vectors $\boldsymbol{u}$ and $\boldsymbol{N}$ are linearly independent in $\mathbb{R}^2$. It is now clear that for any unit vector $\boldsymbol{v}$ we can choose the sign s and a non-zero rational number r, so that $(s\boldsymbol{N} + r\boldsymbol{u})/|s\boldsymbol{N} + r\boldsymbol{u}|$ approximates the unit vector $\boldsymbol{v}$ with any degree of accuracy.

We note that the same construction does not yield a dense subset in $\mathbb{S}^2 \times \mathbb{S}^2$. Therefore, by continuity, we conclude that for any $\{\boldsymbol{q}_1, \boldsymbol{q}_2\} \subset \mathbb{S}^1$ we have

$$\mathsf{K}_2\mathsf{A}_n(\boldsymbol{q}_2)\mathsf{K}_3\mathsf{A}_n(\boldsymbol{q}_1)\mathsf{K}_1 + \mathsf{K}_1\mathsf{A}_n(\boldsymbol{q}_1)\mathsf{K}_3\mathsf{A}_n(\boldsymbol{q}_2)\mathsf{K}_2 \in \Pi$$

for all $\{\mathsf{K}_1, \mathsf{K}_2, \mathsf{K}_3\} \subset \Pi$. Fixing all elements in the above formula except either $\boldsymbol{q}_1$ or $\boldsymbol{q}_2$, and taking linear combinations, we conclude that the three-chain condition has to hold. Theorem 4.12 is proved.

4.5.3 Proof of theorem 4.11

Let us first assume that that $\Pi = \Pi'_{\text{sym}}$, where Π' is the smallest associative $\mathcal{A}$-multialgebra containing Π. Then for every $\{\mathsf{K}_1, \ldots, \mathsf{K}_j\} \subset \Pi$ and every $\{\mathsf{A}_1, \ldots, \mathsf{A}_{j-1}\} \subset \mathcal{A}$ we have

$$\tilde{\mathsf{K}} = \mathsf{K}_1\mathsf{A}_1\mathsf{K}_2\mathsf{A}_2 \ldots \mathsf{K}_{j-1}\mathsf{A}_{j-1}\mathsf{K}_j \in \Pi'.$$

Therefore,

$$\mathsf{K}_1\mathsf{A}_1\mathsf{K}_2\mathsf{A}_2 \ldots \mathsf{K}_{j-1}\mathsf{A}_{j-1}\mathsf{K}_j + \mathsf{K}_j\mathsf{A}_{j-1}\mathsf{K}_{j-1} \ldots \mathsf{A}_2\mathsf{K}_2\mathsf{A}_1\mathsf{K}_1 = \tilde{\mathsf{K}} + \tilde{\mathsf{K}}^T \in \Pi'_{\text{sym}} = \Pi.$$

In particular, three- and four-chain properties will hold.

Conversely, if the three and four-chain properties are satisfied, then by lemma 4.20 the j-chain properties hold for all j. To finish the proof it remains to observe that the smallest associative $\mathcal{A}$-multialgebra Π' containing Π has an explicit characterization:

$$\Pi' = \text{Span}\left\{\mathsf{K}_1\mathsf{A}_1 \ldots \mathsf{A}_{j-1}\mathsf{K}_j : \left\{\mathsf{K}_1, \ldots, \mathsf{K}_j\right\} \subset \Pi, \left\{\mathsf{A}_1, \ldots, \mathsf{A}_{j-1}\right\} \subset \mathcal{A}, j \geqslant 1\right\}.$$

It is now evident that if Π has j-chain properties for all $j \geqslant 1$, then $\tilde{\mathsf{K}} + \tilde{\mathsf{K}}^T \in \Pi$ for any $\tilde{\mathsf{K}} \in \Pi'$. Hence, $\Pi'_{\text{sym}} = \Pi$.

Bibliography

[1] Bergman David J 1982 Self-duality and the low field Hall effect in 2D and 3D metal-insulator composites. *Percolation structures and processes* ed G Deutscher, R Zallen and J Adler (*Annals of the Israel Physical Society* vol 5) (Israel Physical Society) pp 297–321 http://physics.technion.ac.il/~ips/volume5.html

[2] Briane M and Milton G 2009 Homogenization of the three-dimensional hall effect and change of sign of the hall coefficient *Arch. Ration. Mech. Anal.* **193** 715–36

[3] Callen H B 1960 *Thermodynamics* (New York: Wiley)

[4] Clausius R 1879 *Die Mechanische Behandlung der Electricität* (Braunschweig: Vieweg)

[5] Cohn P M 1954 On homomorphic images of special Jordan algebras *Canadian J. Math.* **6** 253–64

[6] Cribb J L 1968 Shrinkage and thermal expansion of a two-phase material *Nature* **220** 576–7

[7] Dvorak G J 1990 On uniform fields in heterogeneous media *Proc. R. Soc. London* A **431** 89–110

[8] Dykhne A M 1971 Conductivity of a two-dimensional two-phase system *Zh. Eksp. Teor. Fiz.* **59** 110–5 (in Russian) English translation is in
Dykhne A M 1971 Conductivity of a two-dimensional two-phase system *Sov. Phys.–JETP* **32** 63–5

[9] Edmonds A R 1996 *Angular Momentum in Quantum Mechanics* (Princeton, NJ: Princeton University Press)

[10] Einstein A 1906 Eine neue Bestimmung der Moleküldimensionen *Ann. Phys.* **324** 289–306

[11] Garnett J C M 1904 Colours in metal glasses and in metallic films *Phil. Trans. R. Soc. Lond.* A **203** 385–420

[12] Grabovsky Y 2017 From microstructure-independent formulas for composite materials to rank-one convex, non-quasiconvex functions *Arch. Ration. Mech. Anal.* to be published

[13] Grabovsky Y 2016 On reflexivity of jordan subalgebras of real symmetric matrices (in preparation)

[14] Grabovsky Y 1998 Exact relations for effective tensors of polycrystals. I: necessary conditions *Arch. Ration. Mech. Anal.* **143** 309–30

[15] Grabovsky Y 2004 Algebra, geometry and computations of exact relations for effective moduli of composites *Advances in Multifield Theories of Continua with Substructure* ed G Capriz and P M Mariano (*Modeling and Simulation in Science, Engineering and Technology*) (Boston, MA: Birkhäuser) pp 167–97

[16] Grabovsky Y and Milton G W 1998 Rank one plus a null-Lagrangian is an inherited property of two-dimensional compliance tensors under homogenization *Proc. R. Soc. Edin.* A **128** 283–99

[17] Grabovsky Y, Milton G W and Sage D S 2000 Exact relations for effective tensors of polycrystals: necessary conditions and sufficient conditions *Commun. Pure. Appl. Math.* **53** 300–53

[18] Haar A 1933 Der Massbegriff in der Theorie der Kontinuierlichen Gruppen *Ann. Math.* **34** 147–69

[19] Hill R 1963 Elastic properties of reinforced solids: some theoretical principles *J. Mech. Phys. Solids* **11** 357–72

[20] Hill R 1964 Theory of mechanical properties of fibre-strengthened materials: I. elastic behaviour *J. Mech. Phys. Solids* **12** 199–212

[21] Jacobson N 1968 *Structure and representations of jordan algebras* (*American Mathematical Society Colloquium Publications* vol 39) (Providence, RI: American Mathematical Society)

[22] Jordan P, Neumann J V and Wigner E 1934 On an algebraic generalization of the quantum mechanical formalism *Ann. Math.* **35** 29–64

[23] Keller J B 1964 A theorem on the conductivity of a composite medium *J. Math. Phys.* **5** 548–9

[24] Landauer R 1978 Electrical conductivity in inhomogeneous media *AIP Conf. Proc.* **40** 2–45

[25] Lurie K A and Cherkaev A V 1986 Effective characteristics of composite materials and the optimal design of structural elements *Usp. Mekh.* **9** 3–81 (in Russian) English translation in Lurie K A and Cherkaev A V 1997 Effective characteristics of composite materials and the optimal design of structural elements *Topics in the Mathematical Modeling of Composite Materials* ed A Cherkaev and R Kohn (*Progress in Nonlinear Differential equations and their Applications* vol 31) (Boston, MA: Birkhäuser) pp 175–258

[26] Lurie K A, Cherkaev A V and Fedorov A V 1984 On the existence of solutions to some problems of optimal design for bars and plates *J. Optim. Theory Appl.* **42** 247–81

[27] Maxwell J C 1881 *A Treatise on Electricity and Magnetism* vol 1, 2nd edn (Oxford: Clarendon)

[28] McCrimmon K 2004 *A Taste of Jordan Algebras* (Berlin: Springer)

[29] Mendelson K S 1975 A theorem on the conductivity of two-dimensional heterogeneous medium *J. Appl. Phys.* **46** 4740–1

[30] Milgrom M 1997 Some more exact results concerning multifield moduli of two-phase composites *J. Mech. Phys. Solids* **45** 399–404

[31] Milgrom M and Shtrikman S 1989 Linear response of polycrystals to coupled fields: exact relations among the coefficients *Phys. Rev.* B **40** 5991–4

[32] Milgrom M and Shtrikman S 1989 Linear response of two-phase composites with cross moduli: exact universal relations *Phys. Rev.* A **40** 1568–75

[33] Mossotti O F 1850 Discussione analitica sull'influenza che l'azione di un mezzo dielettrico ha sulla distribuzione dell'elettricità alla superficie di più corpi elettrici disseminati in esso *Mem. Mat. Fis. Soc. Ital. Sci. Residente Modena* **24** Part II 49–74

[34] Nevard J and Keller J B 1985 Reciprocal relations for effective conductivities of anisotropic media *J. Math. Phys.* **26** 2761–5

[35] Rudin W 1973 *Functional Analysis* (*McGraw-Hill Series in Higher Mathematics*) (New York: McGraw-Hill)

IOP Publishing

Chapter 5

Links

5.1 Links as exact relations

In addition to exact relations, the literature on composite media contains many microstructure-independent formulas that do not quite fit into the setting of exact relations. These formulas do not say anything about the effective properties of a composite, but rather relate effective tensors of two composites that have the same microstructure, but differ in the materials used in their construction. Perhaps the earliest link is due to Mendelson [6]. It says that if a periodic composite medium with local conductivity tensor $\boldsymbol{\sigma}(z)$ has an effective conductivity $\boldsymbol{\sigma}_*$, then the periodic composite with the local tensor

$$\boldsymbol{\sigma}'(z) = \frac{\boldsymbol{\sigma}(z)}{\det \boldsymbol{\sigma}(z)} \tag{5.1}$$

has the effective conductivity tensor $\boldsymbol{\sigma}_*'$ given by

$$\boldsymbol{\sigma}_*' = \frac{\boldsymbol{\sigma}_*}{\det \boldsymbol{\sigma}_*}. \tag{5.2}$$

For example, if we have a two-phase composite made with two isotropic materials with conductivities $\alpha\boldsymbol{I}$ and $\beta\boldsymbol{I}$, then $\boldsymbol{\sigma}' = (1/\alpha)\boldsymbol{I}$, wherever $\boldsymbol{\sigma} = \alpha\boldsymbol{I}$ and $\boldsymbol{\sigma}' = (1/\beta)\boldsymbol{I}$, wherever $\boldsymbol{\sigma} = \beta\boldsymbol{I}$. Observing that $\alpha\beta\boldsymbol{\sigma}' = \alpha\boldsymbol{I}$, wherever $\boldsymbol{\sigma} = \beta\boldsymbol{I}$, and $\beta\boldsymbol{I}$, wherever $\boldsymbol{\sigma} = \alpha\boldsymbol{I}$, we obtain the 'phase-interchange' formula [4, 5]:

$$\boldsymbol{\sigma}^*(\beta, \alpha) = \frac{\alpha\beta\boldsymbol{\sigma}^*(\alpha, \beta)}{\det \boldsymbol{\sigma}^*(\alpha, \beta)}.$$

If the composite is isotropic, i.e. $\boldsymbol{\sigma}^*(\alpha, \beta) = \sigma^*(\alpha, \beta)\boldsymbol{I}$, then the composite with interchanged materials is also isotropic, satisfying

doi:10.1088/978-0-7503-1048-2ch5

$$\sigma^*(\alpha, \beta)\sigma^*(\beta, \alpha) = \alpha\beta.$$

If the microstructure is insensitive to phase interchange, such as the checkerboard geometry [1], then

$$\sigma^*(\alpha, \beta) = \sigma^*(\beta, \alpha) = \sqrt{\alpha\beta}.$$

If we apply Mendelson's formula (5.2) to a polycrystal $\boldsymbol{\sigma}(z) = \boldsymbol{R}(z)\boldsymbol{\sigma}_0\boldsymbol{R}(z)^T$, where $\boldsymbol{R}(z) \in SO(3)$ describes the polycrystalline texture, we obtain

$$\boldsymbol{\sigma}'(z) = \frac{\boldsymbol{\sigma}(z)}{\det \boldsymbol{\sigma}_0}.$$

Hence,

$$\boldsymbol{\sigma}_*' = \frac{\boldsymbol{\sigma}_*}{\det \boldsymbol{\sigma}_0}.$$

However, according to (5.2) we also have

$$\boldsymbol{\sigma}_*' = \frac{\boldsymbol{\sigma}_*}{\det \boldsymbol{\sigma}_*}. \tag{5.2}$$

The two expressions for $\boldsymbol{\sigma}_*'$ agree, only if $\det \boldsymbol{\sigma}_* = \det \boldsymbol{\sigma}_0$, which is an exact relation for two-dimensional conducting composites, discovered first by Keller [4], and developed in [1].

The aim of the foregoing discussion was to show that links are more general and deeper statements about the effective behavior of composites than exact relations. However, their theory is more complicated than the theory of exact relations, and delves deeper into the algebraic structure of Jordan multialgebras.

Links do not have to be limited to a single physical context. For example, we may be interested in links between the effective elastic and conducting properties of one and the same composite. The idea is to regard the pair of tensors $\mathsf{L}_1(z)$, $\mathsf{L}_2(z)$ of material properties at $z \in Q$ in a composites, as a single tensor

$$\hat{\mathsf{L}}(z) = \begin{bmatrix} \mathsf{L}_1(z) & 0 \\ 0 & \mathsf{L}_2(z) \end{bmatrix} \tag{5.3}$$

of an imaginary material, whose properties are described by block-diagonal tensors $\hat{\mathsf{L}}$. To save space, instead of (5.3) we will use a more compact notation $\hat{\mathsf{L}}(z) = [\mathsf{L}_1(z), \mathsf{L}_2(z)]$. In this formalism links become exact relations in $\mathbb{V} = \mathrm{End}(\mathcal{T}_1) \oplus \mathrm{End}(\mathcal{T}_2)$, and hence, can be described by Theorem 4.4, as submanifolds

$$\hat{\mathbb{M}} = \left\{ \hat{\mathsf{L}} \in \mathrm{End}^+(\mathcal{T}_1) \oplus \mathrm{End}^+(\mathcal{T}_2) : \hat{W}_{n_0}(\hat{\mathsf{L}}) \in \hat{\Pi} \right\}, \tag{5.4}$$

independent of the choice of the unit vector $\boldsymbol{n}_0$. In this formula $\hat{\mathsf{L}}$ is given by (5.3) and

$$\hat{W}_{n0}(\hat{\mathsf{L}}) = \left[W_{n0}^{(1)}(\mathsf{L}_1),\ W_{n0}^{(2)}(\mathsf{L}_2)\right],$$

where

$$W_{n}^{(i)}(\mathsf{L}) = \left(\mathsf{I} + \left(\mathsf{L} - \mathsf{L}_0^{(i)}\right)\boldsymbol{\Gamma}_0^{(i)}(\boldsymbol{n})\right)^{-1}\left(\mathsf{L} - \mathsf{L}_0^{(i)}\right), \quad i = 1, 2.$$

The subspace $\hat{\Pi} \subset \mathbb{V}$ has the structure of Jordan $\hat{\mathcal{A}}$-multialgebra, where

$$\hat{\mathcal{A}} = \mathrm{Span}\left\{\left[\boldsymbol{\Gamma}_0^{(1)}(\boldsymbol{n}) - \boldsymbol{\Gamma}_0^{(1)}(\boldsymbol{n}_0),\ \boldsymbol{\Gamma}_0^{(2)}(\boldsymbol{n}) - \boldsymbol{\Gamma}_0^{(2)}(\boldsymbol{n}_0)\right] : |\boldsymbol{n}| = 1\right\}. \tag{5.5}$$

5.2 Algebraic structure of links

5.2.1 The general case

On the most basic level, a precondition for existence of any non-trivial links is the non-trivial structure of the subspace $\hat{\mathcal{A}} \subset \mathcal{A}_1 \oplus \mathcal{A}_2$, where

$$\mathcal{A}_i = \mathrm{Span}\left\{\boldsymbol{\Gamma}_0^{(i)}(\boldsymbol{n}) - \boldsymbol{\Gamma}_0^{(i)}(\boldsymbol{n}_0)\right\}, \quad i = 1, 2. \tag{5.6}$$

If $\hat{\mathcal{A}} = \mathcal{A}_1 \oplus \mathcal{A}_2$, then there are no relations between the two components of elements of $\hat{\mathcal{A}}$. It follows that there are no non-trivial links between $\mathcal{A}_1$ and $\mathcal{A}_2$ multialgebras. The degree of triviality of the structure of $\hat{\mathcal{A}}$ can be quantified by the subspaces

$$\mathfrak{A}_1 = \left\{\mathsf{A}_1 \in \mathcal{A}_1 : [\mathsf{A}_1, 0] \in \hat{\mathcal{A}}\right\}, \qquad \mathfrak{A}_2 = \left\{\mathsf{A}_2 \in \mathcal{A}_2 : [0, \mathsf{A}_2] \in \hat{\mathcal{A}}\right\}. \tag{5.7}$$

The smaller the subspaces $\mathfrak{A}_i$, the more non-trivial links we expect in a given context. The subspace $\hat{\mathcal{A}} \subset \mathcal{A}_1 \oplus \mathcal{A}_2$ can be described by subspaces $\mathfrak{A}_1$, $\mathfrak{A}_2$ and a linear isomorphism $\Xi: \mathcal{A}_1/\mathfrak{A}_1 \to \mathcal{A}_2/\mathfrak{A}_2$, defined by the rule $\bar{\mathsf{A}}_2 = \Xi(\bar{\mathsf{A}}_1)$, if $[\mathsf{A}_1, \mathsf{A}_2] \in \hat{\mathcal{A}}$, where $\bar{\mathsf{A}}_i$ denotes the equivalence class of A_i in $\mathcal{A}_i/\mathfrak{A}_i$, $i = 1, 2$:

$$\hat{\mathcal{A}} = \left\{[\mathsf{A}_1, \mathsf{A}_2] \in \mathcal{A}_1 \oplus \mathcal{A}_2 : \bar{\mathsf{A}}_2 = \Xi\left(\bar{\mathsf{A}}_1\right)\right\}. \tag{5.8}$$

For example, it was shown in [3] that for conductivity $\hat{\mathcal{A}} = \mathcal{A}_1 \oplus \mathcal{A}_2$, if and only if the reference media $\boldsymbol{\sigma}_0^{(1)}$, $\boldsymbol{\sigma}_0^{(2)}$ are not scalar multiples of one another. At the same time, it is easy to see that if $\boldsymbol{\sigma}_0^{(1)} = s\boldsymbol{\sigma}_0^{(2)}$, then $\mathcal{A}_1 = \mathcal{A}_2 = \mathcal{A}$ and $\hat{\mathcal{A}} = \{[A, sA] : A \in \mathcal{A}\}$. In particular,

$$\mathfrak{A}_1 = \mathfrak{A}_2 = \{0\}. \tag{5.9}$$

In this case the link between the first and the second component in the structure of $\hat{\mathcal{A}}$ leads to the presence of non-trivial links, which were computed in [2, 3].

Our goal is to relate Jordan $\hat{\mathcal{A}}$-multialgebras to Jordan $\mathcal{A}$-multialgebras. For a Jordan $\hat{\mathcal{A}}$-multialgebra $\hat{\Pi}$ in $\mathbb{V} = \mathrm{End}(\mathcal{T}_1) \oplus \mathrm{End}(\mathcal{T}_2)$ we define

$$\Pi_1 = \left\{\mathsf{K}_1 \in \mathrm{End}(\mathcal{T}_1) : [\mathsf{K}_1, \mathsf{K}_2] \in \hat{\Pi} \text{ for some } \mathsf{K}_2 \in \mathrm{End}(\mathcal{T}_2)\right\}, \tag{5.10}$$

$$\Pi_2 = \left\{\mathsf{K}_2 \in \mathrm{End}(\mathcal{T}_2) : [\mathsf{K}_1, \mathsf{K}_2] \in \hat{\Pi} \text{ for some } \mathsf{K}_1 \in \mathrm{End}(\mathcal{T}_1)\right\}, \tag{5.11}$$

$$\mathcal{I}_1 = \left\{\mathsf{K}_1 \in \mathrm{End}(\mathcal{T}_1) : [\mathsf{K}_1, 0] \in \hat{\Pi}\right\} \subset \Pi_1, \tag{5.12}$$

$$\mathcal{I}_2 = \left\{\mathsf{K}_2 \in \mathrm{End}(\mathcal{T}_2) : [0, \mathsf{K}_2] \in \hat{\Pi}\right\} \subset \Pi_2. \tag{5.13}$$

Theorem 5.1. *The subspace* $\hat{\Pi} \subset \mathbb{V}$ *is a Jordan* $\hat{\mathcal{A}}$*-multialgebra if and only if*

(i) *the subspaces* Π_j *of* $\mathrm{End}(\mathcal{T}_j)$ *are Jordan* $\mathcal{A}_j$*-multialgebras,* $j = 1, 2$

(ii) *The subspaces* $\mathcal{I}_j \subset \Pi_j$, $j = 1, 2$, *are Jordan ideals, i.e.*

$$\mathsf{J} *_{\mathsf{A}_j} \mathsf{K} \in \mathcal{I}_j \textit{ for all } \mathsf{J} \in \mathcal{I}_j,\ \ \mathsf{K} \in \Pi_j,\ \mathsf{A}_j \in \mathcal{A}_j \quad j = 1, 2.$$

Moreover, $\mathcal{I}_j$ *must contain minimal ideals*

$$\mathfrak{J}_j = \mathrm{Span}\left\{\mathsf{KAK} : \mathsf{K} \in \Pi_j,\ \ \mathsf{A} \in \mathfrak{A}_j\right\}, \quad j = 1, 2.$$

(iii) *The linear isomorphism* $\Phi : \mathcal{F}_1 = \Pi_1/\mathcal{I}_1 \rightarrow \Pi_2/\mathcal{I}_2 = \mathcal{F}_2$, *defined by the rule* $\bar{\mathsf{K}}_2 = \Phi(\bar{\mathsf{K}}_1)$, *whenever* $[\mathsf{K}_1, \mathsf{K}_2] \in \hat{\Pi}$, *satisfies*

$$\Phi(\bar{\mathsf{K}}\bar{\mathsf{A}}\bar{\mathsf{K}}) = \Phi(\bar{\mathsf{K}})\Xi(\bar{\mathsf{A}})\Phi(\bar{\mathsf{K}}), \qquad \bar{\mathsf{K}} \in \Pi_1/\mathcal{I}_1,\ \bar{\mathsf{A}} \in \mathcal{A}_1/\mathfrak{A}_1, \tag{5.14}$$

Where $\Xi : \mathcal{A}_1/\mathfrak{A}_1 \rightarrow \mathcal{A}_2/\mathfrak{A}_2$ *is a linear isomorphism that describes* $\hat{\mathcal{A}}$ *by* (5.8).

Specifically, given the algebraic data Π_j, $\mathcal{I}_j$, $j = 1, 2$ *and* Φ *satisfying properties (i)–(iii) above, the Jordan* $\hat{\mathcal{A}}$*-multialgebra* $\hat{\Pi}$ *can be reconstructed by the formula*

$$\hat{\Pi} = \left\{[\mathsf{K}_1, \mathsf{K}_2] \in \Pi_1 \oplus \Pi_2 : \bar{\mathsf{K}}_2 = \Phi(\bar{\mathsf{K}}_1)\right\}. \tag{5.15}$$

Proof. **Sufficiency.** Suppose that a subspace $\hat{\Pi} \subset \mathbb{V}$ is given by (5.15). Let us show that it is a Jordan $\hat{\mathcal{A}}$-multialgebra. Let $\hat{\mathsf{K}} = [\mathsf{K}_1, \mathsf{K}_2] \in \hat{\Pi}$ and $\hat{\mathsf{A}} = [\mathsf{A}_1, \mathsf{A}_2] \in \hat{\mathcal{A}}$. Then

$$\hat{\mathsf{K}}\hat{\mathsf{A}}\hat{\mathsf{K}} = [\mathsf{K}_1\mathsf{A}_1\mathsf{K}_1, \mathsf{K}_2\mathsf{A}_2\mathsf{K}_2] \in \Pi_1 \oplus \Pi_2,$$

since $\Pi_1 \oplus \Pi_2$ is a Jordan $\hat{\mathcal{A}}$-multialgebra. We observe that for any $\mathsf{J} \in \mathcal{I}_1$ and $\mathsf{A}_0 \in \mathfrak{A}_1$

$$(\mathsf{K}_1 + \mathsf{J})(\mathsf{A}_1 + \mathsf{A}_0)(\mathsf{K}_1 + \mathsf{J}) - \mathsf{K}_1\mathsf{A}_1\mathsf{K}_1 = 2\mathsf{J}*_{\mathsf{A}_1+\mathsf{A}_0}\mathsf{K}_1 + \mathsf{K}_1\mathsf{A}_0\mathsf{K}_1.$$

The first term on the right-hand side is in $\mathcal{I}_1$, since $\mathcal{I}_1$ is an Jordan ideal, while the second term is in $\mathfrak{J}_1 \subset \mathcal{I}_1$. Thus, the equivalence class of $\mathsf{K}_1\mathsf{A}_1\mathsf{K}_1$ in $\mathcal{F}_1$ is independent of the choice of representatives $\mathsf{K}_1 \in \bar{\mathsf{K}}_1$ and $\mathsf{A}_1 \in \bar{\mathsf{A}}_1$. Similarly, the equivalence class of $\mathsf{K}_2\mathsf{A}_2\mathsf{K}_2$ in $\mathcal{F}_2$ is independent of the choice of representatives $\mathsf{K}_2 \in \bar{\mathsf{K}}_2$ and $\mathsf{A}_2 \in \bar{\mathsf{A}}_2$. In other words,

$$\overline{\mathsf{K}_j\mathsf{A}_j\mathsf{K}_j} = \bar{\mathsf{K}}_j\, \bar{\mathsf{A}}_j\, \bar{\mathsf{K}}_j, \quad j = 1, 2. \tag{5.16}$$

By the property (5.14) we get

$$\Phi\left(\overline{\mathsf{K}_1\mathsf{A}_1\mathsf{K}_1}\right) = \Phi\left(\bar{\mathsf{K}}_1\right)\Xi\left(\bar{\mathsf{A}}_1\right)\Phi\left(\bar{\mathsf{K}}_1\right).$$

By definition of the subspace $\hat{\Pi}$ the assumption that $[\mathsf{K}_1, \mathsf{K}_2] \in \hat{\Pi}$ implies that $\bar{\mathsf{K}}_2 = \Phi(\bar{\mathsf{K}}_1)$. Similarly, the assumption that $[\mathsf{A}_1, \mathsf{A}_2] \in \hat{\mathcal{A}}$ implies that $\bar{\mathsf{A}}_2 = \Xi(\bar{\mathsf{A}}_1)$. Thus,

$$\Phi\left(\overline{\mathsf{K}_1\mathsf{A}_1\mathsf{K}_1}\right) = \bar{\mathsf{K}}_2\bar{\mathsf{A}}_2\,\bar{\mathsf{K}}_2 = \overline{\mathsf{K}_2\mathsf{A}_2\mathsf{K}_2}.$$

Hence, $[\mathsf{K}_1\mathsf{A}_1\mathsf{K}_1, \mathsf{K}_2\mathsf{A}_2\mathsf{K}_2] \in \hat{\Pi}$ by definition of $\hat{\Pi}$. The sufficiency is proved now.

Necessity. Part (i). Suppose $\mathsf{K}_1 \in \Pi_1$ and $\mathsf{A}_1 \in \mathcal{A}_1$. Then, by definitions of Π_1 and $\mathcal{A}_1$ there exist $\mathsf{K}_2 \in \Pi_2$ and $\mathsf{A}_2 \in \mathcal{A}_2$, such that $\hat{\mathsf{K}} = [\mathsf{K}_1, \mathsf{K}_2] \in \hat{\Pi}$ and $\hat{\mathsf{A}} = [\mathsf{A}_1, \mathsf{A}_2] \in \hat{\mathcal{A}}$. Thus, $\hat{\mathsf{K}}\hat{\mathsf{A}}\hat{\mathsf{K}} \in \hat{\Pi}$, which implies that $\mathsf{K}_1\mathsf{A}_1\mathsf{K}_1 \in \Pi_1$. The assertion about Π_2 is proved in the same way.

Part (ii). Let $\mathsf{J} \in \mathcal{I}_1$, $\mathsf{K}_1 \in \Pi_1$ and $\mathsf{A}_1 \in \mathcal{A}_1$. Let $\mathsf{K}_2 \in \Pi_2$ and $\mathsf{A}_2 \in \mathcal{A}_2$ be as in the proof of Part (i). Then $[\mathsf{J}, 0] \in \hat{\Pi}$, by definition of $\mathcal{I}_1$. Thus,

$$[\mathsf{J}, 0][\mathsf{A}_1, \mathsf{A}_2][\mathsf{K}_1, \mathsf{K}_2] + [\mathsf{K}_1, \mathsf{K}_2][\mathsf{A}_1, \mathsf{A}_2][\mathsf{J}, 0] \in \hat{\Pi}.$$

It follows that $[\mathsf{J}\mathsf{A}_1\mathsf{K}_1 + \mathsf{K}_1\mathsf{A}_1\mathsf{J}, 0] \in \hat{\Pi}$. We conclude that, $\mathsf{J}*_{\mathsf{A}}\mathsf{K} \in \mathcal{I}_1$ for any $\mathsf{J} \in \mathcal{I}_1$, $\mathsf{A} \in \mathcal{A}_1$ and $\mathsf{K} \in \Pi_1$, proving that $\mathcal{I}_1$ is an ideal. To finish the proof of Part (ii) we take any $\mathsf{K}_1 \in \Pi_1$ and $\mathsf{A} \in \mathfrak{A}_1$. Then, $[\mathsf{A}, 0] \in \hat{\mathcal{A}}$ and there exist $\mathsf{K}_2 \in \Pi_2$, such that $[\mathsf{K}_1, \mathsf{K}_2] \in \hat{\Pi}$. Then

$$[\mathsf{K}_1\mathsf{A}\mathsf{K}_1, 0] = [\mathsf{K}_1, \mathsf{K}_2][\mathsf{A}, 0][\mathsf{K}_1, \mathsf{K}_2] \in \hat{\Pi}.$$

This proves that $\mathfrak{J}_1 \subset \mathcal{I}_1$.

Part (iii). The representation (5.15) follows from definitions of Π_j and $\mathcal{I}_j$, $j = 1, 2$ and Φ. In other words, $[\mathsf{K}_1, \mathsf{K}_2] \in \hat{\Pi}$ if and only if $\bar{\mathsf{K}}_2 = \Phi(\bar{\mathsf{K}}_1)$, and $[\mathsf{A}_1, \mathsf{A}_2] \in \hat{\mathcal{A}}$ if and only if $\bar{\mathsf{A}}_2 = \Xi(\bar{\mathsf{A}}_1)$. Thus,

$$\left[\mathsf{K}_1\mathsf{A}_1\mathsf{K}_1, \mathsf{K}_2\mathsf{A}_2\mathsf{K}_2\right] \in \hat{\Pi} \iff \overline{\mathsf{K}_2\mathsf{A}_2\mathsf{K}_2} = \Phi\left(\overline{\mathsf{K}_1\mathsf{A}_1\mathsf{K}_1}\right).$$

Formula (5.14) follows from (5.16), which is a consequence of part (ii), as we have shown in the proof of sufficiency. □

In order to ensure validity of a link corresponding to $\hat{\Pi} \subset \mathbb{V}$ for all microstructures, we apply Theorem 4.8 to Jordan multialgebras in $\mathbb{V}$. If we rewrite the corresponding three- and four-chain properties for $\hat{\Pi}$ in terms of the algebraic data Π_j, $\mathcal{I}_j$, $j = 1, 2$, and Φ, we will obtain the following theorem.

Theorem 5.2. *The Jordan $\hat{\mathcal{A}}$-multialgebra $\hat{\Pi}$ has three-and four-chain properties if and only if all 3 conditions below are satisfied.*

(i) *Jordan multialgebras Π_j, $j = 1, 2$, have three-and four-chain properties*

(ii) *The ideals $\mathcal{I}_j$, $j = 1, 2$ have the ideal three- and four-chain properties:*

$$\mathsf{J}\mathsf{A}_1\mathsf{K}_1\mathsf{A}_2\mathsf{K}_2 + \mathsf{K}_2\mathsf{A}_2\mathsf{K}_1\mathsf{A}_1\mathsf{J} \in \mathcal{I}_j, \tag{5.17}$$

$$\mathsf{J}\mathsf{A}_1\mathsf{K}_1\mathsf{A}_2\mathsf{K}_2\mathsf{A}_3\mathsf{K}_3 + \mathsf{K}_3\mathsf{A}_3\mathsf{K}_2\mathsf{A}_2\mathsf{K}_1\mathsf{A}_1\mathsf{J} \in \mathcal{I}_j \tag{5.18}$$

for all $\mathsf{J} \in \mathcal{I}_j$, $\{\mathsf{K}_1, \mathsf{K}_2, \mathsf{K}_3\} \subset \Pi_j$, $\{\mathsf{A}_1, \mathsf{A}_2, \mathsf{A}_3\} \subset \mathcal{A}_j$, $j = 1, 2$

(iii) *The Jordan factor multialgebra isomorphism Φ has the isomorphism three- and four-chain properties:*

$$\begin{aligned}\Phi\left(\overline{\mathsf{K}_0\mathsf{A}_1\mathsf{K}_1\mathsf{A}_1'\mathsf{K}_2 + \mathsf{K}_2\mathsf{A}_1'\mathsf{K}_1\mathsf{A}_1\mathsf{K}_0}\right) &= \Phi\left(\bar{\mathsf{K}}_0\right)\mathsf{A}_2\Phi\left(\bar{\mathsf{K}}_1\right)\mathsf{A}_2'\Phi\left(\bar{\mathsf{K}}_2\right)\\ &\quad + \Phi\left(\bar{\mathsf{K}}_2\right)\mathsf{A}_2'\Phi\left(\bar{\mathsf{K}}_1\right)\mathsf{A}_2\Phi\left(\bar{\mathsf{K}}_0\right)\end{aligned} \tag{5.19}$$

$$\begin{aligned}\Phi\left(\overline{\mathsf{K}_0\mathsf{A}_1\mathsf{K}_1\mathsf{A}_1'\mathsf{K}_2\mathsf{A}_1''\mathsf{K}_3 + \mathsf{K}_3\mathsf{A}_1''\mathsf{K}_2\mathsf{A}_1'\mathsf{K}_1\mathsf{A}_1\mathsf{K}_0}\right) &= \Phi\left(\bar{\mathsf{K}}_0\right)\mathsf{A}_2\Phi\left(\bar{\mathsf{K}}_1\right)\mathsf{A}_2'\Phi\left(\bar{\mathsf{K}}_2\right)\mathsf{A}_2''\Phi\left(\bar{\mathsf{K}}_3\right)\\ &\quad + \Phi\left(\bar{\mathsf{K}}_3\right)\mathsf{A}_2''\Phi\left(\bar{\mathsf{K}}_2\right)\mathsf{A}_2'\Phi\left(\bar{\mathsf{K}}_1\right)\mathsf{A}_2\Phi\left(\bar{\mathsf{K}}_0\right)\end{aligned} \tag{5.20}$$

for all $\{\mathsf{K}_0, \mathsf{K}_1, \mathsf{K}_2, \mathsf{K}_3\} \subset \Pi_1$, *and all* $\{[\mathsf{A}_1, \mathsf{A}_2], [\mathsf{A}_1', \mathsf{A}_2'], [\mathsf{A}_1'', \mathsf{A}_2'']\} \subset \hat{\mathcal{A}}$

Proof. **Necessity. Part (i).** Let π_i: $\mathbb{V} \to \mathrm{End}(\mathcal{T}_i)$, $i = 1, 2$, be the projection onto the ith component:

$$\pi_1[\mathsf{K}_1, \mathsf{K}_2] = \mathsf{K}_1, \qquad \pi_2[\mathsf{K}_1, \mathsf{K}_2] = \mathsf{K}_2.$$

By construction of the subspace $\hat{\mathcal{A}}$ it is clear that $\pi_i(\hat{\mathcal{A}}) = \mathcal{A}_i$, $i = 1, 2$. Thus, for any $\{\mathsf{A}_1, \mathsf{A}_1'\} \subset \mathcal{A}_1$ and any $\{\mathsf{K}_1, \mathsf{K}_1', \mathsf{K}_1''\} \subset \Pi_1$ there exist $\{\mathsf{A}_2, \mathsf{A}_2'\} \subset \mathcal{A}_2$ and any $\{\mathsf{K}_2, \mathsf{K}_2', \mathsf{K}_2''\} \subset \Pi_2$, such that

$$\left\{[\mathsf{A}_1, \mathsf{A}_2], [\mathsf{A}_1', \mathsf{A}_2']\right\} \subset \hat{\mathcal{A}}, \qquad \left\{[\mathsf{K}_1, \mathsf{K}_2], [\mathsf{K}_1', \mathsf{K}_2'], [\mathsf{K}_1'', \mathsf{K}_2'']\right\} \subset \hat{\Pi}.$$

Then, by the assumed three-chain property of $\hat{\Pi}$, we conclude that Π_1 must also have the three-chain property. The three-chain property for Π_2 and the four-chain property are proved in the same way.

Part (ii). Suppose now that $\mathsf{J} \in \mathcal{I}_1$. Then $[\mathsf{J}, 0] \in \hat{\Pi}$. Using the three- and four-chain properties of $\hat{\Pi}$ with $[\mathsf{K}_1, \mathsf{K}_2]$ replaced by $[\mathsf{J}, 0]$ we obtain the statement of part (ii) for $i = 1$. The statement for $i = 2$ is proved in the same way.

Part (iii). This is an immediate consequence of the three and four-chain properties of $\hat{\Pi}$.

Sufficiency. We will prove the three-chain property of $\hat{\Pi}$. The four-chain property is proved in the same way. Let us take arbitrary

$$\left\{[\mathsf{A}_1, \mathsf{A}_2], [\mathsf{A}_1', \mathsf{A}_2']\right\} \subset \hat{\mathcal{A}}, \qquad \left\{[\mathsf{K}_1, \mathsf{K}_2], [\mathsf{K}_1', \mathsf{K}_2'], [\mathsf{K}_1'', \mathsf{K}_2'']\right\} \subset \hat{\Pi}.$$

We observe that

$$\mathsf{K}_1\mathsf{A}_1\mathsf{J}\mathsf{A}_1'\mathsf{K}_1' = 2\left(\mathsf{K}_1 *_{\mathsf{A}_1} \mathsf{J}\right)\mathsf{A}_1'\mathsf{K}_1' - \mathsf{J}\mathsf{A}_1\mathsf{K}_1\mathsf{A}_1'\mathsf{K}_1'.$$

Therefore,

$$\mathsf{K}_1\mathsf{A}_1\mathsf{J}\mathsf{A}_1'\mathsf{K}_1' + \mathsf{K}_1'\mathsf{A}_1'\mathsf{J}\mathsf{A}_1\mathsf{K}_1 = 4\left(\mathsf{K}_1 *_{\mathsf{A}_1} \mathsf{J}\right) *_{\mathsf{A}_1'} \mathsf{K}_1' - \mathsf{J}\mathsf{A}_1\mathsf{K}_1\mathsf{A}_1'\mathsf{K}_1' - \mathsf{K}_1'\mathsf{A}_1'\mathsf{K}_1\mathsf{A}_1\mathsf{J} \in \mathcal{I}_1.$$

The same property is true for four-chains as well, since they can be expressed in terms of three-chains with one of the factors in $\mathcal{I}_1$ and a four-chain (5.18). It follows that the equivalence classes in $\mathcal{F}_1 = \Pi_1/\mathcal{I}_1$ of three- and four-chains in (5.19) and (5.20) do not depend on the choices of representatives of elements of Π_1 in $\mathcal{F}_1$. It is now clear that part (iii) in the theorem implies validity of three- and four-chain properties of $\hat{\Pi}$. □

5.2.2 Polycrystalline links

In this section we assume that the rotational equivariance postulate 2.2 holds and that all composites are polycrystalline. The following theorem gives explicit formulations of rotational invariance (or equivariance) of relevant objects.

Theorem 5.3. *Suppose that the reference medium $\hat{\mathsf{L}}_0 = [\mathsf{L}_0^{(1)}, \mathsf{L}_0^{(2)}]$ is isotropic ($\boldsymbol{R} \cdot \hat{\mathsf{L}}_0 = \hat{\mathsf{L}}_0$ for every $\boldsymbol{R} \in SO(d)$). Suppose that $\hat{\Pi}$ is a rotationally invariant Jordan $\hat{\mathcal{A}}$-multialgebra. Then*

(i) *the subspaces $\mathcal{A}_j$, Π_j, $\mathcal{I}_j$, $j = 1, 2$ are rotationally invariant*

(ii) *The maps Ξ and Φ, defined in (5.8) and (5.14), respectively, are rotationally equivariant, i.e.*

$$\Xi(\boldsymbol{R} \cdot \mathsf{A}) = \boldsymbol{R} \cdot \Xi(\mathsf{A}), \quad \Phi\left(\boldsymbol{R} \cdot \overline{\mathsf{K}}\right) = \boldsymbol{R} \cdot \Phi\left(\overline{\mathsf{K}}\right) \quad \boldsymbol{R} \in \mathrm{SO}(d), \mathsf{A} \in \mathcal{A}_1, \mathsf{K} \in \Pi_1,$$

where the action of rotations on elements of the factor-space is defined in a natural way:

$$\boldsymbol{R} \cdot \overline{\mathsf{K}} = \overline{\boldsymbol{R} \cdot \mathsf{K}}, \qquad \boldsymbol{R} \in \mathrm{SO}(d), \mathsf{K} \in \Pi_1.$$

The proof consists of straightforward verification of the required properties.

In this section we focus on the most common case of links between polycrystalline composites within a single physical context, i.e. $\mathcal{T}_1 = \mathcal{T}_2 = \mathcal{T}$. The rotational invariance and equivariance properties listed in Theorem 5.3 allows us to use a wealth of information and powerful tools of representation theory available for rotation groups SO(2) and SO(3). In particular we can show (see lemma 6.9 in section 6.4) that the subspace $\hat{\mathcal{A}}$ satisfies (5.9).

In addition to property (5.9), essential for existence of non-trivial links, there is another, more subtle, algebraic property that signals possible presence of a large number of links. In a sense, links reflect algebraic similarities between Jordan $\mathcal{A}_1$ and $\mathcal{A}_2$ multialgebras. Mathematically, such similarities are enabled by a linear isomorphism (if it exists) $\Phi_0 : \mathrm{End}(\mathcal{T}) \to \mathrm{End}(\mathcal{T})$ satisfying

$$\Phi_0(\mathsf{KAK}) = \Phi_0(\mathsf{K})\Xi(\mathsf{A})\Phi_0(\mathsf{K}), \qquad \forall\, \mathsf{K} \in \mathrm{End}(\mathcal{T}), \forall\, \mathsf{A} \in \mathcal{A}_1, \tag{5.21}$$

where the map Ξ, defined by (5.8) is now a linear transformation $\mathcal{A}_1 \to \mathcal{A}_2$. Then, the transformation

$$\mathfrak{F}_0 : \mathrm{End}(\mathcal{T}) \oplus \mathrm{End}(\mathcal{T}) \to \mathrm{End}(\mathcal{T}) \oplus \mathrm{End}(\mathcal{T}), \quad \mathfrak{F}_0([\mathsf{K}_1, \mathsf{K}_2]) = \left[\mathsf{K}_1, \Phi_0^{-1}(\mathsf{K}_2)\right]$$

has the property

$$\mathfrak{F}_0\left(\hat{\mathsf{K}}\left[\mathsf{A}, \Xi(\mathsf{A})\right]\hat{\mathsf{K}}\right) = \mathfrak{F}_0\left(\hat{\mathsf{K}}\right)[\mathsf{A}, \mathsf{A}]\mathfrak{F}_0\left(\hat{\mathsf{K}}\right), \quad \hat{\mathsf{K}} \in \mathrm{End}(\mathcal{T}) \oplus \mathrm{End}(\mathcal{T}), \mathsf{A} \in \mathcal{A}_1.$$

In other words, $\mathfrak{F}_0$ establishes a one-to-one correspondence between Jordan $\hat{\mathcal{A}}$-multialgebras $\hat{\Pi}$ and Jordan $\hat{\mathcal{A}}_0$-multialgebras $\hat{\Pi}^* = \mathfrak{F}_0(\hat{\Pi})$, where

$$\hat{\mathcal{A}}_0 = \left\{[\mathsf{A}, \mathsf{A}] : \mathsf{A} \in \mathcal{A}_1\right\}.$$

The property (5.14) becomes

$$\Phi\left(\bar{\mathsf{K}}\mathsf{A}\bar{\mathsf{K}}\right) = \Phi\left(\bar{\mathsf{K}}\right)\mathsf{A}\Phi\left(\bar{\mathsf{K}}\right), \quad \forall\, \bar{\mathsf{K}} \in \Pi_1/\mathcal{I}_1, \forall\, \mathsf{A} \in \mathcal{A}_1, \tag{5.22}$$

making Φ a Jordan factor multialgebra isomorphism. Thus, when the isomorphism Φ_0 exists, the computation of all links requires us to classify ideals and corresponding factor multialgebras, up to an SO(d)-equivariant Jordan factor multialgebra isomorphism. If such isomorphism classes contain more than one factor multialgebra, then we only need to find one (simplest) isomorphism between the factor multialgebras $\Pi_1/\mathcal{I}_1$ and $\Pi'_1/\mathcal{I}'_1$ in the same isomorphism class. Once these simple isomorphisms are established the entire collection of links will be described by single representative factor multialgebras $\Pi_1/\mathcal{I}_1$ and their Jordan automorphisms.

Example 5.4. Let us illustrate some of the abstract concepts with the example of of two and three-dimensional conductivity. Let us start in three space dimensions. The only 'non-trivial' polycrystalline link corresponds to $\Pi_1 = \Pi_2 = \mathrm{Sym}(\mathbb{R}^3)$, $\mathcal{I}_1 = \mathcal{I}_2 = \{0\}$ and

$$\Phi(\boldsymbol{K}) = \left(\sigma_2^0/\sigma_1^0\right)\boldsymbol{K}, \tag{5.23}$$

corresponding to the family of links

$$\hat{\mathbb{M}}_\alpha = \left\{[\boldsymbol{\sigma}, \alpha\boldsymbol{\sigma}] : \boldsymbol{\sigma} \in \mathrm{Sym}^+(\mathbb{R}^3)\right\}, \quad \alpha > 0.$$

This is a well-known homogeneity property of homogenization that is usually stated as $(\alpha\boldsymbol{\sigma})^* = \alpha\boldsymbol{\sigma}^*$, and which we have already used (as obvious) in the derivation of the Keller–Dykhne exact relation (4.3) from Mendelson's link (5.1).

By contrast, there are non-trivial links for two-dimensional conductivity. One of them corresponds to $\Pi_1 = \Pi_2 = \mathrm{Sym}(\mathbb{R}^2)$, $\mathcal{I}_1 = \mathcal{I}_2 = \{0\}$ and the map

$$\Phi(\boldsymbol{K}) = \frac{\sigma_2^0}{\sigma_1^0}(\boldsymbol{K} - \mathrm{Tr}\,(\boldsymbol{K})\boldsymbol{I}). \tag{5.24}$$

This is the algebraic representation of Mendelson's theorem (5.1), (5.2).

In order to be thorough, we also need to examine the possibility, where $\Pi_1 = \Pi_2 = \mathrm{Sym}_0(\mathbb{R}^2)$, $\mathcal{I}_1 = \mathcal{I}_2 = \{0\}$. A short calculation shows that all maps Φ satisfying the Jordan automorphism property (5.14) are just restrictions of the maps (5.23) or (5.24) to the subspace $\mathrm{Sym}_0(\mathbb{R}^2)$. We have now proved that aside from the Keller–Dykhne exact relation (4.3) and Mendelson's link (5.1) there are no other microstructure-independent formulas for two-dimensional conducting polycrystals.

5.3 Volume fraction formulas as links

In this section our goal is to develop the theory of formulas of the form

$$f(\mathsf{L}^*) = \langle f(\mathsf{L}(z))\rangle, \tag{5.25}$$

for some $f : \mathrm{End}^+(\mathcal{T}) \to \mathbb{R}^m$, generalizing Hill's formula (4.1). An important feature of Hill's formula is that it serves as a 'companion' to Hill's exact relation (4.2). The idea is to observe that the effective tensor L^* of a periodic composite with local tensor $\mathsf{L}(z)$ is the volume average, if $\mathcal{E}_{\boldsymbol{n}} = \{\boldsymbol{0}\} \subset \mathcal{T}$ for all unit vectors $\boldsymbol{n}$. This suggests that we should be able to obtain microstructure-independent formulas involving volume averages as links between our original composites and fictitious composites with tensor space $\tilde{\mathcal{T}} = \mathbb{R}^N$, where $N = \dim\mathcal{T}$ and $\tilde{\mathcal{E}}_{\boldsymbol{n}} = \{\boldsymbol{0}\}$. Thus,

$$\hat{\mathcal{T}} = \mathcal{T} \oplus \mathbb{R}^N, \qquad \hat{\mathcal{E}}_{\boldsymbol{n}} = \mathcal{E}_{\boldsymbol{n}} \oplus \{\boldsymbol{0}\} \subset \mathcal{T} \oplus \mathbb{R}^N. \tag{5.26}$$

Hence,

$$\hat{\boldsymbol{\Gamma}}_0(\boldsymbol{n}) = [\boldsymbol{\Gamma}_0(\boldsymbol{n}), \boldsymbol{0}], \qquad \hat{\mathcal{A}} = \{[\mathsf{A}, 0] : \mathsf{A} \in \mathcal{A}\}. \tag{5.27}$$

Let $\hat{\Pi} \subset \mathrm{End}(\mathcal{T}) \oplus \mathrm{End}(\mathbb{R}^N)$ be a Jordan $\hat{\mathcal{A}}$-multialgebra. According to theorem 5.1 $\hat{\Pi}$ is completely determined by the algebraic data

$$\Pi = \left\{ \mathsf{K} \in \mathrm{End}(\mathcal{T}) : [\mathsf{K}, \tilde{\mathsf{K}}] \in \hat{\Pi} \text{ for some } \tilde{\mathsf{K}} \in \mathrm{End}(\mathbb{R}^N) \right\}, \tag{5.28}$$

$$\tilde{\Pi} = \left\{ \tilde{\mathsf{K}} \in \mathrm{End}(\mathbb{R}^N) : [\mathsf{K}, \tilde{\mathsf{K}}] \in \hat{\Pi} \text{ for some } \mathsf{K} \in \mathrm{End}(\mathcal{T}) \right\}, \tag{5.29}$$

$$\mathcal{I} = \left\{ \mathsf{K} \in \mathrm{End}(\mathcal{T}) : [\mathsf{K}, 0] \in \hat{\Pi} \right\}, \tag{5.30}$$

$$\tilde{\mathcal{I}} = \left\{ \tilde{\mathsf{K}} \in \mathrm{End}\,(\mathbb{R}^N) : \left[0, \tilde{\mathsf{K}}\right] \in \hat{\Pi} \right\}. \tag{5.31}$$

Using special form (5.27) of the subspace $\hat{\mathcal{A}}$ we obtain that $[\mathsf{K}, \tilde{\mathsf{K}}] \in \hat{\Pi}$ if and only if $[\mathsf{KAK}, 0] \in \hat{\Pi}$, for any $\mathsf{K} \in \Pi$ and $\mathsf{A} \in \mathcal{A}$. Thus, $\Pi^2 \subset \mathcal{I}$, where

$$\Pi^2 = \mathrm{Span}\left\{ \mathsf{K} *_{\mathsf{A}} \mathsf{K}' : \{\mathsf{K}, \mathsf{K}'\} \subset \Pi, \mathsf{A} \in \mathcal{A} \right\}. \tag{5.32}$$

Obviously, Π^2 is an ideal in Π, and the factor algebra Π/Π^2 has a trivial multiplicative structure, since all Jordan products of elements in Π are in Π^2, which is zero in Π/Π^2. In fact, it is clear that the pair (Π, Π^2) contains all the information about Jordan $\hat{\mathcal{A}}$-multialgebras with $\Pi_1 = \Pi$, where Π_1 is defined by (5.10). Thus, all formulas involving volume averages are determined by Jordan $\hat{\mathcal{A}}$-multialgebras of the form

$$\hat{\Pi} = \left\{ \left[\mathsf{K}, \mathcal{P}_{(\Pi^2)^c} \mathsf{K} \right] : \mathsf{K} \in \Pi \right\}, \tag{5.33}$$

where $\mathcal{P}_{(\Pi^2)^c}$ is a projection onto any complement of Π^2 in Π along Π.

Examining sufficient conditions for stability under homogenization, given in Theorem 5.2, we obtain the corresponding sufficient conditions for the volume fraction corresponding to $\hat{\Pi}$ to be microstructure-independent.

Theorem 5.5. *A Jordan $\hat{\mathcal{A}}$-multialgebra $\hat{\Pi}$ with $\hat{\mathcal{A}}$, given by (5.27) possesses three- and four-chain properties if and only if*

$$\mathsf{K}_1\mathsf{A}_1\mathsf{K}_2\mathsf{A}_2\mathsf{K}_3 + \mathsf{K}_3\mathsf{A}_2\mathsf{K}_2\mathsf{A}_1\mathsf{K}_1 \in \Pi^2 \tag{5.34}$$

and

$$\mathsf{K}_1\mathsf{A}_1\mathsf{K}_2\mathsf{A}_2\mathsf{K}_3\mathsf{A}_3\mathsf{K}_4 + \mathsf{K}_4\mathsf{A}_3\mathsf{K}_3\mathsf{A}_2\mathsf{K}_2\mathsf{A}_1\mathsf{K}_1 \in \Pi^2 \tag{5.35}$$

for any $\{\mathsf{K}_1, \mathsf{K}_2, \mathsf{K}_3, \mathsf{K}_4\} \subset \Pi, \quad \{\mathsf{A}_1, \mathsf{A}_2, \mathsf{A}_3\} \subset \mathcal{A}$.

The link submanifold $\hat{\mathbb{M}}$, passing through the reference medium $[\mathsf{L}_0, \tilde{\mathsf{L}}_0]$ and corresponding to the Jordan $\hat{\mathcal{A}}$-multialgebra (5.33) can be written, according to (5.4), as follows

$$\hat{\mathsf{M}} = \left\{\left[\mathsf{L}, \tilde{\mathsf{L}}\right] \in \mathrm{End}^{+}(\mathcal{T}) \times \mathrm{End}^{+}(\mathbb{R}^N) : \tilde{\mathsf{L}}_0 - \tilde{\mathsf{L}} = \mathcal{P}_{\left(\Pi^2\right)^c} W_n(\mathsf{L})\right\}. \tag{5.36}$$

Thus, if $\mathsf{L}(z)$ is a local tensor of a periodic composite and L^* is its effective tensor, then the effective tensor of $\tilde{\mathsf{L}}(z) = \tilde{\mathsf{L}}_0 - \mathcal{P}_{(\Pi^2)^c} W_n(\mathsf{L}(z))$ is

$$\tilde{\mathsf{L}}^* = \tilde{\mathsf{L}}_0 - \mathcal{P}_{\left(\Pi^2\right)^c} W_n(\mathsf{L}^*).$$

However, our choice of $\tilde{\mathcal{E}}_n = \{\mathbf{0}\}$ implies that

$$\tilde{\mathsf{L}}^* = \left\langle \tilde{\mathsf{L}}(z) \right\rangle = \tilde{\mathsf{L}}_0 - \mathcal{P}_{\left(\Pi^2\right)^c} \langle W_n(\mathsf{L}(z)) \rangle.$$

Hence, we obtain the the generalization of Hills formula (4.1):

$$\mathcal{P}_{\left(\Pi^2\right)^c} W_n(\mathsf{L}^*) = \mathcal{P}_{\left(\Pi^2\right)^c} \langle W_n(\mathsf{L}) \rangle. \tag{5.37}$$

For future reference we summarize our results in the form of a theorem.

Theorem 5.6. *Suppose* Π *is a Jordan* $\mathcal{A}$*-multialgebra, corresponding to exact relation submanifold* M *via (4.4). This exact relation admits further formulas involving volume fractions if and only if* $\Pi^2 \neq \Pi$, *in which case* $\dim \Pi - \dim \Pi^2$ *additional formulas can be written in the form (5.37). These formulas are valid for all microstructures, provided the Jordan* $\mathcal{A}$*-multialgebra* Π *satisfies (5.34) and (5.35).*

Bibliography

[1] Dykhne A M 1971 Conductivity of a two-dimensional two-phase system *Sov. Phys.–JETP* **32** 63–5

Dykhne A M 1970 Conductivity of a two-dimensional two-phase system *Zh. Eksp. Teor. Fiz* **59** 110–5

[2] Grabovsky Y 2004 Algebra, geometry and computations of exact relations for effective moduli of composites *Advances in Multifield Theories of Continua with Substructure (Modeling and Simulation in Science, Engineering and Technology)* ed G Capriz and P M Mariano (Boston, MA: Birkhäuser) pp 167–97

[3] Grabovsky Y 2009 Exact relations for effective conductivity of fiber-reinforced conducting composites with the hall effect via a general theory *SIAM J. Math. Anal.* **41** 973–1024

[4] Keller J B 1964 A theorem on the conductivity of a composite medium *J. Math. Phys.* **5** 548–9

[5] Mendelson K S 1975 Effective conductivities of two-phase material with cylindrical phase boundaries *J. Appl. Phys.* **46** 917–8

[6] Mendelson K S 1975 A theorem on the conductivity of two-dimensional heterogeneous medium *J. Appl. Phys.* **46** 4740–1

IOP Publishing

Composite Materials
Mathematical theory and exact relations
Yury Grabovsky

Chapter 6

Computing exact relations and links†

The general theory of exact relations and links developed in the previous chapters opens up the possibility of computing complete lists of all exact relations and links for all kinds of material properties, such as conductivity, elasticity, piezoelectricity, thermoelectricity and thermoelasticity. However, the practical difficulty of performing the necessary calculations depends very much on the dimensionality of spaces $\mathcal{T}$ and $\mathcal{A}$. In fact, brute force calculations can become prohibitively difficult for anything more complicated than conducting composites. In this chapter we develop tools one can use to derive beautiful formulas such as (4.1), (4.3) and (5.1) from the general theory in a systematic manner. Part 3 of this book lists the results of the application of these methods and years of work by both the author and a large number of graduate and undergraduate students. The distinguishing feature of the results in part 3 is that they are *complete* in the sense that all microstructure-independent formulas for effective behavior of composites are logical consequences of the listed facts.

The process of obtaining all exact relations and links consists of four major steps: finding all Jordan $\mathcal{A}$-multialgebras and their squares, finding all Jordan $\hat{\mathcal{A}}$-multialgebras, computing exact relations, including relations containing volume averages, and computing links. We will discuss methods for completing each of these steps and illustrate them by carrying out all calculations for two- and three-dimensional elastic polycrystals.

6.1 Finding Jordan $\mathcal{A}$-multialgebras

6.1.1 Simplifying the subspace $\mathcal{A}$

The first step in computing all exact relations in any physical context is characterizing all Jordan $\mathcal{A}$-multialgebras. It is evident from formula (4.10) that the subspace $\mathcal{A}$ depends on the reference medium L_0, which complicates matters. In this section we will show that in many cases this dependence can be eliminated, or at least significantly simplified, by means of the Milgrom–Shtrikman 'covariance transformations' [11, 12], given by

$$\mathfrak{F}_{\mathsf{C}_1,\mathsf{C}_2}\colon \mathrm{End}(\mathcal{T}) \to \mathrm{End}(\mathcal{T}), \quad \mathfrak{F}_{\mathsf{C}_1,\mathsf{C}_2}(\mathsf{K}) = \mathsf{C}_1\mathsf{K}\mathsf{C}_2, \tag{6.1}$$

where $\{\mathsf{C}_1, \mathsf{C}_2\} \subset \mathrm{End}(\mathcal{T})$ are constant invertible operators. If we define

$$\mathcal{A}_0 = \mathsf{C}_2^{-1}\mathcal{A}\mathsf{C}_1^{-1}, \qquad \Pi_0 = \mathsf{C}_1\Pi\mathsf{C}_2, \tag{6.2}$$

then Π_0 satisfies (4.9) with $\mathcal{A}$ replaced by $\mathcal{A}_0$. Indeed,

$$\mathsf{C}_1(\mathsf{KAK})\mathsf{C}_2 = (\mathsf{C}_1\mathsf{K}\mathsf{C}_2)\left(\mathsf{C}_2^{-1}\mathsf{A}\mathsf{C}_1^{-1}\right)(\mathsf{C}_1\mathsf{K}\mathsf{C}_2).$$

The idea is to choose operators C_1, C_2, so that $\mathcal{A}_0$, given by (6.2), is as simple as possible. If all material tensors are symmetric, then we restrict the set of covariance transformations to maps $\mathfrak{F}_{\mathsf{C}_0,\mathsf{C}_0^T}$, preserving the symmetry of operators. In the case of polycrystalline composites the tensor C_0 must also be isotropic.

The simplest example, where the method is used, is in the context of two-dimensional and three-dimensional conducting composites with the Hall effect [6, 7] (see chapter 8), where for a reference medium $\boldsymbol{L}_0 \in \mathrm{End}^+(\mathbb{R}^d)$, $d = 2, 3$, we compute

$$\boldsymbol{\Gamma}_0(\boldsymbol{n}) = \frac{\boldsymbol{n} \otimes \boldsymbol{n}}{\boldsymbol{L}_0\boldsymbol{n} \cdot \boldsymbol{n}}, \qquad \mathcal{A} = \left\{A \in \mathrm{Sym}\left(\mathbb{R}^d\right)\colon \mathrm{Tr}\,(A\boldsymbol{L}_0) = 0\right\}. \tag{6.3}$$

It is easy to find the covariance transformation simplifying $\mathcal{A}$. Indeed, let $\boldsymbol{\sigma}_0$ be the symmetric part of $\boldsymbol{L}_0$. Then the desired covariance transformation is $\mathfrak{F}_{\boldsymbol{\sigma}_0^{-1/2},\boldsymbol{\sigma}_0^{-1/2}}$, so that

$$\mathcal{A}_0 = \boldsymbol{\sigma}_0^{1/2}\mathcal{A}\boldsymbol{\sigma}_0^{1/2} = \mathrm{Sym}_0\left(\mathbb{R}^d\right),$$

where $\mathrm{Sym}_0(\mathbb{R}^d)$ denotes the subspace of symmetric trace-free matrices given in (4.21).

We will see that the method of covariance transformations is quite effective for polycrystalline composites, in which case the covariance transformations must have the form $\mathfrak{F}_{\mathsf{C}_0,\mathsf{C}_0^T}$, where C_0 is isotropic. From now on we will focus exclusively on polycrystalline exact relations (see definition 4.13).

6.1.2 Rotationally invariant subspaces

Recall from section 4.4.1 that polycrystalline exact relations are generated by rotationally invariant Jordan $\mathcal{A}$-multialgebras. Rotational invariance leads to tremendous reduction in complexity. For example, for three-dimensional elasticity $\mathcal{T} = \mathrm{Sym}(\mathbb{R}^3)$ and $\mathrm{Sym}(\mathcal{T})$ is a 21-dimensional vector space of elastic tensors. If we ignore rotational invariance, then for each $0 < k < 21$ we must find all k-dimensional subspaces Π that satisfy (4.9). The set of all k-dimensional subspaces in a 21-dimensional space forms a $k(21 - k)$-parametric family. If $k = 10$ or 11, the search for exact relations would have to be performed on a 110-dimensional manifold! When we look for rotationally invariant Jordan multialgebras, our task would be reduced to examining only 16 cases, in which the search manifold has dimensions 2 in 2, 1 in 11, and 0 in 3 of the 16 cases.

In order to illustrate our approach we will focus on the case of elastic composites, that provides an excellent showcase of the general methodology.

The first step in our search process is to identify all rotationally invariant subspaces in $\mathrm{Sym}(\mathcal{T})$. In fact, this problem has already arisen in quantum mechanics, see e.g. [2, 5]. Its solution is now a part of a well-developed *representation theory* of compact Lie groups [14]. For d-dimensional elasticity, $\mathcal{T} = \mathrm{Sym}(\mathbb{R}^d)$, $d = 2, 3$. If $\boldsymbol{\varepsilon} \in \mathrm{Sym}(\mathbb{R}^d)$ and $\boldsymbol{R} \in SO(d)$, then the formula

$$\boldsymbol{R} \cdot \boldsymbol{\varepsilon} = \boldsymbol{R}\boldsymbol{\varepsilon}\boldsymbol{R}^T \tag{6.4}$$

defines the action of rotations on $\mathcal{T} = \mathrm{Sym}(\mathbb{R}^d)$. This action on $\mathcal{T}$ 'lifts' to $\mathrm{Sym}(\mathcal{T})$ in a *canonical way*, according to the 'distributive law'

$$\boldsymbol{R} \cdot (\mathsf{L}e) = (\boldsymbol{R} \cdot \mathsf{L})(\boldsymbol{R} \cdot e), \qquad e \in \mathcal{T}, \quad \boldsymbol{R} \in SO(d), \quad \mathsf{C} \in \mathrm{Sym}(\mathcal{T}), \tag{6.5}$$

which for elasticity translates into the explicit formula

$$(\boldsymbol{R} \cdot \mathsf{C})\boldsymbol{\varepsilon} = \boldsymbol{R}\left(\mathsf{C}\left(\boldsymbol{R}^T\boldsymbol{\varepsilon}\boldsymbol{R}\right)\right)\boldsymbol{R}^T, \quad \boldsymbol{\varepsilon} \in \mathrm{Sym}\left(\mathbb{R}^d\right), \quad \mathsf{C} \in \mathrm{Sym}\left(\mathrm{Sym}\left(\mathbb{R}^d\right)\right). \tag{6.6}$$

Formula (6.6) shows that the action of the rotation group on elasticity tensors is sufficiently complicated to rule out naive attempts to answer our questions directly, starting with definition (6.6). The idea is to use the representation theory of compact Lie groups (see e.g. [3]). The main consequence of compactness is the *complete reducibility* property, valid for all finite-dimensional representations. This property becomes apparent if we construct a group-invariant inner product, making group elements act by orthogonal transformations. To construct such an inner product for elasticity we recall that the Frobenius inner product (2.5) on the strain space $\mathrm{Sym}(\mathbb{R}^d)$ is $SO(d)$-invariant:

$$\mathrm{Tr}\,((\boldsymbol{R} \cdot \boldsymbol{A})(\boldsymbol{R} \cdot \boldsymbol{B})) = \mathrm{Tr}\left(\boldsymbol{R}\boldsymbol{A}\boldsymbol{R}^T\boldsymbol{R}\boldsymbol{B}\boldsymbol{R}^T\right) = \mathrm{Tr}\,(\boldsymbol{A}\boldsymbol{B}).$$

Elasticity tensors are symmetric operators on $\mathrm{Sym}(\mathbb{R}^d)$ with respect to this inner product, and the corresponding Frobenius inner product

$$(\mathsf{C}_1, \mathsf{C}_2)_{\mathrm{F}} = \mathrm{Tr}\,(\mathsf{C}_1\mathsf{C}_2) \tag{6.7}$$

on $\mathrm{Sym}(\mathrm{Sym}(\mathbb{R}^d))$ is therefore also $SO(d)$-invariant. The $SO(d)$-invariance of the inner product implies that each rotation is an orthogonal transformation on the Euclidean vector space $\mathrm{Sym}(\mathrm{Sym}(\mathbb{R}^d))$ with the inner product (6.7). As a consequence, if a subspace $U \subset \mathrm{Sym}(\mathrm{Sym}(\mathbb{R}^d))$ is $SO(d)$-invariant, then so is its orthogonal complement. Thus, $\mathrm{Sym}(\mathrm{Sym}(\mathbb{R}^d))$ can be written as a direct sum of $SO(d)$-invariant subspaces, each of which does not have any proper $SO(d)$-invariant subspaces. Such subspaces are called *irreducible representations* of $SO(d)$. We will call them *irreps* for short. The tools of representation theory of rotation groups SO(2) and SO(3) permit an easy identification of all irreps in $\mathrm{Sym}(\mathrm{Sym}(\mathbb{R}^d))$, $d = 2, 3$, by their isomorphism class. While this information is insufficient for our purposes, it is very useful for the eventual explicit characterization of all rotationally invariant subspaces in $\mathrm{Sym}(\mathcal{T})$.

We say that two irreps U and V are isomorphic if there exists a linear bijective map $f : U \to V$, such that $\boldsymbol{R} \cdot f(\boldsymbol{u}) = f(\boldsymbol{R} \cdot \boldsymbol{u})$ for every $\boldsymbol{u} \in U$ and $\boldsymbol{R} \in \mathrm{SO}(d)$. In representation theory the isomorphism classes of all irreps of rotation groups are well-known. In both cases $d = 2$ and $d = 3$, isomorphism classes of irreps are labeled by non-negative integers $m = 0, 1, 2, \ldots$, called *weights*. The irrep W_m of weight m has a very simple geometric meaning in two dimensions, where every rotation can be identified by the angle θ through which it rotates every vector in a plane. We denote such a rotation by $\boldsymbol{R}_\theta$. The irreps W_m, $m \geqslant 1$ are two-dimensional, and $\boldsymbol{R}_\theta \in \mathrm{SO}(2)$ rotates every vector in W_m through the angle $m\theta$. The irrep W_0 is, by contrast, one-dimensional, and rotations leave every vector in W_0 unchanged. For this reason W_0 is sometimes called the *trivial irrep*.

The weight m of an SO(3) irrep W_m inherits its value from a copy of SO(2) that 'sits' inside SO(3), and represents all rotations around a fixed axis.[1] Such a copy of SO(2) is called a 'maximal torus' in representation theory. By the complete reducibility property we can write an SO(3) irrep as a direct sum of SO(2) irreps. According to the representation theory (see appendix C.1), an irrep W_m of $SO(3)$ is a direct sum of $m + 1$ distinct $SO(2)$ irreps of weights $0, 1, 2, \ldots, m$, respectively, showing, in particular, that $\dim(W_m) = 2m + 1$. Hence, the weight of an SO(3) irrep W_m can be defined to be the highest weight of an SO(2) irrep in W_m. This structure of W_m also indicates that there is a unique one-dimensional weight zero (trivial) representation of SO(2) in W_m. A vector spanning this representation is called the zero weight vector in W_m. It is unique up to a scalar multiple.

We have already seen an example of a decomposition of $\mathrm{Sym}(\mathcal{T})$ into a direct sum of $\mathrm{SO}(d)$ irreps in the context of conductivity, where $\mathcal{T} = \mathbb{R}^d$, $d = 2, 3$,

$$\mathrm{Sym}\left(\mathbb{R}^d\right) = \mathbb{R}\boldsymbol{I}_d \oplus \mathrm{Sym}_0\left(\mathbb{R}^d\right), \tag{6.8}$$

where $\mathrm{Sym}_0(\mathbb{R}^d)$ is defined in (4.21). In this example, the one-dimensional irrep $\mathbb{R}\boldsymbol{I}_d$ belongs to W_0 isomorphism class, while $\mathrm{Sym}_0(\mathbb{R}^d)$ is an irrep of weight 2. This information will also be useful for the decomposition of the space of elastic tensors $\mathbb{E}^d = \mathrm{Sym}(\mathrm{Sym}(\mathbb{R}^d))$ into a direct sum of $\mathrm{SO}(d)$ irreps, $d = 2, 3$.

Starting with the decomposition (6.8) we obtain

$$\mathbb{E}^d = \mathrm{Sym}(\mathbb{R}\boldsymbol{I}_d) \oplus \mathrm{Sym}\left(\mathrm{Sym}_0\left(\mathbb{R}^d\right)\right) \oplus \mathrm{Sym}\left(\mathbb{R}\boldsymbol{I}_d, \mathrm{Sym}_0\left(\mathbb{R}^d\right)\right),$$

where $\mathrm{Sym}(V, W)$ denotes the set of all symmetric operators $T \in \mathrm{Sym}(V \oplus W)$ such that $T : V \to W$ (and consequently $T : W \to V$). The space $\mathrm{Sym}(\mathbb{R}\boldsymbol{I}_d) = \mathrm{End}(\mathbb{R}\boldsymbol{I}_d)$ of all linear maps on multiples of the identity is one-dimensional and

[1] Even though there are infinitely many copies of $SO(2) \subset SO(3)$, all of them are conjugate to one another, and hence, each copy assigns one and the same weight to an irrep. For calculations one usually fixes a specific copy of SO(2), much in the same way as one chooses a specific coordinate system to solve a geometric problem.

its elements are rotationally invariant. Hence, $\mathrm{Sym}(\mathbb{R}\boldsymbol{I}_d) \cong W_0$. It is also fairly obvious[2] that $\mathrm{Sym}(\mathbb{R}\boldsymbol{I}_d, \mathrm{Sym}_0(\mathbb{R}^d))$ is isomorphic to $\mathrm{Sym}_0(\mathbb{R}^d) \cong W_2$. Hence, we denote

$$\mathrm{Sym}(\mathbb{R}\boldsymbol{I}_d) = K_0', \qquad \mathrm{Sym}\left(\mathbb{R}\boldsymbol{I}_d, \mathrm{Sym}_0\left(\mathbb{R}^d\right)\right) = K_2', \tag{6.9}$$

where subscripts indicate the weight of the corresponding irrep. It will be useful to fix explicit isomorphisms between irreps of weight 0 and 2 in $\mathbb{E}^d$ and $\mathbb{R}$ and $\mathrm{Sym}_0(\mathbb{R}^d)$, respectively. These isomorphisms will be conveniently denoted by the same letters as the corresponding subrepresentations of $\mathbb{E}^d$:

$$K_0'(\kappa)\boldsymbol{\varepsilon} = \kappa(\mathrm{Tr}\,\boldsymbol{\varepsilon})\boldsymbol{I}_d, \qquad \boldsymbol{\varepsilon} \in \mathrm{Sym}\left(\mathbb{R}^d\right), \tag{6.10}$$

$$K_2'(\boldsymbol{A})\boldsymbol{\varepsilon} = \mathrm{Tr}\,(\boldsymbol{\varepsilon})\boldsymbol{A} + \mathrm{Tr}\,(\boldsymbol{\varepsilon}\boldsymbol{A})\boldsymbol{I}_d, \qquad \boldsymbol{\varepsilon} \in \mathrm{Sym}\left(\mathbb{R}^d\right),\ \boldsymbol{A} \in \mathrm{Sym}_0\left(\mathbb{R}^d\right). \tag{6.11}$$

It remains to decompose the subrepresentation $\mathrm{Sym}(\mathrm{Sym}_0(\mathbb{R}^d))$ into irreducibles. Two of them (one in two dimensions) we can simply guess by constructing rotationally equivariant maps from W_0 and W_2 to $\mathrm{Sym}(\mathrm{Sym}_0(\mathbb{R}^d))$:

$$K_0(\mu)\boldsymbol{\varepsilon} = 2\mu\boldsymbol{\varepsilon}, \quad K_2(\boldsymbol{A})\boldsymbol{\varepsilon} = \mathrm{dev}(\boldsymbol{\varepsilon}\boldsymbol{A} + \boldsymbol{A}\boldsymbol{\varepsilon}), \qquad \{\boldsymbol{\varepsilon}, \boldsymbol{A}\} \subset \mathrm{Sym}_0\left(\mathbb{R}^d\right),$$

where

$$\mathrm{dev}(\boldsymbol{\varepsilon}) = \boldsymbol{\varepsilon} - \frac{\mathrm{Tr}\,\boldsymbol{\varepsilon}}{d}\boldsymbol{I}_d$$

is the deviatoric part of a symmetric $d \times d$ matrix. We note that $K_2(\boldsymbol{A}) = \{0\}$ in two dimensions, due to the identity

$$\boldsymbol{\varepsilon}\boldsymbol{A} + \boldsymbol{A}\boldsymbol{\varepsilon} = \mathrm{Tr}\,(\boldsymbol{\varepsilon}\boldsymbol{A})\boldsymbol{I}_2, \qquad \{\boldsymbol{\varepsilon}, \boldsymbol{A}\} \subset \mathrm{Sym}_0\left(\mathbb{R}^2\right).$$

In order to view $K_0(\mu)$ and $K_2(\boldsymbol{A})$ as elements of $\mathbb{E}^d$ we write

$$K_0(\mu)\boldsymbol{\varepsilon} = 2\mu\,\mathrm{dev}(\boldsymbol{\varepsilon}), \qquad \boldsymbol{\varepsilon} \in \mathrm{Sym}\left(\mathbb{R}^d\right),\ \mu \in \mathbb{R}, \tag{6.12}$$

$$K_2(\boldsymbol{A})\boldsymbol{\varepsilon} = \mathrm{dev}(\mathrm{dev}(\boldsymbol{\varepsilon})\boldsymbol{A} + \boldsymbol{A}\,\mathrm{dev}(\boldsymbol{\varepsilon})), \quad \boldsymbol{\varepsilon} \in \mathrm{Sym}\left(\mathbb{R}^3\right),\ \boldsymbol{A} \in \mathrm{Sym}_0\left(\mathbb{R}^3\right). \tag{6.13}$$

Hence,

$$K_0 = \left\{K_0(\mu) : \mu \in \mathbb{R}\right\}, \tag{6.14}$$

[2] To state the obvious explicitly, we identify an operator $T \in \mathrm{Sym}(\mathbb{R}\boldsymbol{I}_d, \mathrm{Sym}_0(\mathbb{R}^d))$ with $T\boldsymbol{I}_d \in \mathrm{Sym}_0(\mathbb{R}^d)$ and then show that this correspondence is an $SO(d)$ isomorphism.

$$K_2 = \left\{K_2(A) : A \in \mathrm{Sym}_0\left(\mathbb{R}^3\right)\right\}. \tag{6.15}$$

It turns out that the orthogonal complement of $K_0 \oplus K_2$ (K_0 in 2D) in $\mathrm{Sym}(\mathrm{Sym}_0(\mathbb{R}^d))$ is isomorphic to the irrep W_4 of $SO(d)$. We denote this subspace $K_4 \subset \mathbb{E}^d$. For our purposes we need a more direct and explicit characterization of K_4. This has been accomplished in [1]. Let us summarize the results.

Let $\mathsf{C} \in \mathbb{E}^d$ and $\boldsymbol{n} \in \mathbb{R}^d$ be a unit vector. Then there is a unique symmetric $d \times d$ matrix $A_{\mathsf{C}}(\boldsymbol{n})$, such that for every $\boldsymbol{a} \in \mathbb{R}^d$

$$A_{\mathsf{C}}(\boldsymbol{n})\boldsymbol{a} \cdot \boldsymbol{a} = (\mathsf{C}(\boldsymbol{a} \otimes \boldsymbol{n}))\boldsymbol{n} \cdot \boldsymbol{a}.$$

The matrix $A_{\mathsf{C}}(\boldsymbol{n})$ is called the *acoustic tensor* of C. The name comes from the fact that the eigenvalues of $A_{\mathsf{C}}(\boldsymbol{n})$ are squares of the speeds of d different sound waves (a compression wave and $d - 1$ shear waves) propagating through the anisotropic elastic material in the direction $\boldsymbol{n}$. Observe that $\mathrm{Tr}\, A_{\mathsf{C}}(\boldsymbol{n})$ is quadratic in $\boldsymbol{n} \in \mathbb{R}^d$. Thus, there is a unique symmetric $d \times d$ matrix $\boldsymbol{M}_{\mathsf{C}}$, such that

$$\boldsymbol{M}_{\mathsf{C}}\, \boldsymbol{n} \cdot \boldsymbol{n} = \mathrm{Tr}\, A_{\mathsf{C}}(\boldsymbol{n}).$$

The subspace $K_4 \subset \mathbb{E}^d$ can now be described as follows.

$$K_4 = \left\{\mathsf{C} \in \mathbb{E}^d : \boldsymbol{M}_{\mathsf{C}} = 0, \mathsf{C}\boldsymbol{I}_d = 0\right\}. \tag{6.16}$$

To summarize,

$$\mathbb{E}^d = \mathrm{Sym}\left(\mathrm{Sym}\left(\mathbb{R}^d\right)\right) = \begin{cases} K_0' \oplus K_2' \oplus K_0 \oplus K_4, & d = 2, \\ K_0' \oplus K_2' \oplus K_0 \oplus K_2 \oplus K_4, & d = 3. \end{cases} \tag{6.17}$$

$$\mathsf{C} = \begin{cases} K_0'(\kappa) + K_0(\mu) + K_2'(A) + \mathsf{K}_4(\mathsf{C}), & d = 2, \\ K_0'(\kappa) + K_0(\mu) + K_2'(A) + K_2(\boldsymbol{B}) + \mathsf{K}_4(\mathsf{C}), & d = 3, \end{cases} \tag{6.18}$$

where $\{A, \boldsymbol{B}\} \subset \mathrm{Sym}_0(\mathbb{R}^d)$ and $\mathsf{K}_4(\mathsf{C}) \in K_4$ is the orthogonal projection of $\mathsf{C} \in \mathbb{E}^d$ onto K_4. Moreover, for a given $\mathsf{C} \in \mathbb{E}^d$ we can obtain the explicit expressions for κ, μ, A and $\boldsymbol{B}$, according to the following formulas (see [1]),

$$\kappa = \frac{1}{d^2}\mathrm{Tr}\,(\mathsf{C}\boldsymbol{I}_d), \qquad A = \frac{1}{d}\mathrm{dev}(\mathsf{C}\boldsymbol{I}_d), \tag{6.19}$$

$$\boldsymbol{B} = \frac{2\mathrm{dev}(d\boldsymbol{M}_{\mathsf{C}} - 2\mathsf{C}\boldsymbol{I}_d)}{(d-2)(d+4)}, \qquad \mu = \frac{\mathrm{Tr}\,(d\boldsymbol{M}_{\mathsf{C}} - \mathsf{C}\boldsymbol{I}_d)}{d(d-1)(d+2)}. \tag{6.20}$$

In fact, these formulas are not hard to derive, if we observe that, due to (6.10), (6.11) we have

$$\mathsf{C}\boldsymbol{I}_d = d(\kappa\boldsymbol{I}_d + A). \tag{6.21}$$

It is also fairly straightforward to compute, using formulas (6.10)–(6.13), that

$$M_{\mathrm{C}} = \kappa \boldsymbol{I}_d + 2\boldsymbol{A} + \mu\left(d + 1 - \frac{2}{d}\right)\boldsymbol{I}_d + \left(\frac{d}{2} + 1 - \frac{4}{d}\right)\boldsymbol{B}. \tag{6.22}$$

Solving the systems (6.21) and (6.22) for κ, μ, $\boldsymbol{A}$ and $\boldsymbol{B}$ we obtain formulas (6.19) and (6.20).

Now the search for all SO(d)-invariant Jordan multialgebras in $\mathbb{E}^d$ is reduced to SO(d)-invariant subspaces of $\mathbb{E}^d$. We note that formula (6.17) does not mean that there are only a small finite number of such subspaces. In fact, the presence of several irreps of the same weight, such as K_0 and K_0', as well as K_2 and K_2' makes it possible to combine them to create infinite families of subrepresentations. For example, the subspaces

$$K_0^\alpha = \left\{K_0(\mu) + K_0'(\alpha\mu) : \mu \in \mathbb{R}\right\}, \tag{6.23}$$

$$K_2^\alpha = \left\{K_2(A) + K_2'(\alpha A) : A \in \mathrm{Sym}_0(\mathbb{R}^3)\right\} \tag{6.24}$$

are also rotationally invariant. The list of all rotationally invariant subspaces in $\mathbb{E}^d$ can thus be described as follows:

Theorem 6.1.

(a) *A subspace* $\Pi \subset \mathrm{Sym}(\mathrm{Sym}(\mathbb{R}^2))$ *is rotationally invariant if and only if* $\Pi = U_0 \oplus U_2 \oplus U_4$, *where* U_0 *is either* $\{0\}$, K_0', K_0^α *for some* $\alpha \in \mathbb{R}$, *or* $K_0 \oplus K_0'$; U_2 *is either* $\{0\}$ *or* K_2'; U_4 *is either* $\{0\}$ *or* K_4.

(b) *A subspace* $\Pi \subset \mathrm{Sym}(\mathrm{Sym}(\mathbb{R}^3))$ *is rotationally invariant* if and only if $\Pi = U_0 \oplus U_2 \oplus U_4$, *where* U_0 *is either* $\{0\}$, K_0', K_0^α *or* $K_0 \oplus K_0'$; U_2 *is either* $\{0\}$, K_2', K_2^β *or* $K_2 \oplus K_2'$; U_4 *is either* $\{0\}$ *or* K_4.

6.1.3 Computing the subspace $\mathcal{A}$

We continue to use two- and three-dimensional elasticity to illustrate the process of computation of all exact relations and links listed in part 3 of the book. Once all rotationally invariant subspaces in Sym($\mathcal{T}$) have been identified,[3] the subspace $\mathcal{A}$, given by (4.10) can be computed with little extra effort. Let $\mathsf{C}_0 \in \mathbb{E}_+^d$ be an arbitrary isotropic positive definite (this is what subscript '+' in $\mathbb{E}_+^d$ indicates) reference medium. Then C_0 has the form[4]

$$\mathsf{C}_0\boldsymbol{\varepsilon} = \kappa_0 \,\mathrm{Tr}\,(\boldsymbol{\varepsilon})\boldsymbol{I} + 2\mu_0 \mathrm{dev}(\boldsymbol{\varepsilon}).$$

[3] It is a curious observation that only irreps of weights 0–4 occur in End($\mathcal{T}$) for every known physical property of materials.

[4] An arbitrary isotropic tensor for any three-dimensional material property is a positive linear combination of orthogonal projections onto the SO(3)-irreps in $\mathcal{T}$.

It is not difficult to compute [13] that

$$\boldsymbol{\Gamma}_0(\boldsymbol{n})\boldsymbol{\varepsilon} = \frac{1}{\mu_0}\big((\boldsymbol{\varepsilon n}) \odot \boldsymbol{n} - \rho_0(\boldsymbol{\varepsilon n} \cdot \boldsymbol{n})\boldsymbol{n} \otimes \boldsymbol{n}\big), \tag{6.25}$$

where $\boldsymbol{a} \odot \boldsymbol{n} = \frac{1}{2}(\boldsymbol{a} \otimes \boldsymbol{n} + \boldsymbol{n} \otimes \boldsymbol{a})$, and

$$\rho_0 = \frac{\lambda_0 + \mu_0}{\lambda_0 + 2\mu_0} = \frac{d\kappa_0 + (d-2)\mu_0}{d\kappa_0 + 2(d-1)\mu_0}, \quad d = 2, 3. \tag{6.26}$$

Formula (4.17) shows that in order to compute the subspace $\mathcal{A}$ we need to find the projections of $\boldsymbol{\Gamma}_0(\boldsymbol{n})$ onto K_2', K_2 and K_4. Using explicit formulas (6.19), (6.20) we obtain

$$\boldsymbol{A} = \frac{\mathrm{dev}(\boldsymbol{n} \otimes \boldsymbol{n})}{d(\lambda_0 + 2\mu_0)}, \qquad \boldsymbol{B} = \frac{7 - 4\rho_0}{14\mu_0}\mathrm{dev}(\boldsymbol{n} \otimes \boldsymbol{n}). \tag{6.27}$$

We also verify that the projection of $\boldsymbol{\Gamma}_0(\boldsymbol{n})$ onto K_4 is non-zero (for any $|\boldsymbol{n}| = 1$), if $\rho_0 > 0$, so that $\mathcal{A}_{2d} = K_2' \oplus K_4$, and

$$\mathcal{A}_{3d} = K_2^{\alpha_0} \oplus K_4, \qquad \alpha_0 = \frac{14(1 - \rho_0)}{3(7 - 4\rho_0)} = \frac{14\mu_0}{3(3\kappa_0 + 8\mu_0)},$$

where K_2^{α} is given by (6.24). We see that in three-dimensional elasticity the subspace $\mathcal{A}_{3d}$ depends on C_0, while, the subspace $\mathcal{A}_{2d}$ does not. However, we can transform $\mathcal{A}_{3d}$ into a subspace $\mathcal{A}_0$, which is independent of C_0, by means of a covariance transformation $\mathfrak{F}_{\mathsf{C}_1,\mathsf{C}_2}$, defined in (6.1). In order to preserve both the SO(3)-invariance of all subspaces and the symmetry of tensors, we can only use the transformation $\mathfrak{F}_{\overline{\mathsf{C}},\overline{\mathsf{C}}}$, where $\overline{\mathsf{C}}$ has to be isotropic:

$$\overline{\mathsf{C}}\boldsymbol{\varepsilon} = \overline{\kappa}\,\mathrm{Tr}\,(\boldsymbol{\varepsilon})\boldsymbol{I}_3 + 2\overline{\mu}\,\mathrm{dev}(\boldsymbol{\varepsilon}), \qquad \boldsymbol{\varepsilon} \in \mathrm{Sym}(\mathbb{R}^3).$$

In that case, it is easy to check that

$$\overline{\mathsf{C}}K_2'(\boldsymbol{A})\overline{\mathsf{C}} = 6\overline{\kappa}\overline{\mu}K_2'(\boldsymbol{A}), \qquad \overline{\mathsf{C}}K_2(\boldsymbol{B})\overline{\mathsf{C}} = 4\overline{\mu}^2K_2(\boldsymbol{B}).$$

Hence,

$$\overline{\mathsf{C}}K_2^{\beta}\overline{\mathsf{C}} = K_2^{\gamma}, \qquad \gamma = \frac{3\overline{\kappa}}{2\overline{\mu}}\beta.$$

This shows that in order to eliminate the dependence of $\mathcal{A}_0$ on C_0 we may choose $\overline{\kappa}/\overline{\mu} = \alpha_0$, in which case

$$\mathcal{A}_0 = \overline{\mathsf{C}}^{-1}\mathcal{A}_{3d}\overline{\mathsf{C}}^{-1} = K_4 \oplus K_2^{2/3}. \tag{6.28}$$

The choice of the 'slope' 2/3 in $\mathcal{A}_0$ is convenient because

$$\left(K_2(\boldsymbol{A}) + \frac{2}{3}K_2'(\boldsymbol{A})\right)\boldsymbol{\varepsilon} = \boldsymbol{\varepsilon A} + \boldsymbol{A\varepsilon}, \quad \boldsymbol{\varepsilon} \in \mathrm{Sym}(\mathbb{R}^3),\ \boldsymbol{A} \in \mathrm{Sym}_0(\mathbb{R}^3). \tag{6.29}$$

6.1.4 Subspace multiplication

We are now ready to discuss the algebraic problem of finding all rotationally invariant Jordan $\mathcal{A}_0$-multialgebras. For three-dimensional elastic composites the straightforward approach would be to test each of the 32 possibilities for Π listed in theorem 6.1, on whether it satisfies (4.9). Instead of verifying (4.9) directly, we observe that it can be rewritten in the language of subspace multiplication. For two subspaces U_1 and U_2 of $\mathrm{End}(\mathcal{T})$ we define

$$U_1U_2 = \mathrm{Span}\left\{\mathsf{K}_1\mathsf{K}_2 : \mathsf{K}_1 \in U_1, \mathsf{K}_2 \in U_2\right\}. \tag{6.30}$$

With this notation, a subspace $\Pi \subset \mathrm{Sym}(\mathcal{T})$ is a Jordan $\mathcal{A}$-multialgebra if and only if

$$(\Pi\mathcal{A}\Pi)_{\mathrm{sym}} \subset \Pi, \tag{6.31}$$

where

$$U_{\mathrm{sym}} = U \cap \mathrm{Sym}(\mathcal{T}).$$

We remark that that for $\Pi \subset \mathrm{Sym}(\mathcal{T})$ and $\mathcal{A} \subset \mathrm{Sym}(\mathcal{T})$ the subspace $U = \Pi\mathcal{A}\Pi$ is *symmetric*, i.e. $\mathsf{K} \in U \Rightarrow \mathsf{K}^T \in U$. In that case

$$U_{\mathrm{sym}} = \left\{\mathsf{K} + \mathsf{K}^T : \mathsf{K} \in U\right\}.$$

It is easy to check that if U_1 and U_2 are rotationally invariant, then so is U_1U_2. Moreover, the usual space addition and space multiplication (6.30) satisfy the distributive laws

$$(U_1 + U_2)U_3 = U_1U_3 + U_2U_3, \qquad U_3(U_1 + U_2) = U_3U_1 + U_3U_2.$$

Hence, in order to compute the product U_1U_2 of any two rotationally invariant subspaces we only need to know the products of all of the irreps in $\mathrm{End}(\mathcal{T})$. The representation theory of SO(3) permits us to derive explicit formulas for such products which, in turn, makes it possible to use computer algebra software to compute all the SO(3)-invariant Jordan $\mathcal{A}$-multialgebras.

We begin the analysis by noting the similarity between a product of subspaces in $\mathrm{End}(\mathcal{T})$, given by (6.30), and a *tensor product* of the same subspaces. Specifically, the map

$$\Phi : U_1 \otimes U_2 \to U_1U_2, \qquad \Phi(\mathsf{K}_1 \otimes \mathsf{K}_2) = \mathsf{K}_1\mathsf{K}_2 \tag{6.32}$$

is SO(d) equivariant, i.e.

$$\boldsymbol{R} \cdot \Phi(\mathsf{K}_1 \otimes \mathsf{K}_2) = \Phi((\boldsymbol{R} \cdot \mathsf{K}_1) \otimes (\boldsymbol{R} \cdot \mathsf{K}_2)).$$

That means that Φ will either 'transfer' an irrep in $U_1 \otimes U_2$ to an isomorphic irrep in U_1U_2, or it will 'kill' it. The decomposition of a tensor product of irreps into a direct sum of irreps is one of the basic problems studied in representation theory. In the case of SO(3) this problem has arisen in quantum mechanics in order to understand angular momentum in a hydrogen atom [15]. The decomposition of a tensor product of irreps into a direct sum of irreps is given by the following formulas, called the Clebsch–Gordan formulas,

$$W_m \otimes W_0 \cong W_m, \quad m \geqslant 0 \tag{6.33}$$

$$W_m \otimes W_m \cong \begin{cases} W_0 \oplus W_0 \oplus W_{2m}, & d = 2, \\ W_0 \oplus W_1 \oplus \cdots \oplus W_{2m}, & d = 3, \end{cases} \quad m \geqslant 1 \tag{6.34}$$

$$W_m \otimes W_n \cong \begin{cases} W_{|m-n|} \oplus W_{m+n}, & d = 2, \\ W_{|m-n|} \oplus \cdots \oplus W_{m+n}, & d = 3, \end{cases} \quad m \neq n, m \geqslant 1, n \geqslant 1. \tag{6.35}$$

We can use these formulas to identify all of the irreps in $\mathrm{End}(\mathcal{T}) \cong \mathcal{T} \otimes \mathcal{T}$. According to (6.8), for elasticity

$$\mathcal{T} = \mathrm{Sym}(\mathbb{R}^d) = \mathbb{R}\boldsymbol{I}_d \oplus \mathrm{Sym}_0(\mathbb{R}^d) \cong W_0 \oplus W_2.$$

Then

$$\mathrm{End}(\mathcal{T}) \cong \mathrm{End}(W_0) \oplus \mathrm{Hom}(W_0, W_2) \oplus \mathrm{Hom}(W_2, W_0) \oplus \mathrm{End}(W_2),$$

where $\mathrm{Hom}(U, V)$ denotes the space of linear maps from U to V. It is well-known that $\mathrm{Hom}(U, V) \cong U \otimes V$, as a representation of $SO(d)$. Therefore, according to (6.33), (6.34)

$$\mathrm{End}\big(\mathrm{Sym}(\mathbb{R}^d)\big) \cong$$
$$\begin{cases} W_0 \oplus W_2 \oplus W_2 \oplus W_0 \oplus W_4, & d = 2, \\ W_0 \oplus W_2 \oplus W_2 \oplus W_0 \oplus W_1 \oplus W_2 \oplus W_3 \oplus W_4, & d = 3. \end{cases} \tag{6.36}$$

We also recall from section 6.1.2 that

$$\mathrm{Sym}(\mathrm{Sym}(\mathbb{R}^d)) = \begin{cases} K_0' \oplus K_2' \oplus K_0 \oplus K_4, & d = 2, \\ K_0' \oplus K_2' \oplus K_0 \oplus K_2 \oplus K_4, & d = 3. \end{cases} \tag{6.37}$$

This shows that for $d = 3$, the terms in (6.36) with odd weights comprise the space of all antisymmetric operators on $\mathrm{Sym}_0(\mathbb{R}^3)$. For $d = 2$, we see that all antisymmetric operators on $\mathrm{Sym}_0(\mathbb{R}^2)$ are rotationally invariant (isotropic).

Surprisingly the theory of invariant subspace multiplication is simplest when $d = 3$. Hence, in the following discussion we will be dealing only with SO(3)-invariant subspaces. In the most general case $\mathcal{T}$ can be written as a direct sum of k_0 irreps of weight 0, k_1 irreps of weight 1, and so on:

$$\mathcal{T} \cong \mathbb{R}^{k_0} \otimes W_0 \oplus \mathbb{R}^{k_1} \otimes W_1 \oplus \cdots \oplus \mathbb{R}^{k_n} \otimes W_n.$$

It is convenient to think of $\mathcal{T}$ as a subspace in $\mathbb{R}^p \otimes \mathcal{W}$, where

$$\mathcal{W} = W_0 \oplus \cdots \oplus W_n.$$

So that

$$\mathrm{End}(\mathcal{T}) \subset \mathrm{End}\big(\mathbb{R}^p \otimes \mathcal{W}\big) = \mathrm{End}(\mathbb{R}^p) \otimes \mathrm{End}(\mathcal{W}).$$

Our first task is to identify (coordinatize) all irreps in

$$\mathrm{End}(\mathbb{R}^p) \otimes \mathrm{End}(\mathcal{W}) = \mathrm{End}(\mathbb{R}^p) \otimes \left(\bigoplus_{j,k=0}^{n} \mathrm{Hom}\big(W_j,\ W_k\big) \right).$$

It is a special property of SO(3) that, according to (6.35),

$$\mathrm{Hom}\big(W_j,\ W_k\big) \cong W_j \otimes W_k \cong \bigoplus_{\alpha=|j-k|}^{j+k} W_\alpha,$$

so that *there is at most one irrep of each weight in* $\mathrm{Hom}(W_j,\ W_k)$. We still need to deal with the fact that there could be multiple irreps of the same weight in various $\mathrm{Hom}(W_j,\ W_k)$ blocks. For that purpose we select specific zero weight vectors $\mathsf{E}_{jk\alpha}$ (fixed points of a fixed copy of SO(2) inside SO(3)) in each irrep $W_\alpha \subset \mathrm{Hom}(W_j,\ W_k)$. We may think of this choice as the choice of a basis in a vector space for solving linear equations numerically. Choosing specific zero weight vectors $\mathsf{E}_{jk\alpha}$ becomes essential only if there are multiple irreps of weight α in $\mathrm{Sym}(\mathcal{T})$. What makes the SO(3) theory particularly simple is the fact that all SO(3) irreps are of real type. That means that, if W_α is an irrep in $\mathrm{End}(\mathbb{R}^p) \otimes \mathrm{End}(\mathcal{W})$, then its zero weight vector can be written as

$$\sum_{j,k=0}^{n} \boldsymbol{X}_{jk} \otimes \mathsf{E}_{jk\alpha},$$

where $\boldsymbol{X}_{jk} \in \mathrm{End}(\mathbb{R}^p)$ for each j and k. We will denote such zero weight vector $\mathsf{E}(\mathbb{X})$, where $\mathbb{X}$ is the $(n+1) \times (n+1)$ block-matrix, whose (jk)th block is the $p \times p$ matrix $\boldsymbol{X}_{jk}$, where block numbering runs from 0 to n. The unique irrep of weight α containing $\mathsf{E}(\mathbb{X})$ will be accordingly denoted by $W_\alpha(\mathbb{X})$. We call the block-matrix $\mathbb{X}$ the homogeneous coordinates of $W_\alpha \subset \mathrm{End}(\mathbb{R}^p) \otimes \mathrm{End}(\mathcal{W})$, since $\lambda\mathbb{X}$, $\lambda \neq 0$, defines the same irrep. The fundamental question is to identify the product $W_\beta(\mathbb{Y})W_\alpha(\mathbb{X})$ as a direct sum of irreps in $\mathrm{End}(\mathcal{T})$ with specified homogeneous coordinates. The map (6.32) is SO(3)-invariant, and hence, there are block-matrices $\mathbb{Z}^\gamma$, such that

$$W_\beta(\mathbb{Y})W_\alpha(\mathbb{X}) = \bigoplus_{\gamma=|\alpha-\beta|}^{\alpha+\beta} W_\gamma(\mathbb{Z}^\gamma),$$

with the convention that $\mathbb{Z}^\gamma = 0$, if there is no weight γ irrep in $W_\beta(\mathbb{Y})W_\alpha(\mathbb{X})$. In appendix C we prove the explicit formula for block-matrices $\mathbb{Z}^\gamma$:

$$\boldsymbol{Z}_{jr}^{\gamma} = \frac{1}{\hat{K}_{jr\gamma}} \sum_{k=0}^{n} \boldsymbol{Y}_{kr}\boldsymbol{X}_{jk}\hat{R}_{jkr}^{\alpha\beta\gamma}\hat{K}_{kr\beta}\hat{K}_{jk\alpha}, \tag{6.38}$$

where the Racah coefficients $\hat{R}_{jkr}^{\alpha\beta\gamma}$, given by (C.10), are absolute numbers, while the numbers $\hat{K}_{jr\gamma}$, given by (C.9) involve the choice of specific zero weight vectors $w_j \in W_j \subset \mathcal{W}$, as well as a choice of an SO(3)-invariant inner product on $\mathcal{W}$, even though the resulting matrices $\mathbb{Z}^\gamma$ are independent of these choices. We set $\boldsymbol{Z}_{jr}^{\gamma} = 0$, whenever $\hat{K}_{jr\gamma} = 0$. Formula (6.38) permits us to compute a product of any two SO(3)-invariant subspaces. In particular, we can use it to build a 'Jordan $\mathcal{A}_0$-multiplication table'. The Maple™ implementation of subspace multiplication, as well as the Jordan $\mathcal{A}_0$-multiplication table for piezoelectricity ($n = 2, p = 1$) which includes elasticity, as a particular case, was built by an undergraduate researcher Daniel Lapsley, working under the author's supervision.

For elasticity, $\mathcal{A}_0$ is given by (6.28). Tables 6.1 and 6.2 show the 'Jordan products' $(U_1\mathcal{A}_0U_2)_{\rm sym}$ for almost all pairs of irreps U_1 and U_2 in Sym(Sym($\mathbb{R}^d$)), $d = 2, 3$, Since Jordan multiplication is commutative, we do not need to list the product $(U_2\mathcal{A}_0U_1)_{\rm sym}$ if the product $(U_1\mathcal{A}_0U_2)_{\rm sym}$ has already been listed. We note that table 6.2 omits K_2^α (or rather contains the results for $\alpha = 0$ only). For the purposes of identifying all SO(3)-invariant Jordan $\mathcal{A}_0$-multialgebras it will be sufficient to have the following results:

$$K_4 \subset \left(K_2^\alpha \mathcal{A}_0 K_2^\beta\right)_{\rm sym}, \qquad K_4 \subset \left(K_2^\alpha \mathcal{A}_0 K_0^\beta\right)_{\rm sym}. \tag{6.39}$$

For the sake of completeness, using Daniel Lapsley's Maple™ program we give the missing formulas

$$\left(K_2^\alpha \mathcal{A}_0 K_0'\right)_{\rm sym} = \alpha K_0' \oplus K_2', \qquad \left(K_2^\alpha \mathcal{A}_0 K_2'\right)_{\rm sym} = \alpha K_0' \oplus K_2' \oplus K_0 \oplus K_2 \oplus K_4,$$

$$\left(K_2^\alpha \mathcal{A}_0 K_0^\beta\right)_{\rm sym} = K_4 \oplus \begin{cases} K_0' \oplus K_2^{-49/24}, & \alpha = -\dfrac{7}{6}, \beta = \dfrac{49}{8}, \\ \beta K_0' \oplus K_2' \oplus K_2, & \alpha = -\dfrac{7}{6}, \beta \neq \dfrac{49}{8}, \\ K_0^{30\alpha\beta/(6\alpha+7)} \oplus K_2^{7\alpha/4}, & \alpha \neq -\dfrac{7}{6}, \beta = 3\alpha^2 - \dfrac{7\alpha}{4}, \\ K_0^{30\alpha\beta/(6\alpha+7)} \oplus K_2' \oplus K_2, & \alpha \neq -\dfrac{7}{6}, \beta \neq 3\alpha^2 - \dfrac{7\alpha}{4}, \end{cases}$$

$$\left(K_2^\alpha \mathcal{A}_0 K_2^\beta\right)_{\rm sym} = \left(\alpha^2 + \beta^2\right)K_2' \oplus K_2 \oplus K_4 \oplus \begin{cases} K_0^{15\alpha\beta}, & \alpha + \beta = \dfrac{7}{12}, \\ K_0 \oplus \alpha\beta K_0' & \alpha + \beta \neq \dfrac{7}{12}, \end{cases}$$

$$\left(K_2^\alpha \mathcal{A}_0 K_4\right)_{\rm sym} = K_0 \oplus \alpha K_2' \oplus K_2 \oplus K_4.$$

Table 6.1. Two-dimensional Jordan $\mathcal{A}_{2d}$-multiplication table.

	K_0'	K_2'	K_0^α	K_4
K_0'	0	K_0'	K_2'	K_2'
K_2'		$K_0' \oplus K_2'$	$K_2' \oplus K_0^{2\alpha} \oplus K_4$	$K_2' \oplus K_0 \oplus K_4$
K_0^β			$K_4 \oplus (\alpha+\beta)K_2'$	$K_0 + \beta K_2'$
K_4				K_4

Table 6.2. Three-dimensional Jordan $\mathcal{A}_0$-multiplication table.

	K_0'	K_2'	K_0^α	K_2	K_4
K_0'	0	K_0'	K_2'	K_2'	K_2'
K_2'		$K_0' \oplus K_2'$	$K_2' \oplus K_0^{5\alpha} \oplus K_2 \oplus K_4$	$K_2' \oplus K_0 \oplus K_2 \oplus K_4$	$K_2' \oplus K_0 \oplus K_2 \oplus K_4$
K_0^β			$K_2^{(\alpha+\beta)/2} \oplus K_4$	$K_0 \oplus K_2 \oplus K_4 \oplus \beta K_2'$	$K_0 \oplus K_2 \oplus K_4 \oplus \beta K_2'$
K_2				$K_0 \oplus K_2 \oplus K_4$	$K_0 \oplus K_2 \oplus K_4$
K_4					$K_0 \oplus K_2 \oplus K_4$

6.1.5 Computing all multialgebras

Once $\mathcal{A}$-products of irreps have been computed, it is then a fairly simple matter, in most cases, to characterize all SO(d)-invariant Jordan $\mathcal{A}$-multialgebras. Let us again turn to the cases of two-dimensional and three-dimensional elasticity for illustration. The main observation is that in addition to 'coordinate irreps' denoted by K_j and K_j' we also have 'askew' irreps K_0^α and K_2^α. The most difficult part of the analysis is to identify all 'askew' SO(d)-invariant Jordan $\mathcal{A}$-multialgebras, i.e. multialgebras containing K_0^α or K_2^α, but not $K_0 \oplus K_0'$ or $K_2 \oplus K_2'$. The following theorem gives an answer to this problem.

Theorem 6.2. *There are no proper rotationally invariant Jordan $\mathcal{A}$-multialgebras containing either K_0^α or K_2^α for $\alpha \neq 0$.*

Proof. First, consider two-dimensional elasticity. According to the multiplication table 6.1, if $\alpha \neq 0$ and $K_0^\alpha \subset \Pi$, then $K_4 \oplus K_2' \subset \Pi$. But then $K_0' \subset \Pi$ and $K_0 \subset \Pi$. This proves that $\Pi = \mathbb{E}^2$.

Next, consider three-dimensional elasticity. In that case table 6.2 shows that if $K_0^\alpha \subset \Pi$, $\alpha \neq 0$, then $K_2^\alpha \subset \Pi$. The result would follow, when we prove that $K_2^\alpha \subset \Pi$ implies that $\Pi = \mathbb{E}^3$.

According to (6.39), if $K_2^\alpha \subset \Pi$, $\alpha \neq 0$, then $K_4 \subset \Pi$. This in turn implies that $K_0 \oplus K_2 \subset \Pi$. In particular, $K_2^\alpha + K_2 \subset \Pi$. But, obviously,

$$K_2^\alpha + K_2 = \begin{cases} K_2, & \alpha = 0, \\ K_2 \oplus K_2', & \alpha \neq 0. \end{cases}$$

Finally, $K_2' \subset \Pi$ implies that $K_0' \subset \Pi$ and so, $\Pi = \mathbb{E}^3$. □

It remains now to examine only a small finite number of cases. The results are given in the following complete lists of SO(d)-invariant Jordan $\mathcal{A}$-multialgebras.

1. $\Pi = \mathrm{Sym}(\mathrm{Sym}_0(\mathbb{R}^d))$. We observe that $\Pi = \mathrm{Ann}(\mathbb{R}\boldsymbol{I}_d) \cap \mathbb{E}^d$. Hence, this solution corresponds to a uniform field relation (UFR) via (4.23). As such, it can be written as a set of all elastic tensors C with the property $\mathsf{C}\boldsymbol{I}_d = d\kappa_0\boldsymbol{I}_d$ for some constant bulk modulus $\kappa_0 > 0$. This exact relation was discovered in [10].
2. $\Pi = K_0' \oplus K_2'$. We will see that this Jordan multialgebra corresponds to the exact relation (9.13) in section 9.2. This is the first exact relation obtained in [8] by the application of the general theory of exact relations. In fact, no other proof exists in the three-dimensional case.
3. $\Pi = K_0'$. We observe that $\Pi = \mathrm{Ann}(\mathrm{Sym}_0(\mathbb{R}^d))$. Hence, this solution corresponds to a UFR via (4.23). This corresponds to the Hill exact relation (4.2).
4. If $d = 2$, then $\Pi = K_4$. This solution corresponds to the Lurie–Cherkaev exact relation (9.7).

Tables 6.1 and 6.2 allow fast calculations of Jordan 'squares' (5.32). We observe that $\Pi^2 = \Pi$, except for $\Pi = K_0'$, in which case $\Pi^2 = \{0\}$. According to the theory of section 5.3, $\Pi = K_0'$ must have an accompanying volume fraction relation. In the next section we will show that this volume fraction relation is the famous Hill formula (4.1).

In order to verify that the manifolds $\mathbb{M}$ corresponding to the computed Jordan multialgebras are exact relations, we need to show that they are reflexive in the sense of definition 4.10. The reflexivity of UFRs is automatic (see theorem 4.15). In order to establish reflexivity of the remaining multialgebras we construct explicit associative multialgebras, whose set of symmetric operators coincides with the given Jordan multialgebra. It is easy to verify that when $d = 2$, $\Pi = K_4$ is also an associative $\mathcal{A}$-multialgebra. For $\Pi = K_0' \oplus K_2'$, the subspace

$$\Pi' = \left\{\mathsf{C} \in \mathrm{End}\left(\mathrm{Sym}\left(\mathbb{R}^d\right)\right): \mathsf{C}\mathrm{Sym}_0\left(\mathbb{R}^d\right) \subset \mathbb{R}\boldsymbol{I}_d\right\},$$

is an associative $\mathcal{A}$-multialgebra, whose symmetric elements comprise Π. This becomes apparent, if we recall that all elements of $\mathcal{A}$ map $\boldsymbol{I}_d$ into symmetric trace-free matrices, so that $\mathsf{C}_1\mathsf{A}\mathsf{C}_2 \in \Pi'$, whenever $\{\mathsf{C}_1, \mathsf{C}_2\} \subset \Pi'$ and $\mathsf{A} \in \mathcal{A}$.

6.2 Computing exact relations

The final task is to compute explicit representations of the manifold $\mathbb{M}$, given by (4.4), for each Jordan $\mathcal{A}$-multialgebra. Using formula (4.4) for this purpose is inconvenient, since it involves a unit vector $\boldsymbol{n}_0$, which, according to the general theory, the manifold $\mathbb{M}$ does not depend on. In this section we will show that we can compute the explicit representation of $\mathbb{M}$ via (4.4), where instead of the transformation $W_{\boldsymbol{n}}$, given by (3.26) and (3.41), we use a simplified transformation W_{M}, given by (3.26), provided M has the property

$$\mathsf{K}(\overline{\mathbf{\Gamma}} - \mathsf{M})\mathsf{K} \in \Pi, \quad \forall \mathsf{K} \in \Pi, \tag{6.40}$$

where the fixed tensor $\overline{\mathbf{\Gamma}}$ can be chosen arbitrarily among the points on the affine hull $\mathcal{G}$ of $\{\mathbf{\Gamma}_0(\boldsymbol{n}) : |\boldsymbol{n}| = 1\}$. In particular, $\overline{\mathbf{\Gamma}}$, given by (4.19), or $\mathbf{\Gamma}_0(\boldsymbol{n})$, for any $|\boldsymbol{n}| = 1$ are all valid choices.

Definition 6.3. *A tensor* M *satisfying (6.40) is called an inversion key for the Jordan multialgebra* Π.

Choosing the inversion key is the same as recognizing that the Jordan $\mathcal{A}$-multialgebra Π is also a Jordan $\mathcal{A}'$-multialgebra, where

$$\mathcal{A}' = \mathrm{Span}\left\{\mathcal{A}, \overline{\mathbf{\Gamma}} - \mathsf{M}\right\}. \tag{6.41}$$

It is easy to see that if M satisfies (6.40) with one choice of $\overline{\mathbf{\Gamma}} \in \mathcal{G}$ it will also satisfy (6.40) with any other choice of $\overline{\mathbf{\Gamma}} \in \mathcal{G}$. Indeed, if $\overline{\mathbf{\Gamma}}' \in \mathcal{G}$, then

$$\mathsf{K}\left(\overline{\mathbf{\Gamma}}' - \mathsf{M}\right)\mathsf{K} = \mathsf{K}\left(\overline{\mathbf{\Gamma}} - \mathsf{M}\right)\mathsf{K} + \mathsf{K}\left(\overline{\mathbf{\Gamma}}' - \overline{\mathbf{\Gamma}}\right)\mathsf{K} \in \Pi,$$

since $\overline{\mathbf{\Gamma}}' - \overline{\mathbf{\Gamma}} \in \mathcal{A}$.

To choose the simplest inversion key we first check if $\mathsf{M} = 0$ satisfies (6.40). If not, then other possibilities become apparent if we use isotropic $\overline{\mathbf{\Gamma}}$, given by (4.19). In this case $\overline{\mathbf{\Gamma}}$ is a sum of scalar multiples of orthogonal projections onto $SO(d)$-invariant subspaces of $\mathcal{T}$. If $\mathsf{M} = 0$ does not satisfy (6.40), we can then try M to be the sum of one or more of the terms in the expression for $\overline{\mathbf{\Gamma}}$. In applications we were always able to find a much simpler choice for M than $\overline{\mathbf{\Gamma}}$.

Let us prove now that we can compute an exact relation submanifold $\mathbb{M}$ using the simplified transformation W_{M}. It is therefore necessary that the operator $\mathsf{I} + (\mathsf{L} - \mathsf{L}_0)\mathsf{M}$ be invertible. However, our criterion (6.40) does not guarantee that M will have the universal invertibility property (see definition 3.17). For this reason we denote

$$\mathbb{P}_{\mathsf{M}} = \left\{\mathsf{L} \in \mathrm{End}^{+}(\mathcal{T}) : \mathsf{I} + (\mathsf{L} - \mathsf{L}_0)\mathsf{M} \text{ is invertible}\right\}.$$

We also define

$$\mathbb{M}' = \mathbb{P}_{\mathsf{M}} \bigcap \mathbb{M}. \tag{6.42}$$

Lemma 6.4. *Suppose that* $\mathsf{L}_0 \in \mathbb{M}$, *where* $\mathbb{M}$ *is given by (4.4) in terms of a Jordan* $\mathcal{A}$*-multialgebra* Π. *Then the set* $\mathbb{M}'$ *is an open and dense subset of* $\mathbb{M}$.

Proof. The idea is to obtain an explicit parametrization of $\mathbb{M}$ by $m = \dim\mathbb{M}$ real parameters. Recall that $\mathbb{K} = W_{\boldsymbol{n}_0}(\mathbb{M})$ is an open subset of a Jordan $\mathcal{A}$-multialgebra Π, containing $0 \in \mathrm{End}(\mathcal{T})$. Let $\{\mathsf{K}_1, \ldots, \mathsf{K}_m\}$ be a basis of Π. For any $\boldsymbol{u} \in \mathbb{R}^m$ we denote $\mathsf{K}(\boldsymbol{u}) = u_1\mathsf{K}_1 + \ldots + u_m\mathsf{K}_m$. Then, $\mathcal{O} = \{\boldsymbol{u} \in \mathbb{R}^m : \mathsf{K}(\boldsymbol{u}) \in \mathbb{K}\}$ is an open subset of $\mathbb{R}^m$, containing $0 \in \mathbb{R}^m$. The inverse of W_{n_0}, can be computed by (3.46), where $\mathsf{M} = \mathbf{\Gamma}_0(\boldsymbol{n})$ has the universal invertibility property, according to corollary 3.19. This formula defines a diffeomorphism $\Phi : \mathcal{O} \to \mathbb{M}$

$$\Phi(\boldsymbol{u}) = \mathsf{L}_0 + \left(\mathsf{I} - \mathsf{K}(\boldsymbol{u})\mathbf{\Gamma}_0(\boldsymbol{n}_0)\right)^{-1}\mathsf{K}(\boldsymbol{u}), \quad \boldsymbol{u} \in \mathcal{O}.$$

Thus, in order to prove the lemma we need to prove that the operator

$$\mathsf{P}(\boldsymbol{u}) = \mathsf{I} + (\Phi(\boldsymbol{u}) - \mathsf{L}_0)\mathsf{M}$$

is invertible on an open and dense subset of $\mathcal{O}$. Substituting the expression for $\Phi(\boldsymbol{u})$ into the formula for $\mathsf{P}(\boldsymbol{u})$ we obtain

$$\mathsf{P}(\boldsymbol{u}) = \left(\mathsf{I} - \mathsf{K}(\boldsymbol{u})\mathbf{\Gamma}_0(\boldsymbol{n}_0)\right)^{-1}(\mathsf{I} - \mathsf{K}(\boldsymbol{u})(\mathbf{\Gamma}_0(\boldsymbol{n}_0) - \mathsf{M})).$$

It remains to observe that the polynomial

$$p(\boldsymbol{u}) = \det(\mathsf{I} - \mathsf{K}(\boldsymbol{u})(\mathbf{\Gamma}_0(\boldsymbol{n}_0) - \mathsf{M}))$$

is non-zero on a sufficiently small neighborhood of $0 \in \mathbb{R}^m$, since $\mathsf{K}(0) = 0$. Hence, $p(\boldsymbol{u})$ is a non-zero polynomial on $\mathbb{R}^m$. We conclude that $p(\boldsymbol{u}) \neq 0$ on an open and dense subset $\mathcal{O}' \subset \mathcal{O}$. Thus, $\mathsf{P}(\boldsymbol{u})$ must be invertible for all $\boldsymbol{u} \in \mathcal{O}'$. The lemma is proved now. □

Lemma 6.4 implies that the subset $\mathbb{M}' \subset \mathbb{M}$, given by (6.42), describes the submanifold $\mathbb{M}$ completely, since $\overline{\mathbb{M}'} = \mathbb{M}$, where the closure is taken in the relative topology of $\mathrm{End}^+(\mathcal{T})$. We can now prove our main inversion theorem.

Theorem 6.5. *Suppose* M *is an inversion key for the Jordan multialgebra* Π *in the sense of definition 6.3. Then*

$$\mathbb{M}' = \{\mathsf{L} \in \mathbb{P}_{\mathsf{M}} : W_{\mathsf{M}}(\mathsf{L}) \in \Pi\}, \tag{6.43}$$

where the submanifold $\mathbb{M}'$ *is given by (6.42) and* W_{M}*, by (3.26).*

Proof. Let $\tilde{\mathbb{M}}$ denote the right-hand side of (6.43). To establish the theorem we will show that $\mathbb{M}' \subset \tilde{\mathbb{M}}$ and $\tilde{\mathbb{M}} \subset \mathbb{M}'$.

To prove the first inclusion we assume that $\mathsf{L} \in \mathbb{M}'$ and prove that $\mathsf{L} \in \tilde{\mathbb{M}}$. We have

$$\mathsf{I} + (\mathsf{L} - \mathsf{L}_0)\mathsf{M} = \mathsf{I} + (\mathsf{L} - \mathsf{L}_0)\mathbf{\Gamma}_0(\boldsymbol{n}_0) - (\mathsf{L} - \mathsf{L}_0)\mathsf{A}', \quad \mathsf{A}' = \mathbf{\Gamma}_0(\boldsymbol{n}_0) - \mathsf{M}.$$

Multiplying by $(\mathsf{I} + (\mathsf{L} - \mathsf{L}_0)\mathbf{\Gamma}_0(\boldsymbol{n}_0))^{-1}$ on the left, and using definition (3.41) of $W_{\boldsymbol{n}_0}(\mathsf{L})$, we obtain

$$\left(\mathsf{I} + (\mathsf{L} - \mathsf{L}_0)\mathbf{\Gamma}_0(\boldsymbol{n}_0)\right)^{-1}(\mathsf{I} + (\mathsf{L} - \mathsf{L}_0)\mathsf{M}) = \mathsf{I} - W_{\boldsymbol{n}_0}(\mathsf{L})\mathsf{A}'. \tag{6.44}$$

The left-hand side of (6.44) is invertible, since $\mathsf{L} \in \mathbb{P}_{\mathsf{M}}$. If we take inverse of both sides in (6.44) and then multiply by $W_{\boldsymbol{n}_0}(\mathsf{L})$ on the right we obtain

$$W_{\mathsf{M}}(\mathsf{L}) = \left(\mathsf{I} - W_{n_0}(\mathsf{L})\mathsf{A}'\right)^{-1} W_{n_0}(\mathsf{L}) \in \Pi,$$

by lemma 4.5, which applies, since $W_{n_0}(\mathsf{L}) \in \Pi$ and $\mathsf{A}' \in \mathcal{A}'$, where $\mathcal{A}'$ is defined in (6.41). The fact that M is an inversion key implies that Π is a Jordan $\mathcal{A}'$-multialgebra. This proves the inclusion $\mathbb{M}' \subset \tilde{\mathbb{M}}$. The other inclusion is proved in a symmetric way by computing that

$$W_{n_0}(\mathsf{L}) = \left(\mathsf{I} + W_{\mathsf{M}}(\mathsf{L})\mathsf{A}'\right)^{-1} W_{\mathsf{M}}(\mathsf{L}) \in \Pi.$$

□

The simplification afforded by a choice of the inversion key can also be combined with the covariance transformation that was used to simplify the subspace $\mathcal{A}$.

Theorem 6.6. *Suppose $\mathcal{A}_0$ is given by (6.2) and Π_0 is a Jordan $\mathcal{A}_0$-multialgebra. Suppose that the operator M_0 satisfies*

$$\mathsf{K}\left(\overline{\boldsymbol{\Gamma}}^0 - \mathsf{M}_0\right)\mathsf{K} \in \Pi_0 \quad \forall \mathsf{K} \in \Pi_0, \tag{6.45}$$

where $\overline{\boldsymbol{\Gamma}}^0 = \mathsf{C}_2^{-1}\overline{\boldsymbol{\Gamma}}\mathsf{C}_1^{-1}$. Then the exact relation corresponding to Π_0 can be computed by the formula

$$\mathbb{M}' = \left\{\mathsf{L} \in \mathrm{End}^+(\mathcal{T}) : \mathsf{L} = \mathsf{L}_0 + \mathsf{C}_1^{-1}(\mathsf{I} - \mathsf{K}\mathsf{M}_0)^{-1}\mathsf{K}\mathsf{C}_2^{-1},\ \ \mathsf{K} \in \Pi_0\right\}. \tag{6.46}$$

Example 6.7.

Let us compute the exact relation corresponding to the SO(d)-invariant Jordan $\mathcal{A}_0$-multialgebra $\Pi_0 = K_0' \oplus K_2'$, from section 6.1.5. Recall that for elasticity $\boldsymbol{\Gamma}_0(\boldsymbol{n})$ is given by (6.25). Then we easily compute, using (6.19) and (6.20), that $\overline{\boldsymbol{\Gamma}}$ is an isotropic elastic tensor with bulk and shear moduli, given, respectively, by

$$\kappa_\Gamma = \begin{cases} \dfrac{1}{4(\kappa_0 + \mu_0)}, & d = 2, \\[2ex] \dfrac{1}{3(3\kappa_0 + 4\mu_0)}, & d = 3, \end{cases} \qquad \mu_\Gamma = \begin{cases} \dfrac{\kappa_0 + 2\mu_0}{8\mu_0(\kappa_0 + \mu_0)}, & d = 2, \\[2ex] \dfrac{3(\kappa_0 + 2\mu_0)}{10\mu_0(3\kappa_0 + 4\mu_0)}, & d = 3. \end{cases} \tag{6.47}$$

Recall that we have simplified the subspace $\mathcal{A}$ by (6.28). That means that in theorem 6.6 we need to set $\mathsf{C}_1 = \mathsf{C}_2 = \overline{\mathsf{C}}$, where the bulk and shear moduli $\bar{\kappa}$ and $\bar{\mu}$, respectively, of $\overline{\mathsf{C}}$ can be chosen arbitrarily for two-dimensional elasticity, and are constrained by

$$\frac{\overline{\kappa}}{\overline{\mu}} = \alpha_0 = \frac{14\mu_0}{3(3\kappa_0 + 8\mu_0)}, \tag{6.48}$$

for three-dimensional elasticity. This gives us some freedom to simplify $\overline{\mathbf{\Gamma}}^0$ in (6.45). In two space dimensions we can choose

$$\overline{\kappa} = \frac{1}{2}\sqrt{\kappa_\Gamma}, \qquad \overline{\mu} = \frac{1}{2}\sqrt{\mu_\Gamma},$$

so that both the bulk and the shear moduli of $\overline{\mathbf{\Gamma}}^0 = \overline{\mathsf{C}}^{-1}\mathbf{\Gamma}\overline{\mathsf{C}}^{-1}$ are 1. For three-dimensional elasticity we may only choose

$$\overline{\kappa} = \frac{1}{3}\sqrt{\kappa_\Gamma},$$

making the bulk modulus of $\overline{\mathbf{\Gamma}}^0$ equal to 1. The shear modulus $\overline{\mu}$ is then uniquely determined by (6.48).

Let us show that for $\Pi = K_0' \oplus K_2'$ we can choose $\mathsf{M}_0 = \boldsymbol{I}_d \otimes \boldsymbol{I}_d$, $d = 2, 3$, i.e.,

$$\mathsf{M}_0\boldsymbol{\varepsilon} = (\operatorname{Tr}\boldsymbol{\varepsilon})\boldsymbol{I}_d, \qquad \boldsymbol{\varepsilon} \in \operatorname{Sym}(\mathbb{R}^d).$$

Indeed, this choice implies that $(\overline{\mathbf{\Gamma}}^0 - \mathsf{M}_0)\boldsymbol{I}_d = 0$, since in both two- and three-dimensional elasticity the bulk modulus of $\overline{\mathbf{\Gamma}}^0$ is 1. Observing that every $\mathsf{K} \in \Pi$ maps $\operatorname{Sym}_0(\mathbb{R}^d)$ into multiples of the identity, it follows that $(\overline{\mathbf{\Gamma}}^0 - \mathsf{M}_0)\mathsf{K}$, and consequently $\mathsf{K}(\overline{\mathbf{\Gamma}}^0 - \mathsf{M}_0)\mathsf{K}$ annihilate $\operatorname{Sym}_0(\mathbb{R}^d)$. Thus,

$$\mathsf{K}(\overline{\mathbf{\Gamma}}^0 - \mathsf{M}_0)\mathsf{K} \subset K_0' \subset \Pi.$$

We can now use the inversion key M_0 to compute the corresponding exact relation. We begin by parametrizing Π using the maps (6.10)–(6.13). So, if $\mathsf{K} \in \Pi$, then $\mathsf{K} = K_0'(\theta) + K_2'(\boldsymbol{A})$, $\theta \in \mathbb{R}$, $\boldsymbol{A} \in \operatorname{Sym}_0(\mathbb{R}^d)$. Observe that

$$\mathsf{K}\boldsymbol{I}_d = d\theta\boldsymbol{I}_d + d\boldsymbol{A} = \boldsymbol{B} \in \operatorname{Sym}(\mathbb{R}^d).$$

We note that $\boldsymbol{B}$ also uniquely determines parameters θ and $\boldsymbol{A}$. In fact,

$$\mathsf{K} = \frac{1}{d}\left(\boldsymbol{I}_d \otimes \boldsymbol{B} + \boldsymbol{B} \otimes \boldsymbol{I}_d - \frac{1}{d}(\operatorname{Tr}\boldsymbol{B})\boldsymbol{I}_d \otimes \boldsymbol{I}_d\right). \tag{6.49}$$

We compute

$$\left(\mathsf{I} - \mathsf{K}\boldsymbol{I}_d \otimes \boldsymbol{I}_d\right)^{-1} = \left(\mathsf{I} - \boldsymbol{B} \otimes \boldsymbol{I}_d\right)^{-1} = \mathsf{I} + \frac{\boldsymbol{B} \otimes \boldsymbol{I}_d}{1 - \operatorname{Tr}\boldsymbol{B}}.$$

Hence,

$$\left(\mathsf{I} - \mathsf{K}\boldsymbol{I}_d \otimes \boldsymbol{I}_d\right)^{-1}\mathsf{K} = \mathsf{K} + \frac{\boldsymbol{B} \otimes \boldsymbol{B}}{1 - \operatorname{Tr}\boldsymbol{B}}.$$

Using (6.49) we obtain

$$(\mathsf{I} - \mathsf{K}\mathsf{M}_0)^{-1}\mathsf{K} = \frac{1}{d}\left(\boldsymbol{I}_d \otimes \boldsymbol{B} + \boldsymbol{B} \otimes \boldsymbol{I}_d - \frac{1}{d}(\operatorname{Tr} \boldsymbol{B})\boldsymbol{I}_d \otimes \boldsymbol{I}_d\right) + \frac{\boldsymbol{B} \otimes \boldsymbol{B}}{1 - \operatorname{Tr} \boldsymbol{B}}.$$

Substituting this expression into (6.46) gives a (rather ugly) parametrization of $\mathbb{M}$. It is a remarkable property of exact relations that they invariably possess beautiful representations. Indeed, the ugly expression above can also be written as

$$(\mathsf{I} - \mathsf{K}\mathsf{M}_0)^{-1}\mathsf{K} = -\frac{\boldsymbol{I}_d \otimes \boldsymbol{I}_d}{d^2} + \boldsymbol{C} \otimes \boldsymbol{C}, \qquad \boldsymbol{C} = \frac{\boldsymbol{B} + (1 - \operatorname{Tr} \boldsymbol{B})\boldsymbol{I}_d/d}{\sqrt{1 - \operatorname{Tr} \boldsymbol{B}}}. \tag{6.50}$$

Thus,

$$\mathsf{C} = \mathsf{C}_0 - \frac{1}{d^4\overline{\kappa}^2}\boldsymbol{I}_d \otimes \boldsymbol{I}_d + \overline{\mathsf{C}}^{-1}\boldsymbol{C} \otimes \overline{\mathsf{C}}^{-1}\boldsymbol{C}. \tag{6.51}$$

We see now that $\mathbb{M}$ is parametrized by $\boldsymbol{D} = \overline{\mathsf{C}}^{-1}\boldsymbol{C} \in \operatorname{Sym}(\mathbb{R}^d)$ and has the form

$$\mathsf{C} = \tilde{\mathsf{C}}_0 + \boldsymbol{D} \otimes \boldsymbol{D}, \tag{6.52}$$

where

$$\tilde{\mathsf{C}}_0 = \mathsf{C}_0 - \frac{1}{d^2\kappa_\Gamma}\boldsymbol{I}_d \otimes \boldsymbol{I}_d$$

is an isotropic elastic tensor with shear modulus μ_0 and bulk modulus

$$\tilde{\kappa} = \kappa_0 - \frac{1}{d^2\kappa_\Gamma} = \begin{cases} -\mu_0, & d = 2, \\ -4\mu_0/3, & d = 3. \end{cases}$$

Thus, $\tilde{\mathsf{C}}_0 = 2\mu_0\mathsf{T}$, where T is the isotropic 'null-Lagrangian', given by the formula

$$\mathsf{T}\boldsymbol{\varepsilon} = \boldsymbol{\varepsilon} - (\operatorname{Tr} \boldsymbol{\varepsilon})\boldsymbol{I}_d, \qquad \boldsymbol{\varepsilon} \in \operatorname{Sym}\left(\mathbb{R}^d\right). \tag{6.53}$$

Thus, we have proved that all elastic tensors of materials in the exact relation corresponding to $\Pi = K_0' \oplus K_2'$ have a 'rank-one plus a null-Lagrangian' structure, given by

$$\mathsf{C} = 2\mu_0\mathsf{T} + \boldsymbol{D} \otimes \boldsymbol{D} : \boldsymbol{D} \in \operatorname{Sym}\left(\mathbb{R}^3\right). \tag{6.54}$$

6.3 Computing volume fraction relations

Recall that additional formulas involving volume averages accompany an exact relation, if the Jordan 'square' (5.32) is strictly smaller that the original Jordan multialgebra. In order to compute such formulas explicitly we derive the rule for finding the inversion key M, so that the transformation W_{M} can be used in (5.37) instead of W_n.

Theorem 6.8. *Let* Π *be a Jordan* $\mathcal{A}$*-multialgebra, and let* $\mathcal{P}_{(\Pi^2)^c}$ *denote the orthogonal projection onto the orthogonal complement of* Π^2 *in* Π. *Then*

$$\mathcal{P}_{(\Pi^2)^c} W_{\mathsf{M}}(\mathsf{L}^*) = \mathcal{P}_{(\Pi^2)^c}\left\langle W_{\mathsf{M}}(\mathsf{L}(z))\right\rangle, \tag{6.55}$$

where W_{M} *is given by* (3.26) *and* M *satisfies*

$$\mathsf{K}\left(\overline{\boldsymbol{\Gamma}} - \mathsf{M}\right)\mathsf{K} \in \Pi^2 \quad \forall \mathsf{K} \in \Pi. \tag{6.56}$$

Whether or not M *satisfies (6.56) does not depend on the choice of* $\overline{\boldsymbol{\Gamma}} \in \mathcal{G}$*—the affine hull of* $\{\boldsymbol{\Gamma}_0(\boldsymbol{n}) : |\boldsymbol{n}| = 1\}$.

Proof. In our theory the formulas involving volume averages are particular instances of links, as was discussed in section 5.3. The links, according to section 5.1, can be interpreted as exact relations. Thus, formulas (6.45) and (6.46) must apply to volume fraction relations. Therefore, in order to prove the theorem we need to show that these formulas reduce to (6.56) and (6.55), respectively, for Jordan $\hat{\mathcal{A}}$-multialgebra $\hat{\Pi}$, defined in (5.33), where $\hat{\mathcal{A}}$ is given by (5.27). Now, let $\hat{\mathsf{M}} = [\mathsf{M}, 0]$ be such that (6.45) is satisfied for $\hat{\Pi}$. That means that

$$\left[\mathsf{K}\left(\overline{\boldsymbol{\Gamma}} - \mathsf{M}\right)\mathsf{K}, 0\right] \in \hat{\Pi}, \quad \forall \mathsf{K} \in \Pi.$$

Thus, according to (5.33) we conclude that (6.56) must hold. Then, by theorem 6.5, $\hat{\mathcal{M}}$, given by (5.36) can also be described as the closure of

$$\hat{\mathbb{M}}' = \left\{\left[\mathsf{L}, \tilde{\mathsf{L}}\right] \in \mathbb{M}' \times \mathrm{End}^+(\mathbb{R}^N) : \tilde{\mathsf{L}}_0 - \tilde{\mathsf{L}} = \mathcal{P}_{(\Pi^2)^c} W_{\mathsf{M}}(\mathsf{L})\right\}.$$

In other words, formula (6.55) is valid. □

According to (6.56) the volume fraction inversion key can also serve as the inversion key for the corresponding multialgebra Π. The converse is not always true. It is important to emphasize that formulas (6.55) and (6.56) use the original (unsimplified) versions of $\boldsymbol{\Gamma}_0(\boldsymbol{n})$ and $\mathcal{A}$. If we apply the covariance transformation (6.1) and use $\mathcal{A}_0$, given by (6.2), instead of $\mathcal{A}$, then formula (6.55) becomes

$$\mathcal{P}_{(\Pi_0^2)^c}\mathsf{K}(\mathsf{L}^*) = \mathcal{P}_{(\Pi_0^2)^c}\langle \mathsf{K}(\mathsf{L}(z))\rangle, \tag{6.57}$$

where

$$\mathsf{K}(\mathsf{L}) = \mathsf{C}_1 W_{\mathsf{C}_2\mathsf{M}_0\mathsf{C}_1}(\mathsf{L})\mathsf{C}_2$$

is the inverse of the transformation in (6.46), and Π_0 is a Jordan $\mathcal{A}_0$-multialgebra. However, the inversion key M_0 is required to satisfy

$$\mathsf{K}\left(\overline{\boldsymbol{\Gamma}}^0 - \mathsf{M}_0\right)\mathsf{K} \in \Pi_0^2 \quad \forall \mathsf{K} \in \Pi_0. \tag{6.58}$$

We note that in cases when the simplest inversion key M_0 for Π_0 also satisfies (6.58), the calculations in (6.46) usually also yield the explicit formulas for $\mathsf{K} = \mathsf{K}(\mathsf{L})$. In this case the seemingly more complicated formula (6.57) is more convenient than the more compact (6.55).

Let us now apply this theory for computing the volume fraction relation associated with the Jordan $\mathcal{A}_0$-multialgebra $\Pi = K_0'$, obtained in section 6.1.5. It has the property that $\Pi^2 = \{0\}$. The analysis in the example in section 6.2 shows that we should choose $\mathsf{M}_0 = \boldsymbol{I}_d \otimes \boldsymbol{I}_d$, since $(\overline{\boldsymbol{\Gamma}}^0 - \mathsf{M}_0)\boldsymbol{I}_d = 0$. Indeed, all operators in K_0' map $\mathrm{Sym}(\mathbb{R}^d)$ to multiples of the identity, and hence, $(\overline{\boldsymbol{\Gamma}}^0 - \mathsf{M}_0)\mathsf{K} = 0$. Consequently, $\mathsf{K}(\overline{\boldsymbol{\Gamma}}^0 - \mathsf{M}_0)\mathsf{K} = 0$, as required by (6.58). Thus, all calculations from section 6.2 apply with $\boldsymbol{B} = d\theta\boldsymbol{I}_d$. This gives us

$$\left(\mathsf{I} - \mathsf{K}\mathsf{M}_0\right)^{-1}\mathsf{K} = \mathsf{K} + \frac{\boldsymbol{B} \otimes \boldsymbol{B}}{1 - \mathrm{Tr}\,\boldsymbol{B}} = \frac{\theta\boldsymbol{I}_d \otimes \boldsymbol{I}_d}{1 - d^2\theta}$$

Hence, C is isotropic with the shear modulus μ_0 and the bulk modulus

$$\kappa = \frac{1}{d}\mathsf{C}\boldsymbol{I}_d = \kappa_0 + \frac{\theta}{d\overline{\kappa}^2\left(1 - d^2\theta\right)} = \kappa_0 + \frac{\theta}{\kappa_\Gamma\left(1 - d^2\theta\right)}.$$

Solving for θ we obtain

$$\theta = \theta(\kappa) = \frac{\kappa_\Gamma(\kappa - \kappa_0)}{1 + d^2\kappa_\Gamma(\kappa - \kappa_0)}.$$

According to (6.57) the volume fraction relation reads $\theta(\kappa^*) = \langle\theta(\kappa)\rangle$. Hence, after some simple algebra, we conclude that

$$\frac{1}{\kappa_\Gamma^{-1} - d^2\kappa_0 + d^2\kappa^*} = \left\langle \frac{1}{\kappa_\Gamma^{-1} - d^2\kappa_0 + d^2\kappa} \right\rangle.$$

This gives

$$\begin{cases} \dfrac{1}{\mu_0 + \kappa^*} = \left\langle \dfrac{1}{\mu_0 + \kappa} \right\rangle, & d = 2, \\[2ex] \dfrac{1}{4\mu_0 + 3\kappa^*} = \left\langle \dfrac{1}{4\mu_0 + 3\kappa} \right\rangle, & d = 3. \end{cases} \tag{6.59}$$

This is the well-known Hill's formula for the effective bulk modulus of composites made of isotropic materials with constant shear modulus.

6.4 Finding Jordan $\hat{\mathcal{A}}$-multialgebras

In section 5.1 we showed that links can be regarded as exact relations in general. In this section we will focus on the most important subcase of polycrystalline links when $\mathcal{T}_1 = \mathcal{T}_2 = \mathcal{T}$. In this case we can say a bit more about the structure of the subspace $\hat{\mathcal{A}}$.

Lemma 6.9. *Suppose $\mathcal{T}_1 = \mathcal{T}_2 = \mathcal{T}$ and $\hat{\mathsf{L}}_0 = [\mathsf{L}_0^{(1)}, \mathsf{L}_0^{(1)}]$ is isotropic. Assume that $\mathcal{A}_1$ and $\mathcal{A}_2$ are isomorphic as representations of* SO(d), *$d = 2$ or 3. Then the subspace $\hat{\mathcal{A}}$ satisfies (5.9).*

Proof. According to lemma 4.14 the subspace $\hat{\mathcal{A}}$ is generated by a single vector $\tilde{\mathbf{\Gamma}} = \mathbf{\Gamma}_0(\boldsymbol{n}_0) - \overline{\mathbf{\Gamma}}$ via (4.17), where $\mathbf{\Gamma}$ stands for $\hat{\mathbf{\Gamma}}$ in this proof. This means that in three dimensions $\tilde{\mathbf{\Gamma}}$ must be a zero weight vector (the fixed point of a copy of $SO(2) \subset SO(3)$ of all rotations around $\boldsymbol{n}_0$). A single zero weight vector would generate a subrepresentation containing at most one copy of each irrep of SO(3) (via the action of raising and lowering operators, described in appendix C). It follows that in three dimensions the subspace $\hat{\mathcal{A}}$ must be a direct sum of non-isomorphic irreps of SO(3). The same is true in two dimensions, but for a different reason. If an SO(2) representation V is a direct sum of n isomorphic irreps of weight $k > 0$, then V, as a representation of SO(2), is isomorphic to $\mathbb{C}^n$ (as a real vector space of dimension $2n$) with the action of the counterclockwise rotation $\boldsymbol{R}_\theta$ through angle θ corresponding to the multiplication of elements of $\mathbb{C}^n$ by $e^{ik\theta}$. Any SO(2) subrepresentation of V is then identified with a complex subspace of $\mathbb{C}^n$ (regarded as a real subspace of twice the dimension). It is now obvious that a single non-zero vector in V spans the unique one-dimensional complex subspace of $\mathbb{C}^n$. This subspace is isomorphic to a single copy of an SO(2) irrep of weight k. Thus, in two dimensions the subspace $\hat{\mathcal{A}}$ must also be a direct sum of non-isomorphic irreps of SO(2). It is easy to check that subspaces $\mathfrak{A}_1$ and $\mathfrak{A}_2$, defined in (5.7), are rotationally invariant. According to (5.8), $\hat{\mathcal{A}}$ is isomorphic to $\mathfrak{A}_1 \oplus (\mathcal{A}_1/\mathfrak{A}_1) \oplus \mathfrak{A}_2$ as a representation of SO(d). It is also isomorphic to $\mathfrak{A}_1 \oplus (\mathcal{A}_2/\mathfrak{A}_2) \oplus \mathfrak{A}_2$. Now, the assumption that $\mathcal{A}_1$ and $\mathcal{A}_2$ are isomorphic as representations of SO(d) combined with

$$\mathfrak{A}_1 \oplus (\mathcal{A}_1/\mathfrak{A}_1) \oplus \mathfrak{A}_2 \cong \hat{\mathcal{A}} \cong \mathfrak{A}_1 \oplus (\mathcal{A}_2/\mathfrak{A}_2) \oplus \mathfrak{A}_2 \tag{6.60}$$

implies that $\mathfrak{A}_1 \cong \mathfrak{A}_2$ as representations of SO(d). But this contradicts the fact that $\hat{\mathcal{A}}$ contains no more than one copy of each irrep, unless $\mathfrak{A}_1 = \mathfrak{A}_2 = \{0\}$, since, according to (6.60) there is a subrepresentation of $\hat{\mathcal{A}}$, containing $\mathfrak{A}_1 \oplus \mathfrak{A}_2$. □

We remark that the assumption in lemma 6.9 that $\mathcal{A}_1$ and $\mathcal{A}_2$ are isomorphic as representations of SO(d) is a natural one, since, the subspaces $\mathcal{A}$ depends on the reference medium L_0 continuously, while the set of isomorphism classes is discreet. It follows that in general $\mathcal{A}_1 \cong \mathcal{A}_2$ on an open dense subset of isotropic tensors L_0. In fact, in most examples the isomorphism class of $\mathcal{A}$ is independent of L_0.

As we have remarked in section 5.2.2, lemma 6.9 implies that

$$\hat{\mathcal{A}} = \left\{[\mathsf{A}, \Xi(\mathsf{A})] \in \mathcal{A}_1 \oplus \mathcal{A}_2 : \mathsf{A} \in \mathcal{A}_1\right\}, \tag{6.61}$$

where $\Xi : \mathcal{A}_1 \mapsto \mathcal{A}_2$ is a linear SO(d)-equivariant isomorphism, i.e.

$$\boldsymbol{R} \cdot \Xi(\mathsf{A}) = \Xi(\boldsymbol{R} \cdot \mathsf{A}), \qquad \boldsymbol{R} \in SO(d).$$

Then, the key factor determining the number and structure of links is the existence or non-existence of the map (5.21). We note that in all of our case studies we could use the Milgrom–Shtrikman 'covariance transformations' (6.1) discussed in section 6.1.1 to map $\mathcal{A}_1$ and $\mathcal{A}_2$ into one and the same simplified subspace $\mathcal{A}_0$. In other words, we could always find isotropic invertible transformations $\{\mathsf{C}_1, \mathsf{C}_2\} \subset \mathrm{End}_{SO(d)}(\mathcal{T})$, such that

$$\mathsf{C}_1^{-T}\mathcal{A}_1\mathsf{C}_1^{-1} = \mathcal{A}_0 = \mathsf{C}_2^{-T}\mathcal{A}_2\mathsf{C}_2^{-1}. \tag{6.62}$$

However, the ability to transform $\mathcal{A}_1$ and $\mathcal{A}_2$ into one and the same subspace $\mathcal{A}_0$ is insufficient to ensure the existence of the isomorphism (5.21). After the application of the Milgrom–Shtrikman covariance transformation (6.62) we obtain

$$\hat{\mathcal{A}}_0 = \left[\mathsf{C}_1^{-T}, \mathsf{C}_2^{-T}\right]\hat{\mathcal{A}}\left[\mathsf{C}_1^{-1}, \mathsf{C}_2^{-1}\right] = \left\{[\mathsf{A}, \Xi_0(\mathsf{A})] \in \mathcal{A}_0 \oplus \mathcal{A}_0 : \mathsf{A} \in \mathcal{A}_0\right\}, \tag{6.63}$$

where

$$\Xi_0(\mathsf{A}) = \mathsf{C}_2^{-T}\Xi\left(\mathsf{C}_1^{T}\mathsf{A}\mathsf{C}_1\right)\mathsf{C}_2^{-1}, \qquad \mathsf{A} \in \mathcal{A}_0.$$

As we will see in the examples below, it is not always possible to find a transformation Φ_0, such that

$$\Phi_0(\mathsf{KAK}) = \Phi_0(\mathsf{K})\Xi_0(\mathsf{A})\Phi_0(\mathsf{K}), \qquad \forall \mathsf{K} \in \mathrm{Sym}(\mathcal{T}), \forall \mathsf{A} \in \mathcal{A}_0. \tag{6.64}$$

We now apply theorems 5.1 and 5.3 to obtain the following characterization of Jordan $\hat{\mathcal{A}}_0$-multialgebras $\hat{\Pi}$

$$\hat{\Pi} = \left\{\mathsf{K}_1, \mathsf{K}_2\right] : \overline{\mathsf{K}}_j \in \mathcal{F}_j, \overline{\mathsf{K}}_2 = \Phi\left(\overline{\mathsf{K}}_2\right)\right\}, \qquad \mathcal{F}_j = \Pi_j/\mathcal{I}_j, j = 1, 2, \tag{6.65}$$

where Π_j, $j = 1, 2$, are Jordan $\mathcal{A}_0$-multialgebras, $\mathcal{I}_j \subset \Pi_j$ are ideals and $\Phi : \mathcal{F}_1 \to \mathcal{F}_2$ is an SO(d)-invariant linear isomorphism satisfying

$$\Phi\left(\overline{\mathsf{K}}\mathsf{A}\overline{\mathsf{K}}\right) = \Phi\left(\overline{\mathsf{K}}\right)\Xi_0(\mathsf{A})\Phi\left(\overline{\mathsf{K}}\right), \qquad \forall \overline{\mathsf{K}} \in \mathcal{F}_1, \forall \mathsf{A} \in \mathcal{A}_0. \tag{6.66}$$

In case, when the global isomorphism (6.64) does not exist, it is theoretically possible that such an isomorphism exists on a subalgebra Π:

$$\Phi(\mathsf{KAK}) = \Phi(\mathsf{K})\Xi_0(\mathsf{A})\Phi(\mathsf{K}), \qquad \forall \mathsf{K} \in \Pi, \forall \mathsf{A} \in \mathcal{A}_0, \tag{6.67}$$

where $\Phi : \Pi \to \Pi$ is an SO(d)-invariant linear isomorphism.

As before, we will use the case of elastic composites for the purposes of illustrating the algorithm of obtaining all Jordan $\hat{\mathcal{A}}$-multialgebras, corresponding to links between effective tensors of composites with the same microstructure. Remarkably, the cases of two- and three-dimensional elasticity are essentially different, since the global isomorphism (6.64) exists in two and does not exist in three space dimensions. As a consequence we find that there are three non-trivial links in two-dimensional elasticity and only one in three dimensions.

6.4.1 Computing $\hat{\mathcal{A}}$ in elasticity

Let us compute the subspace $\hat{\mathcal{A}}$ for two- and three-dimensional elasticity. Using formulas (6.27) and noting that the first term in (6.25) has no component isomorphic to W_4 we obtain

$$\boldsymbol{\Gamma}_0(\boldsymbol{n}) - \boldsymbol{\Gamma}_0(\boldsymbol{n}_0) = \begin{cases} aK_2''(\boldsymbol{G}) + \dfrac{\rho_0}{\mu_0}\mathsf{K}_4, & d = 2, \\ aK_2''(\boldsymbol{G}) + bK_2(\boldsymbol{G}) + \dfrac{\rho_0}{\mu_0}\mathsf{K}_4, & d = 3, \end{cases}$$

where $\boldsymbol{G} = \boldsymbol{n} \otimes \boldsymbol{n} - \boldsymbol{n}_0 \otimes \boldsymbol{n}_0$, ρ_0 is defined in (6.26) and

$$a = \frac{1 - \rho_0}{\mu_0 d} = d\kappa_\Gamma, \qquad b = \frac{7 - 4\rho_0}{14\mu_0} \tag{6.68}$$

are the coefficients in (6.27). The precise expression of the tensor $\mathsf{K}_4 \in K_4$ is unimportant for computing $\hat{\mathcal{A}}$. It is only important that K_4 is independent of the elastic moduli κ_0, μ_0. We can simplify $\boldsymbol{\Gamma}_0(\boldsymbol{n}) - \boldsymbol{\Gamma}_0(\boldsymbol{n}_0)$ by a covariance transformation $\boldsymbol{\Gamma}_0 \mapsto \overline{\mathsf{C}}^{-1}\boldsymbol{\Gamma}_0\overline{\mathsf{C}}^{-1}$, where $\overline{\mathsf{C}}$ is an isotropic elastic tensor with bulk and shear moduli $\overline{\kappa}$, $\overline{\mu}$, respectively. Using the formulas

$$\overline{\mathsf{C}}^{-1}K_2'(A)\overline{\mathsf{C}}^{-1} = \frac{K_2'(A)}{2d\overline{\kappa}\overline{\mu}}, \quad \overline{\mathsf{C}}^{-1}K_2(\boldsymbol{B})\overline{\mathsf{C}}^{-1} = \frac{K_2(\boldsymbol{B})}{4\overline{\mu}^2}, \quad \overline{\mathsf{C}}^{-1}\mathsf{K}_4\overline{\mathsf{C}}^{-1} = \frac{\mathsf{K}_4}{4\overline{\mu}^2}, \tag{6.69}$$

we choose, for isotropic reference media $\hat{\mathsf{C}}_0 = [\mathsf{C}_1^0, \mathsf{C}_2^0]$ with bulk and shear moduli κ_j and μ_j, respectively, $j = 1, 2$,

$$\overline{\mu}_j = \begin{cases} m_0\sqrt{\dfrac{\rho_j}{\mu_j}}, & d = 2, \\ k_0\sqrt{b_j}, & d = 3, \end{cases} \qquad \overline{\kappa}_j = \begin{cases} k_0 a_j\sqrt{\dfrac{\mu_j}{\rho_j}}, & d = 2, \\ \dfrac{k_0 a_j}{\sqrt{b_j}}, & d = 3, \end{cases} \qquad j = 1, 2. \tag{6.70}$$

Here m_0 and k_0 are arbitrary positive constants that do not affect the resulting subspace

$$\hat{\mathcal{A}}_0 = \begin{cases} \{[\mathsf{A}, \mathsf{A}] : \mathsf{A} \in \mathcal{A}_{2d}\}, & d = 2, \\ \{[\mathsf{A}, \Xi_0(\mathsf{A})] : \mathsf{A} \in \mathcal{A}_0\}, & d = 3, \end{cases} \quad \mathcal{A}_{2d} = K_2' \oplus K_4, \tag{6.71}$$

where $\mathcal{A}_0$ is given by (6.28). The map $\Xi_0 : \mathcal{A}_0 \to \mathcal{A}_0$ is defined in terms of the unique representation $\mathsf{A} = \mathsf{A}_2 + \mathsf{A}_4$, where $\mathsf{A}_2 \in K_2^{2/3}$, and $\mathsf{A}_4 \in K_4$. Then

$$\Xi_0(\mathsf{A}) = \mathsf{A}_2 + \xi \mathsf{A}_4, \qquad \xi = \frac{(3\kappa_1 + 8\mu_1)(3\kappa_2 + \mu_2)}{(3\kappa_2 + 8\mu_2)(3\kappa_1 + \mu_1)}.$$

It is easy to see that $\xi = 1$ if and only if $\mathsf{L}_2 = s\mathsf{L}_1$ for some scalar $s > 0$. We note that in three space dimensions the relation (6.48) holds for each $j = 1, 2$.

6.4.2 Ideals and factor multialgebras in elasticity

The next step in our program of obtaining a complete list of all links is to identify all Jordan multialgebra–ideal pairs. In the case of elasticity, there are so few Jordan multialgebras that it is a simple matter (using multiplication tables 6.1 and 6.2) to produce the following list of all Jordan multialgebra–ideal pairs. In this regard we make two simple observations. The first, is that $\mathcal{I} = \{0\}$ is an ideal in every Jordan multialgebra. The second, is that any subspace $\mathcal{I} \subset \Pi$, such that $\Pi^2 \subset \mathcal{I}$ is an ideal. It is then easy to verify that in the setting of elasticity, the only multialgebra–ideal pair, not covered by the above observations, is $\Pi = K_0' \oplus K_2'$, $\mathcal{I} = K_0'$. Hence, the complete list of multialgebra–ideal pairs for elastic composites is

1. $\Pi = \mathbb{E}^d = \mathrm{Sym}(\mathrm{Sym}(\mathbb{R}^d))$, $\mathcal{I} = \{0\}$.
2. $\Pi = \mathrm{Sym}(\mathrm{Sym}_0(\mathbb{R}^d))$, $\mathcal{I} = \{0\}$.
3. $\Pi = K_0' \oplus K_2'$, $\mathcal{I} = K_0'$ and $\mathcal{I} = \{0\}$.
4. $\Pi = K_0'$, $\mathcal{I} = \{0\}$.
5. If $d = 2$, then $\Pi = K_4$, $\mathcal{I} = \{0\}$.

A simple inspection shows that different factor multialgebras are not isomorphic even as representations of $SO(d)$, let alone Jordan factor multialgebras.

6.4.3 Jordan isomorphisms in three-dimensional elasticity

Now that all the factor multialgebras are computed, it remains to find all $SO(3)$-invariant Jordan factor multialgebra isomorphisms Φ described in theorem 5.1. It is here that the different structures of subspaces $\hat{\mathcal{A}}$ for the two- and three-dimensional elasticity will result in a very different set of links for the two cases.

Theorem 6.10. *Suppose that $\xi \neq 1$. Then*

(i) *There are no $SO(3)$-equivariant linear isomorphisms Φ of $\mathbb{E}^3 = \mathrm{Sym}(\mathrm{Sym}(\mathbb{R}^3))$, satisfying (6.67).*

(ii) *There are no $SO(3)$-equivariant linear isomorphisms Φ of $K_0 \oplus K_2 \oplus K_4$, satisfying (6.67).*

(iii) *There are no SO(3)-equivariant linear isomorphisms* Φ *of* $K_0' \oplus K_2'$, *satisfying (6.67).*

(iv) *The only SO(3)-equivariant linear isomorphism of* $\mathcal{F} = (K_0' \oplus K_2')/K_0' \cong K_2'$, *satisfying (6.66) is* $\Phi(\overline{\mathsf{K}}) = \overline{\mathsf{K}}$.

(v) *All maps* $\Phi(\mathsf{K}) = \lambda\mathsf{K}$, $\mathsf{K} \in K_0'$ *satisfy (6.67).*

Proof. Before we prove (i), let us first prove (ii)–(v). In each case (ii)–(v) an SO(3)-equivariant linear isomorphism Φ must be a multiple of the identity on each irrep. (This is known as Schur's lemma in representation theory.) Then, for the map Φ in (ii), there exist non-zero scalars λ_0, λ_2, λ_4, such that $\Phi(\mathsf{K}) = \lambda_i\mathsf{K}$ for every $\mathsf{K} \in K_i$, $i = 0, 2, 4$. Recall that $\mathcal{A}_0 = K_2^{2/3} \oplus K_4$. Let us test (6.67) by setting $\mathsf{K} = K_0(\mu)$, $\mathsf{A} = (2/3)K_2'(\boldsymbol{A}) + K_2(\boldsymbol{A}) + \mathsf{A}_4$. Then, according to (6.69) (see also table 6.2)

$$\mathsf{KAK} = 4\mu^2(K_2(\boldsymbol{A}) + \mathsf{A}_4).$$

Thus,

$$\Phi(\mathsf{KAK}) = 4\mu^2(\lambda_2K_2(\boldsymbol{A}) + \lambda_4\mathsf{A}_4).$$

We also have

$$\Phi(\mathsf{K})\Xi_0(\mathsf{A})\Phi(\mathsf{K}) = 4\mu^2\lambda_0^2(K_2(\boldsymbol{A}) + \xi\mathsf{A}_4).$$

It follows that

$$\lambda_2 = \lambda_0^2, \qquad \lambda_4 = \xi\lambda_0^2.$$

Now let us take $\mathsf{K} \in K_j$, $j = 2, 4$ and $\mathsf{A} \in K_2^{3/2}$, such that

$$\mathsf{KAK} = \mathsf{B}_0 + \mathsf{B}_2 + \mathsf{B}_4, \qquad \mathsf{B}_i \in K_i\backslash\{0\},\ i = 0, 2, 4.$$

This is possible, because $(K_jK_2^{3/2}K_j)_{\text{sym}} = K_0 \oplus K_2 \oplus K_4$, $j = 2, 4$, and the sets $\{(\mathsf{K}, \mathsf{A}) \in K_j \oplus K_2^{3/2} : \mathsf{KAK} = \mathsf{B}_0 + \mathsf{B}_2 + \mathsf{B}_4, \mathsf{B}_k \neq 0\}$, $k = 0, 2, 4$ are open and dense in $K_j \oplus K_2^{3/2}$, $j = 2, 4$. Then, we must have

$$\Phi(\mathsf{KAK}) = \lambda_0\mathsf{B}_0 + \lambda_2\mathsf{B}_2 + \lambda_4\mathsf{B}_4.$$

We also have

$$\Phi(\mathsf{K})\Xi_0(\mathsf{A})\Phi(\mathsf{K}) = \lambda_j^2\mathsf{KAK} = \lambda_j^2(\mathsf{B}_0 + \mathsf{B}_2 + \mathsf{B}_4), \quad j = 2, 4,$$

since $\Xi_0(\mathsf{A}) = \mathsf{A}$ for $\mathsf{A} \in K_2^{3/2}$. This implies that $\lambda_0 = \lambda_2 = \lambda_4 = 1$. It is now clear that (6.67) cannot hold, unless $\xi = 1$.

Now let us prove (iii). We first note that for $\mathsf{K}_0 \in K_0'$, $\mathsf{K}_2 \in K_2'$, $\mathsf{A}_2 \in K_2^{3/2}$, $\mathsf{A}_4 \in K_4$ we have

$$\mathsf{K}_0\mathsf{A}_4\mathsf{K}_2 = \mathsf{K}_2\mathsf{A}_4\mathsf{K}_0 = 0, \qquad \mathsf{K}_0\mathsf{A}_2\mathsf{K}_2 + \mathsf{K}_2\mathsf{A}_2\mathsf{K}_0 = \mathsf{B}_0 \in K_0'.$$

Thus, (6.67) implies

$$\lambda_0 \mathsf{B}_0 = \lambda_0 \lambda_2 \mathsf{B}_0.$$

It follows that $\lambda_2 = 1$. Now let $\mathsf{K} \in K_2'$. Then

$$\mathsf{K}\mathsf{A}_4\mathsf{K} = \mathsf{B}_0 \in K_0'.$$

Hence, (6.67) implies

$$\lambda_0 \mathsf{B}_0 = \lambda_2^2 \xi \mathsf{B}_0 = \xi \mathsf{B}_0.$$

This implies that $\lambda_0 = \xi$. To obtain the contradiction we compute for $\mathsf{K} = K_2'(\boldsymbol{B})$ and $\mathsf{A} \in K_2^{2/3}$, given by (6.29)

$$K_2'(\boldsymbol{B})\mathsf{A}K_2'(\boldsymbol{B})\boldsymbol{\varepsilon} = 2\mathrm{Tr}\,(\boldsymbol{AB})K_2'(\boldsymbol{B})\boldsymbol{\varepsilon} + K_0'(2\,\mathrm{Tr}\,(\boldsymbol{BAB}))\boldsymbol{\varepsilon}. \tag{6.72}$$

Choosing non-singular $\boldsymbol{A} = \boldsymbol{B} \in \mathrm{Sym}_0(\mathbb{R}^3)$, we obtain

$$\mathsf{KAK} = \mathsf{B}_0 + \mathsf{B}_2, \qquad \mathsf{B}_i \in K_i' \backslash \{0\},\ i = 0, 2.$$

Then, for (6.67) to hold we must have

$$\lambda_0 \mathsf{B}_0 + \lambda_2 \mathsf{B}_2 = \lambda_2^2 (\mathsf{B}_0 + \mathsf{B}_2), \tag{6.73}$$

which cannot hold for $\lambda_0 = \xi \neq 1$, $\lambda_2 = 1$.

Part (iv) is very simple, since we only need to test the maps $\Phi(\overline{\mathsf{K}}) = \lambda\overline{\mathsf{K}}$. Factoring (6.73) by K_0' we conclude that we must have $\lambda^2 = \lambda$, while $\Phi(\overline{\mathsf{K}}) = \overline{\mathsf{K}}$ is easily seen to satisfy (6.66), since

$$\overline{\mathsf{K}}\mathsf{A}_4\overline{\mathsf{K}} = \overline{0}, \quad \forall\, \overline{\mathsf{K}} \in K_2'.$$

Part (v) is obvious, since $\mathsf{KAK} = 0$ for all $\mathsf{K} \in K_0'$ and all $\mathsf{A} \in \mathcal{A}_0$.

We are now ready to prove part (i). Let us assume that such Φ exists. Then for $\Pi = K_0 \oplus K_2 \oplus K_4$ its Φ image $\Phi(\Pi)$ must be isomorphic to Π as a representation of SO(3). It also has to be a Jordan $\mathcal{A}_0$-multialgebra. But Π is the only SO(3)-invariant $\mathcal{A}_0$-multialgebra isomorphic to Π as a representation of SO(3). This implies that $\Phi(\Pi) = \Pi$, which does not exist due to part (ii). □

We conclude that in the case $\xi \neq 1$ there is only one non-trivial link, corresponding to

$$\hat{\Pi} = \left\{[\mathsf{K}, \mathsf{K} + \mathsf{J}] : \mathsf{J} \in K_0', \mathsf{K} \in K_0' \oplus K_2'\right\}. \tag{6.74}$$

Moreover, when $\xi = 1$, the arguments in the proof of theorem 6.10 show that $\Phi(\mathsf{K}) = \mathsf{K}$ are the only automorphisms in cases (i)–(iii). We also remark that the link in case (v) does not bring any new information, since we already have fully explicit formulas (4.1) for the effective elastic tensor.

Finally, we need to verify that link (6.74) satisfies sufficient conditions from theorem 5.2. We note that the map Φ in (6.74) is trivial, i.e. $\Phi(\overline{\mathsf{K}}) = \overline{\mathsf{K}}$, so that conditions (5.19), (5.20) follow from (5.17), (5.18) and the already verified three- and four-chain properties for $\Pi = K_0' \oplus K_4'$. Conditions (5.17), (5.18) for (6.74) follow from a simple verification that

$$(K_0'\mathcal{A}_0K_2')_{\text{sym}} = K_0', \qquad (K_2'\mathcal{A}_0K_0')_{\text{sym}} = K_0'.$$

6.4.4 Jordan isomorphisms in two-dimensional elasticity

Recall that Ξ_0 is the identity transformation in two-dimensional elasticity. Thus, the goal is to compute all Jordan automorphisms of the factor multialgebras listed in section 6.4.2. Let us first consider the case $\mathcal{I} = \{0\}$. Using the fact than none of the Jordan $\mathcal{A}_{2d}$-multialgebras are isomorphic as representations of SO(2), we conclude that any Jordan automorphism Φ of $\mathbb{E}^2$ must map $K_0' \oplus K_2'$ into itself and $K_0 \oplus K_4$ into itself. It follows that

$$\Phi(K_0) = K_0, \quad \Phi(K_4) = K_4, \quad \Phi\left(K_0'\right) = K_0', \quad \Phi\left(K_2'\right) = K_2'.$$

Let $\mathsf{K}'_0 = K_0'(1)$, $\mathsf{K}_0 = K_0(1)$. Then

$$\Phi(\mathsf{K}_0') = \lambda_0'\mathsf{K}_0', \qquad \Phi(\mathsf{K}_0) = \lambda_0\mathsf{K}_0.$$

According to the multiplication table 6.1

$$\mathsf{K}_0\mathsf{A}\mathsf{K}_0 = \mathsf{K}_4(\mathsf{A}) \in K_4, \qquad \mathsf{A} \in \mathcal{A}_{2d}.$$

Applying Φ and using (6.67) we obtain

$$\Phi(\mathsf{K}_4(\mathsf{A})) = \lambda_0^2\mathsf{K}_4(\mathsf{A}).$$

As A ranges over $\mathcal{A}_{2d}$, $\mathsf{K}_4(\mathsf{A})$ must range over K_4 and thus,

$$\Phi(\mathsf{K}) = \lambda_0^2\mathsf{K}, \quad \mathsf{K} \in K_4.$$

Similarly,

$$\mathsf{K}_0'\mathsf{A}\mathsf{K}_0 + \mathsf{K}_0\mathsf{A}\mathsf{K}_0' = \mathsf{K}_2'(\mathsf{A}) \in K_2', \qquad \mathsf{A} \in \mathcal{A}_{2d}.$$

Applying Φ we conclude that $\Phi(\mathsf{K}_2'(\mathsf{A})) = \lambda_0\lambda_0'\mathsf{K}_2'(\mathsf{A})$, so that

$$\Phi(\mathsf{K}) = \lambda_0\lambda_0'\mathsf{K}, \quad \mathsf{K} \in K_2'.$$

Using these results we apply Φ to

$$\mathsf{K}_0'\mathsf{A}\mathsf{K} + \mathsf{K}\mathsf{A}\mathsf{K}_0' = \theta(\mathsf{K}, \mathsf{A})\mathsf{K}_0', \qquad \mathsf{A} \in \mathcal{A}_{2d}, \ \mathsf{K} \in K_2'$$

and to

$$\mathsf{K}_0'\mathsf{A}\mathsf{K} + \mathsf{K}\mathsf{A}\mathsf{K}_0' = \mathsf{K}_2'(\mathsf{K}, \mathsf{A}), \qquad \mathsf{A} \in \mathcal{A}_{2d}, \ \mathsf{K} \in K_4,$$

we conclude that we must have $\lambda_0 = \lambda_0' = 1$. Thus, $\Phi(\mathsf{K}) = \mathsf{K}$ for all $\mathsf{K} \in \mathbb{E}^2$.

Let us now compute all Jordan automorphisms of $K_0 \oplus K_4$. As before,

$$\Phi(\mathsf{K}_0) = \lambda_0 \mathsf{K}_0, \qquad \Phi(\mathsf{K}) = \lambda_0^2 \mathsf{K}, \quad \mathsf{K} \in K_4.$$

Applying Φ to the remaining two relations

$$\mathsf{KAK} = \mathsf{P}(\mathsf{K}, \mathsf{A}) \in K_4, \quad \mathsf{K}_0\mathsf{AK} + \mathsf{KAK}_0 = \theta(\mathsf{K}, \mathsf{A})\mathsf{K}_0, \quad \mathsf{A} \in \mathcal{A}_{2d}, \; \mathsf{K} \in K_4$$

we obtain $\lambda_0^2 = 1$, which in addition to the trivial automorphism gives a new one:

$$\Phi(\tau\mathsf{K}_0 + \mathsf{K}_4) = \mathsf{K}_4 - \tau\mathsf{K}_0, \qquad \mathsf{K}_4 \in K_4, \tag{6.75}$$

giving rise to the corresponding Jordan $\hat{\mathcal{A}}_0$-multialgebra

$$\hat{\Pi} = \left\{[\mathsf{K}, \Phi(\mathsf{K})]: \mathsf{K} \in K_0 \oplus K_4\right\}. \tag{6.76}$$

We have already remarked on the fact that sometimes links contain more information than exact relations. For example, (6.76) implies that K_4 must be a Jordan $\mathcal{A}_{2d}$-multialgebra. Indeed,

$$K_4 = \{\mathsf{K} \in K_0 \oplus K_4 : \Phi(\mathsf{K}) = \mathsf{K}\},$$

i.e. K_4 is the space of fixed points of an automorphism Φ. Then, for any fixed point K of a Jordan multialgebra automorphism Φ we have

$$\Phi(\mathsf{KAK}) = \Phi(\mathsf{K})\mathsf{A}\Phi(\mathsf{K}) = \mathsf{KAK},$$

showing that KAK is also a fixed point of Φ. Therefore, the set of fixed points of a Jordan $\mathcal{A}$-multialgebra automorphism is always a Jordan $\mathcal{A}$-multialgebra. We also note that the three- and four-chain properties (see definition 4.7) of the fixed point subspace follow from (5.19) and (5.20) that had to be verified in order to check whether the underlying link $\hat{\Pi}$, given by (6.76), holds for all pairs of composites, regardless of the microstructure.

Next, we compute all automorphisms of $K_0' \oplus K_2'$. As before, $\Phi(\mathsf{K}_0') = \lambda'_0\mathsf{K}_0'$. However, to make progress we need to go a little beyond the multiplication table. We compute for $\mathsf{K} = K_2'(\boldsymbol{B})$ and $\mathsf{A} \in K_2^{2/3}$, given by (6.29),

$$\mathsf{K}_0'\mathsf{A}K_2'(\boldsymbol{B}) + K_2'(\boldsymbol{B})\mathsf{A}\mathsf{K}'_0 = 4\,\mathrm{Tr}\,(\boldsymbol{AB})\mathsf{K}_0'.$$

Suppose that Φ, acting on K_2' transforms $\boldsymbol{B}$ into $\boldsymbol{B}'$, i.e. $\Phi(K_2'(\boldsymbol{B})) = K_2'(\boldsymbol{B}')$. Then, we must have

$$\mathrm{Tr}\,(\boldsymbol{AB}) = \mathrm{Tr}\,(\boldsymbol{AB}'), \qquad \forall \boldsymbol{A} \in \mathrm{Sym}_0\left(\mathbb{R}^2\right).$$

But that implies $\boldsymbol{B}' = \boldsymbol{B}$. To complete the calculation we apply Φ to (6.72) and obtain $\lambda'_0 = 1$, proving that there are no non-trivial automorphisms of $K_0' \oplus K_2'$. Incidentally, the same calculation also proves that there are no non-trivial automorphisms of the factor multialgebra $K_2' \cong (K_0' \oplus K_2')/K_0'$.

Finally, we compute all SO(2)-invariant automorphisms of K_4. In order to do that we need to delve a little deeper into the structure of $\mathbb{E}^2$. We begin by recalling that every $\boldsymbol{\varepsilon} \in \mathrm{Sym}(\mathbb{R}^2)$ can be decomposed as

$$\boldsymbol{\varepsilon} = \frac{1}{2}(\mathrm{Tr}\,\boldsymbol{\varepsilon})\boldsymbol{I}_2 + \mathrm{dev}(\boldsymbol{\varepsilon}).$$

We now make one more step and parametrize $\mathrm{Sym}_0(\mathbb{R}^2)$ by complex numbers as follows

$$\mathbb{C} \ni z = x + \mathrm{i}y \mapsto \psi(z) = \begin{bmatrix} x & y \\ y & -x \end{bmatrix}.$$

Thus, we will write

$$\boldsymbol{\varepsilon} = \frac{1}{2}(\mathrm{Tr}\,\boldsymbol{\varepsilon})\boldsymbol{I}_2 + \psi(e), \quad e \in \mathbb{C}.$$

Then we can parametrize the SO(2) irrep K_4 by complex numbers as follows:

$$\mathbb{C} \ni v \mapsto K_4(v), \qquad K_4(v)\boldsymbol{\varepsilon} = \psi(v\bar{e}).$$

The subspace $\mathcal{A}_0$ can now be parametrized by two complex numbers

$$\mathcal{A}_0 = \{\mathsf{A}(a, b) = \mathsf{A}_2(a) + K_4(b) : \{a, b\} \subset \mathbb{C}\}, \qquad \mathsf{A}_2(a)\boldsymbol{\varepsilon} = \psi(a)\boldsymbol{\varepsilon} + \boldsymbol{\varepsilon}\psi(a).$$

It is now easy to check that

$$K_4(v)\mathsf{A}_2(a)K_4(v) = 0, \qquad K_4(v)K_4(b)K_4(v) = K_4\big(\bar{b}v^2\big). \tag{6.77}$$

Any SO(2)-equivariant map $\Phi : K_4 \to K_4$ must have the form $\Phi(K_4(v)) = K_4(cv)$ for some $c \in \mathbb{C}$. Applying Φ to (6.77) we conclude that $c^2 = c$, i.e. $c = 1$ is the only choice for which Φ satisfies (6.67).

Let us now verify that link (6.76) satisfies sufficient conditions from theorem 5.2. We note that the ideals $\mathcal{I}_1$, $\mathcal{I}_2$ in (6.76) are trivial, so that that conditions (5.17), (5.18) are trivially satisfied. Hence, we only need to verify conditions (5.19), (5.20) for (6.75). To prove (5.19), (5.20) for (6.75) we recall that $K_0 \oplus K_4$ is the set of symmetric operators in the associative $\mathcal{A}_0$-multialgebra $\mathrm{End}(\mathrm{Sym}_0(\mathbb{R}^2))$. For any $\mathsf{K} \in \mathrm{End}(\mathrm{Sym}_0(\mathbb{R}^2))$ there is a unique pair of complex numbers v, w, such that $\mathsf{K} = K_4(v) + P_0(w)$, where $P_0(w)\psi(e) = \psi(we)$. We claim that the transformation

$$\hat{\Phi}(K_4(v) + P_0(w)) = K_4(v) - P_0(w)$$

is the associative $\mathcal{A}_0$-multialgebra automorphism. This claim can be verified directly by applying $\hat{\Phi}$ to

$$\begin{aligned}
&\big(K_4(v) + P_0(w)\big)\big(\mathsf{A}_2(a) + K_4(b)\big)\big(K_4(v') + P_0(w')\big)\psi(e) \\
&= \psi\Big(\big(vv'\bar{b} + w\overline{w'}b\big)\bar{e} + (vw'\bar{b} + w\overline{v'}b)e\Big) \\
&= \Big(K_4\big(vv'\bar{b} + w\overline{w'}b\big) + P_0\big(vw'\bar{b} + w\overline{v'}b\big)\Big)\psi(e).
\end{aligned}$$

Hence, we see that Φ is a restriction of an associative $\mathcal{A}_0$-multialgebra automorphism to the set of symmetric operators there. Hence, it must satisfy (5.19), (5.20).

6.5 Computing links

We continue to view links as exact relations in $\mathrm{End}(\mathcal{T}_1) \oplus \mathrm{End}(\mathcal{T}_2)$. Hence, the passage from the algebraic representation of links to their formulation in physical variables occurs along the lines of section 6.2, where we explain how to find the simplest possible inversion key $\hat{\mathsf{M}} = [\mathsf{M}_1, \mathsf{M}_2]$ to be used in (6.46). In other words, from the general theoretic point of view there is nothing to be added to the methods of section 6.2. Instead we illustrate the procedure using the example of elastic polycrystals. We observe that $\hat{\boldsymbol{\Gamma}} = [\overline{\boldsymbol{\Gamma}}_1, \overline{\boldsymbol{\Gamma}}_2]$, where $\overline{\boldsymbol{\Gamma}}_j$ are isotropic elastic tensors with bulk and shear moduli κ_Γ^j, μ_Γ^j, $i = 1, 2$, given by (6.47) with κ_0 and μ_0, replaced by κ_j, μ_j, respectively. Recall that we have also applied the Milgrom–Shtrikman covariance transformation (6.70) in order to simplify the subspace $\hat{\mathcal{A}}$. Hence, the same covariance transformation, applied to $\hat{\boldsymbol{\Gamma}}$ transforms it into $\hat{\boldsymbol{\Gamma}}^0 = [\overline{\boldsymbol{\Gamma}}_1^0, \overline{\boldsymbol{\Gamma}}_2^0]$, where $\overline{\boldsymbol{\Gamma}}_j^0$, $j = 1, 2$, are isotropic tensors with bulk and shear moduli κ_j^0, μ_j^0, $j = 1, 2$, given by

$$\kappa_j^0 = \frac{\kappa_\Gamma^j}{d^2\overline{\kappa}_j^2}, \qquad \mu_j^0 = \frac{\mu_\Gamma^j}{4\overline{\mu}_j^2}, \qquad j = 1, 2, \tag{6.78}$$

where $\overline{\kappa}_j$ and $\overline{\mu}_j$ are defined in (6.70).

6.5.1 Links in three-dimensional elasticity

In three dimensional elasticity there is a single non-trivial link, corresponding to (6.74). If $\hat{\mathsf{M}}_0 = [\mathsf{M}_1, \mathsf{M}_2]$ is its inversion key, then the defining property (6.45) is equivalent to the following set of conditions on the two isotropic elastic tensors M_1, M_2.

- $\mathsf{K}(\overline{\boldsymbol{\Gamma}}_j^0 - \mathsf{M}_j)\mathsf{K} \in K_0' \oplus K_2'$, for all $\mathsf{K} \in K_0' \oplus K_2'$, $j = 1, 2$.
- $\mathsf{K}(\overline{\boldsymbol{\Gamma}}_j^0 - \mathsf{M}_j)\mathsf{J} + \mathsf{J}(\overline{\boldsymbol{\Gamma}}_j^0 - \mathsf{M}_j)\mathsf{K} \in K_0'$, for all $\mathsf{K} \in K_0' \oplus K_2'$, $\mathsf{J} \in K_0'$, $j = 1, 2$.
- $\mathsf{K}(\overline{\boldsymbol{\Gamma}}_1^0 - \overline{\boldsymbol{\Gamma}}_2^0 - \mathsf{M}_1 + \mathsf{M}_1)\mathsf{K} \in K_0'$, for all $\mathsf{K} \in K_0' \oplus K_2'$.

It is easy to see that $\mathsf{M}_j = \kappa_j^0 \boldsymbol{I}_3 \otimes \boldsymbol{I}_3 = \kappa_j^0 \mathsf{M}_0$ satisfies

$$\mathsf{K}\left(\overline{\boldsymbol{\Gamma}}_j^0 - \mathsf{M}_j\right)\mathsf{K} \in K_0', \quad \forall \mathsf{K} \in K_0' \oplus K_2'.$$

Hence, all the requirements in the list above for the inversion key are satisfied. In order to compute the link corresponding to (6.74) we will use the inversion formula (6.46) for each component:

$$\mathsf{C}_j = \mathsf{C}_j^0 + \overline{\mathsf{C}}_j^{-1}\left(\mathsf{I} - \kappa_j^0 \mathsf{K}_j \mathsf{M}_0\right)^{-1} \mathsf{K}_j \overline{\mathsf{C}}_j^{-1}, \quad j = 1, 2, \tag{6.79}$$

where $\overline{\mathsf{C}}_j$ are isotropic elastic tensors with bulk and shear moduli $\overline{\kappa}_j$, $\overline{\mu}_j$, $j = 1, 2$, respectively, given by (6.70). From (6.54) we already know that

$$\mathsf{C}_j = 2\mu_j \mathsf{T} + \boldsymbol{D}_j \otimes \boldsymbol{D}_j, \quad j = 1, 2. \tag{6.80}$$

The goal is to compute the link between $\boldsymbol{D}_1$ and $\boldsymbol{D}_2$, corresponding to (6.74), which can also be recorded as

$$\operatorname{dev}(\boldsymbol{B}_1) = \operatorname{dev}(\boldsymbol{B}_2), \qquad \boldsymbol{B}_j = \mathsf{K}_j \boldsymbol{I}_d, j = 1, 2. \tag{6.81}$$

We first observe that

$$\left(\mathsf{I} - \kappa_j^0 \mathsf{K}_j \mathsf{M}_0\right)^{-1} \mathsf{K}_j = \frac{1}{\kappa_j^0}\left(\mathsf{I} - \left(\kappa_j^0 \mathsf{K}_j\right)\mathsf{M}_0\right)^{-1}\left(\kappa_j^0 \mathsf{K}_j\right).$$

Then we can use formula (6.50) applied to $\kappa_j^0 \mathsf{K}_j$ to obtain

$$\left(\mathsf{I} - \kappa_j^0 \mathsf{K}_j \mathsf{M}_0\right)^{-1} \mathsf{K}_j = -\frac{\boldsymbol{I}_d \otimes \boldsymbol{I}_d}{d^2 \kappa_j^0} + \boldsymbol{C}_j \otimes \boldsymbol{C}_j,$$

where

$$\boldsymbol{C}_j = \frac{\kappa_j^0 \boldsymbol{B} + \left(1 - \kappa_j^0 \operatorname{Tr} \boldsymbol{B}\right)\boldsymbol{I}_d / d}{\sqrt{\kappa_j^0\left(1 - \kappa_j^0 \operatorname{Tr} \boldsymbol{B}\right)}}.$$

Hence,

$$\boldsymbol{D}_j = \overline{\mathsf{C}}_j^{-1} \boldsymbol{C}_j = \frac{\kappa_j^0 \overline{\mathsf{C}}_j^{-1} \boldsymbol{B}_j + \left(1 - \kappa_j^0 \operatorname{Tr} \boldsymbol{B}_j\right)\boldsymbol{I}_d / \left(d^2 \overline{\kappa}_j\right)}{\sqrt{\kappa_j^0\left(1 - \kappa_j^0 \operatorname{Tr} \boldsymbol{B}\right)}}.$$

Taking a trace we obtain

$$\operatorname{Tr}\left(\boldsymbol{D}_j\right) = \frac{1}{d\overline{\kappa}_j \sqrt{\kappa_j^0\left(1 - \kappa_j^0 \operatorname{Tr} \boldsymbol{B}\right)}}. \tag{6.82}$$

Taking the deviatoric part and using (6.82) and (6.78) we obtain

$$\operatorname{dev}\left(\boldsymbol{B}_j\right) = \frac{2\overline{\mu}_j \operatorname{dev}\left(\boldsymbol{D}_j\right)}{d\overline{\kappa}_j \kappa_j^0 \operatorname{Tr}\left(\boldsymbol{D}_j\right)} = \frac{2d\overline{\kappa}_j \overline{\mu}_j \operatorname{dev}\left(\boldsymbol{D}_j\right)}{\kappa_\Gamma^j \operatorname{Tr}\left(\boldsymbol{D}_j\right)}.$$

By (6.70) and (6.68) we find

$$\frac{\overline{\kappa}_j \overline{\mu}_j}{\kappa_\Gamma^j} = \begin{cases} \dfrac{m_0 k_0 a_j}{\kappa_\Gamma^j}, & d = 2, \\[2ex] \dfrac{k_0^2 a_j}{\kappa_\Gamma^j}, & d = 3, \end{cases} = \begin{cases} 2m_0 k_0, & d = 2, \\ 3k_0^2, & d = 3, \end{cases}$$

which is independent of the moduli of the reference media κ_j and μ_j. We conclude that (6.81) is equivalent to

$$\frac{\mathrm{dev}(\boldsymbol{D}_1)}{\mathrm{Tr}\,(\boldsymbol{D}_1)} = \frac{\mathrm{dev}(\boldsymbol{D}_2)}{\mathrm{Tr}\,(\boldsymbol{D}_2)},$$

which can also be written as

$$\frac{\boldsymbol{D}_1}{\mathrm{Tr}\,(\boldsymbol{D}_1)} = \frac{\boldsymbol{D}_2}{\mathrm{Tr}\,(\boldsymbol{D}_2)}. \tag{6.83}$$

The link (6.83) can also be restated in simpler terms, if we observe that (6.83) is satisfied if and only if $\boldsymbol{D}_2$ is a scalar multiple of $\boldsymbol{D}_1$.

What this means physically, is that if we have two composites $\mathsf{C}_1(z)$ and $\mathsf{C}_2(z)$ that have the form (6.80), where $\boldsymbol{D}_2(z) = \Lambda(z)\boldsymbol{D}_1(z)$, then $\boldsymbol{D}_2^*$ must be a scalar multiple of $\boldsymbol{D}_1^*$

$$\boldsymbol{D}_2^* = \Lambda^*\boldsymbol{D}_1^*,$$

where the scalar Λ^* is microstructure-dependent. We observe that Hill's exact relation (4.2) is obtained from this link, if we choose $\boldsymbol{D}_1 = \boldsymbol{I}_3$.

6.5.2 Links in two-dimensional elasticity

Recall that in addition to (6.83) there are two more links in two dimensional elasticity, corresponding to Jordan $\hat{\mathcal{A}}_0$-multialgebras (6.76) and

$$\hat{\Pi} = \left\{[\mathsf{K}, \mathsf{K}] : \mathsf{K} \in \mathbb{E}^2\right\}. \tag{6.84}$$

We note that the seemingly trivial multialgebra (6.84) may correspond to a non-trivial link since the reference media C_1^0 and C_2^0 are not necessarily scalar multiples of one another. In particular, we cannot choose $\hat{\mathsf{M}}_0 = [0, 0]$ for the inversion key for (6.84), since $\overline{\boldsymbol{\Gamma}}_1^0 \neq \overline{\boldsymbol{\Gamma}}_2^0$, while the inversion key $\hat{\mathsf{M}}_0 = [\mathsf{M}_1, \mathsf{M}_2]$ must satisfy

$$\mathsf{K}\left(\overline{\boldsymbol{\Gamma}}_1^0 - \mathsf{M}_1\right)\mathsf{K} = \mathsf{K}\left(\overline{\boldsymbol{\Gamma}}_2^0 - \mathsf{M}_2\right)\mathsf{K}, \qquad \forall\,\mathsf{K} \in \mathbb{E}^2.$$

This is equivalent to

$$\overline{\boldsymbol{\Gamma}}_1^0 - \mathsf{M}_1 = \overline{\boldsymbol{\Gamma}}_2^0 - \mathsf{M}_2. \tag{6.85}$$

Let us now compute the link corresponding to (6.84). In order to simplify (somewhat) the calculations we observe that the first component of $\hat{\mathsf{M}}_0$ can be chosen arbitrarily. Hence, we choose $\mathsf{M}_1 = 0$. In that case $\mathsf{M}_2 = \overline{\boldsymbol{\Gamma}}_2^0 - \overline{\boldsymbol{\Gamma}}_1^0$ and

$$\mathsf{K} = \overline{\mathsf{C}}_1\left(\mathsf{C}_1 - \mathsf{C}_1^0\right)\overline{\mathsf{C}}_1.$$

Substituting this expression into (6.46) for C_2 we obtain our initial representation of the link corresponding to (6.84):

$$\mathsf{C}_2 = \mathsf{C}_2^0 + \left[\overline{\mathsf{C}}_2 - \overline{\mathsf{C}}_1\left(\mathsf{C}_1 - \mathsf{C}_1^0\right)\overline{\mathsf{C}}_1\left(\overline{\boldsymbol{\Gamma}}_2^0 - \overline{\boldsymbol{\Gamma}}_1^0\right)\overline{\mathsf{C}}_2\right]^{-1}\overline{\mathsf{C}}_1\left(\mathsf{C}_1 - \mathsf{C}_1^0\right)\overline{\mathsf{C}}_1\overline{\mathsf{C}}_2^{-1}.$$

We observe that

$$\overline{\mathsf{C}}_1\left(\overline{\boldsymbol{\Gamma}}_2^0 - \overline{\boldsymbol{\Gamma}}_1^0\right)\overline{\mathsf{C}}_2 = \xi\mathsf{T},$$

where T is given by (6.53) and

$$\xi = \frac{\mu_2\kappa_1 - \mu_1\kappa_2}{2\sqrt{\kappa_1\kappa_2\mu_1\mu_2(\kappa_1 + \mu_1)(\kappa_2 + \mu_2)}}.$$

After some matrix algebra we obtain

$$\mathsf{C}_2 = \xi^{-1}\mathsf{T}\left[\mathsf{C}_1^0 + \xi^{-1}\overline{\mathsf{C}}_1^{-1}\overline{\mathsf{C}}_2\mathsf{T} - \mathsf{C}_1\right]^{-1}\left\{\left(\mathsf{C}_1 - \mathsf{C}_1^0\right)\left(\overline{\mathsf{C}}_1\overline{\mathsf{C}}_2^{-1} - \xi\mathsf{T}\mathsf{C}_2^0\right) + \overline{\mathsf{C}}_1^{-1}\overline{\mathsf{C}}_2\mathsf{C}_2^0\right\}.$$

We next compute

$$\mathsf{C}_1^0 + \xi^{-1}\overline{\mathsf{C}}_1^{-1}\overline{\mathsf{C}}_2\mathsf{T} = \eta\mathsf{T}, \qquad \eta = \frac{2\mu_1\kappa_1(\mu_2 + \kappa_2)}{\mu_2\kappa_1 - \mu_1\kappa_2},$$

$$-\mathsf{C}_1^0\left(\overline{\mathsf{C}}_1\overline{\mathsf{C}}_2^{-1} - \xi\mathsf{T}\mathsf{C}_2^0\right) + \overline{\mathsf{C}}_1^{-1}\overline{\mathsf{C}}_2\mathsf{C}_2^0 = 0.$$

Hence,

$$\overline{\mathsf{C}}_1\overline{\mathsf{C}}_2^{-1} - \xi\mathsf{T}\mathsf{C}_2^0 = \left(\overline{\mathsf{C}}_1\mathsf{C}_1^0\right)^{-1}\overline{\mathsf{C}}_2\mathsf{C}_2^0 = \beta\mathsf{I}, \qquad \beta^2 = \frac{\kappa_2\mu_2(\kappa_1 + \mu_1)}{\kappa_1\mu_1(\kappa_2 + \mu_2)}.$$

Thus, we obtain

$$\mathsf{C}_2 = \frac{\beta}{\xi}\mathsf{T}(\eta\mathsf{T} - \mathsf{C}_1)^{-1}\mathsf{C}_1. \tag{6.86}$$

The most compact form of this two-parameter family of links is obtained if we rewrite (6.86) in terms of compliance tensors $\mathsf{S}_j = \mathsf{C}_j^{-1}$, $j = 1, 2$:

$$\mathsf{S}_2 = b_0(\mathsf{S}_1 - a_0\mathsf{T}). \tag{6.87}$$

where

$$b_0 = \frac{\xi\eta}{\beta} = \frac{\kappa_1\mu_1(\kappa_2 + \mu_2)}{\mu_2\kappa_2(\mu_1 + \kappa_1)}, \qquad a_0 = \frac{1}{\eta} = \frac{\mu_2\kappa_1 - \mu_1\kappa_2}{2\mu_1\kappa_1(\mu_2 + \kappa_2)}.$$

This family of links was discovered in [4].

Finally, we want to compute the link, corresponding to $\hat{\mathcal{A}}_0$-multialgebra (6.76). Before we plunge into computations let us mention another important simplifying device. We have seen how linear independence of isotropic reference media tensors C_1^0 and C_2^0 in $\hat{\mathsf{C}}^0 = [\mathsf{C}_1^0, \mathsf{C}_2^0]$ complicated our calculations. In some sense we have overcome this difficulty when we computed the global link (6.87), since it has the property that for any $\hat{\mathsf{C}}^0 = [\mathsf{C}_1^0, \mathsf{C}_2^0]$ we can choose values of a_0 and b_0, so that the link (6.87) passes through $\hat{\mathsf{C}}^0$. Now it is sufficient to compute only the link corresponding to (6.76) that passes through $\hat{\mathsf{C}}^0 = [\mathsf{I}, \mathsf{I}]$. Indeed, let $\mathsf{C}_2 = \Lambda(\mathsf{C}_1)$, $\mathsf{C}_1 \in \mathbb{M}$, be a link, where $\mathbb{M}$ is the set of all elastic tensors $\mathsf{C} > 0$ with the property $\mathsf{C}\boldsymbol{I}_d = d\kappa_0\boldsymbol{I}_d$ for some constant bulk modulus $\kappa_0 > 0$. Let $\mathsf{C}'_0 = \Lambda(\mathsf{C}_0)$. Let Ψ be the global link (6.87), such that $\Psi(\mathsf{I}) = \mathsf{C}_0$ and let Ψ' be the global link, satisfying $\Psi'(\mathsf{C}'_0) = \mathsf{I}$. Consider now the map

$$\tilde{\Lambda}(\mathsf{C}) = \Psi'(\Lambda(\Psi(\mathsf{C}))).$$

Then, $\tilde{\Lambda}(\mathsf{I}) = \mathsf{I}$ and $\tilde{\Lambda}(\mathsf{C})$ is also a link. Indeed, let $\mathsf{C}_1(z)$ be a composite and $\mathsf{C}_2(z) = \tilde{\Lambda}(\mathsf{C}_1(z))$. Introduce related composites $\mathsf{C}_3(z) = \Psi(\mathsf{C}_1(z))$ and $\mathsf{C}_4(z) = \Lambda(\mathsf{C}_3(z))$. Then $\mathsf{C}_2(z) = \Psi'(\mathsf{C}_4(z))$. If C^*_j, is the effective tensor of $\mathsf{C}_j(z)$, $j = 1, 2, 3, 4$, then, since Ψ and Ψ' are links, $\mathsf{C}^*_3 = \Psi(\mathsf{C}^*_1)$ and $\mathsf{C}^*_2 = \Psi'(\mathsf{C}^*_4)$. By assumption Λ is a link, and therefore $\mathsf{C}^*_4 = \Lambda(\mathsf{C}^*_3)$. But then $\mathsf{C}^*_2 = \tilde{\Lambda}(\mathsf{C}^*_1)$, proving that $\tilde{\Lambda}(\mathsf{C})$ is a link. Conversely, every link Λ can be obtained as a $[\Psi, \Psi']$-image of a link $\tilde{\Lambda}$, passing through $\hat{\mathsf{C}}^0 = [\mathsf{I}, \mathsf{I}]$. In this case

$$\overline{\boldsymbol{\Gamma}}_1^0 = \overline{\boldsymbol{\Gamma}}_2^0 = K'_0\left(\frac{1}{4}\right) + K_0\left(\frac{3}{8}\right).$$

We claim that

$$\hat{\mathsf{M}} = \left[K_0\left(\frac{3}{8}\right), K_0\left(\frac{3}{8}\right)\right]$$

is an inversion key. Indeed, it follows from an obvious relation

$$\mathsf{K}\mathsf{K}'_0\mathsf{K} = 0, \qquad \forall\, \mathsf{K} \in \mathrm{Sym}\big(\mathrm{Sym}_0(\mathbb{R}^2)\big).$$

The inversion formulas (6.46) read

$$\mathsf{C}_1 = \mathsf{I} + \left(\mathsf{I} - \frac{3}{8}\mathsf{K}K_0(1)\right)^{-1}\mathsf{K}, \quad \mathsf{C}_2 = \mathsf{I} + \left(\mathsf{I} - \frac{3}{8}\Phi(\mathsf{K})K_0(1)\right)^{-1}\Phi(\mathsf{K}),$$

where Φ is given by (6.75). It is clear that $\boldsymbol{I}_2$ is an eigenstrain for both C_1 and C_2 (with eigenvalue 1). Therefore, the action needs to be restricted to the two-dimensional space of symmetric, trace-free 2×2 matrices. On that space $K_0(1) = 2\mathsf{I}$ and hence,

$$\mathsf{C}_1 = \mathsf{I} - \left(\mathsf{I} + \frac{3}{4}\mathsf{K}\right)^{-1}\mathsf{K}, \quad \mathsf{C}_2 = \mathsf{I} - \left(\mathsf{I} + \frac{3}{4}\Phi(\mathsf{K})\right)^{-1}\Phi(\mathsf{K}), \tag{6.88}$$

where we have replaced K with $-\mathsf{K}$. This shows that C_1 is a rational function of K on $\mathrm{Sym}_0(\mathbb{R}^2)$. A crucial observation is that the transformation Φ preserves eigenstrains of K. Indeed, on $\mathrm{Sym}_0(\mathbb{R}^2)$

$$\Phi(\mathsf{K}) = \mathsf{K} - (\mathrm{Tr}\,\mathsf{K})\mathsf{I},$$

where the trace is taken on $\mathrm{Sym}_0(\mathbb{R}^2)$. Thus, C_1 and C_2 have the same eigenstrains, while the relation between their eigenvalues can be computed from (6.88). Suppose that K has eigenvalues α_1, α_2. Then the eigenvalues of C_1 are

$$2\mu_j = 1 - \frac{\alpha_j}{1 + 3\alpha_j/4}, \qquad j = 1, 2. \tag{6.89}$$

The eigenvalues of $\Phi(\mathsf{K})$ will be $-\alpha_2$, $-\alpha_1$, i.e. $-\alpha_2$ is the eigenvalue of $\Phi(K)$ corresponding to the α_1 eigenvalue of K. Thus, the eigenvalues of C_2 are

$$2\mu_j' = 1 + \frac{\alpha_{j'}}{1 - 3\alpha_{j'}/4}, \qquad j = 1, 2,$$

where

$$j' = 3 - j = \begin{cases} 2, & j = 1, \\ 1, & j = 2. \end{cases}$$

Equivalently,

$$2\mu_{j'}' = 1 + \frac{\alpha_j}{1 - 3\alpha_j/4}, \qquad j = 1, 2. \tag{6.90}$$

Eliminating α_j from (6.89), (6.90) we obtain

$$\left(\frac{1}{\mu_j} + 2\right)\left(\frac{1}{\mu_{j'}'} + 2\right) = 16, \qquad j = 1, 2.$$

In order to obtain the general form of this link we apply the global link (6.87), and after somewhat tedious calculations obtain

$$\left(\frac{1}{\mu_j} + \frac{1}{\kappa_0}\right)\left(\frac{1}{\mu_{j'}'} + \frac{1}{\kappa_0'}\right) = a_0^2, \qquad j = 1, 2, \tag{6.91}$$

where a_0 is an arbitrary constant and κ_0 and κ'_0 are the constant bulk moduli of $\mathsf{C}_1(z)$ and $\mathsf{C}_2(z)$, respectively. This family of links was discovered in [9].

Bibliography

[1] Auffray N, Kolev B and Petitot M 2014 On anisotropic polynomial relations for the elasticity tensor *J. Elast.* **115** 77–103

[2] Biedenharn L C and Louck J D 1981 *Angular Momentum in Quantum Physics* (*Encyclopedia of Mathematics and its Applications* vol 8) (Reading, MA: Addison-Wesley)

[3] Bröcker T and tom Dieck T 1985 *Representations of Compact Lie Groups* (*Graduate Texts in Mathematics* vol 98) (New York: Springer)

[4] Cherkaev A V, Lurie K A and Milton G W 1992 Invariant properties of the stress in plane elasticity and equivalence classes of composites *Proc. R. Soc. Lond.* A **438** 519–29

[5] Edmonds A R 1996 *Angular Momentum in Quantum Mechanics* (Princeton, NJ: Princeton University Press)

[6] Grabovsky Y 2004 Algebra, geometry and computations of exact relations for effective moduli of composites *Advances in Multifield Theories of Continua with Substructure* ed G Capriz and P M Mariano (*Modeling and Simulation in Science, Engineering and Technology*) (Boston, MA: Birkhäuser) pp 167–97

[7] Grabovsky Y 2009 Exact relations for effective conductivity of fiber-reinforced conducting composites with the Hall effect via a general theory *SIAM J. Math. Anal.* **41** 973–1024

[8] Grabovsky Y, Milton G W and Sage D S 2000 Exact relations for effective tensors of polycrystals: necessary conditions and sufficient conditions *Commun. Pure. Appl. Math.* **53** 300–53

[9] Helsing J, Milton G W and Movchan A B 1997 Duality relations, correspondences and numerical results for planar elastic composites *J. Mech. Phys. Solids* **45** 565–90

[10] Lurie K A, Cherkaev A V and Fedorov A V 1984 On the existence of solutions to some problems of optimal design for bars and plates *J. Optim. Theory Appl.* **42** 247–81

[11] Milgrom M and Shtrikman S 1989 Linear response of polycrystals to coupled fields: exact relations among the coefficients *Phys. Rev.* B **40** 5991–4

[12] Milgrom M and Shtrikman S 1989 Linear response of two-phase composites with cross moduli: exact universal relations *Phys. Rev.* A **40** 1568–75

[13] Milton G W 2002 *The Theory of Composites* (*Cambridge Monographs on Applied and Computational Mathematics* vol 6) (Cambridge: Cambridge University Press)

[14] Weyl H 1931 *Theory of Groups and Quantum Mechanics* (New York: E P Dutton)

[15] Wigner E P 1931 *Gruppentheorie und ihre Anwendung auf die Quantenmechanik der Atomspektren* (Braunschweig: Vieweg)

Part III

Case studies

Chapter 7

Introduction

This part of the book is a compendium of results on exact relations and links for effective properties of conducting, elastic, piezoelectric, thermoelastic and thermoelectric composites, given without proof, but with references to published derivations, when they exist. If no references are given, then the author is not aware of a prior publication where a given formula is derived. Many results in the literature, such as [2, 6, 11, 19], for example, give formulas for effective moduli of two-phase composites. Such formulas will not be listed explicitly among exact relations, because they do not hold for general multiphase composites. Instead, these results can be derived using links that map the two constituents into one of the general exact relations in our lists. The resulting formulas are therefore obtained by a combination of a listed exact relation and a link. This technique is explained and illustrated in [5, 7]. (See also section 8.1.2 for a simple example.)

The results listed here are of fundamental nature. They are valid for *all* composites: multiphase, polycrystalline, periodic, random, functionally graded, etc. Moreover, composites themselves do not have to be homogeneous, as is the case for functionally graded composites. The effective tensors of such composites are also inhomogeneous, i.e. spatially dependent, and all our formulas apply pointwise. The applicability of listed results depends on the physical validity of three assumptions:

- the response of each material in a composite is linear (the linear constitutive law);
- there are no residual stresses, net charges or thermal imbalances;
- the constituents are in perfect mechanical/electric/thermal contact.

Even though the latter two assumptions might not be valid for real composites, one can use formulas in this part of the book to estimate the extent to which the failure of these assumptions has affected the effective behavior of composites, by providing benchmarks for comparison between theory and experiment.

doi:10.1088/978-0-7503-1048-2ch7

Exact relations are sets of explicit formulas satisfied by the effective parameters of a composite involving properties and volume fractions of constituents that are valid for *all microstructures*. By contrast, *links* relate effective moduli of two composites with the same microstructure, but different properties of the constituents.

One of the earliest examples of an exact relation was encountered by Keller in [12], and was proved in full generality by Dykhne [4]. It states that any two-dimensional conducting composite, all of whose constituents have the same determinant d_0 of their conductivity tensors, must have the effective conductivity tensor $\boldsymbol{\sigma}^*$, satisfying $\det \boldsymbol{\sigma}^* = d_0$, *regardless of the microstructure*. In fact, the microstructure, or texture, as it is called for polycrystals, affects all components of the effective conductivity tensor, and only their specific combination, the determinant, remains completely insensitive to the microstructure. In our list of exact relations, this result would be listed as a single equation

$$\det \boldsymbol{\sigma} = d_0, \tag{7.1}$$

to be interpreted as described above.

In order to explain the concept of a link, let us assume that we somehow know the effective tensor L^* of a composite whose microscopic properties are described by the tensor $\mathsf{L}(\boldsymbol{x})$ of material properties at the point $\boldsymbol{x}$ in a composite. Now, let us imagine that at every point $\boldsymbol{x}$ in a composite we have replaced the material $\mathsf{L}(\boldsymbol{x})$ with the material $\mathsf{L}'(\boldsymbol{x}) = \lambda_0 \mathsf{L}(\boldsymbol{x})$, where λ_0 is a positive scalar, independent of $\boldsymbol{x}$. Then the effective tensor L'_* of the new composite will be $\lambda_0 \mathsf{L}^*$. This obvious, for experts, statement is an example of a link, which in our notation will be recorded as $\mathsf{L}' = \lambda_0 \mathsf{L}$. An example of a non-trivial link in the context of two-dimensional conductivity is due to Mendelson [15, 16]. Once again we assume that we somehow know the effective conductivity tensor $\boldsymbol{\sigma}^*$ of a two-dimensional conducting composite whose microscopic properties are described by the local tensor field $\boldsymbol{\sigma}(\boldsymbol{x})$. Now, let us imagine that at every point $\boldsymbol{x}$ in this composite we have replaced the material $\boldsymbol{\sigma}(\boldsymbol{x})$ with the material

$$\boldsymbol{\sigma}'(\boldsymbol{x}) = \frac{\boldsymbol{\sigma}(\boldsymbol{x})}{\det \boldsymbol{\sigma}(\boldsymbol{x})}. \tag{7.2}$$

Then, according to [15, 16], the effective tensor $\boldsymbol{\sigma}'_*$ of the new composite must be

$$\boldsymbol{\sigma}'_* = \frac{\boldsymbol{\sigma}^*}{\det \boldsymbol{\sigma}^*}. \tag{7.3}$$

We emphasize that in this result, no constraints whatsoever were placed on the microstructure of the original composite $\boldsymbol{\sigma}(\boldsymbol{x})$. In our notation, this result will be recorded as a single equation

$$\boldsymbol{\sigma}' = \frac{\boldsymbol{\sigma}}{\det \boldsymbol{\sigma}}. \tag{7.4}$$

In general links contain a wealth of information. For example, they can simplify and even obviate specific exact relations. For instance, we can use the 'trivial' link

$\sigma' = \lambda_0\sigma$ applied to the two-dimensional conductivity to simplify the Keller–Dykhne exact relation (7.1), by eliminating the parameter d_0:

$$\det\sigma = 1. \tag{7.5}$$

That means that we can always recover (7.1) by combining (7.5) and the link $\sigma' = \lambda_0\sigma$. In turn, the simplified exact relation (7.5) is a consequence of (7.4). To see this, we observe that the set of all tensors left invariant by any link must be an exact relation. Clearly, the set of conductivity tensors σ left invariant by (7.4) is (7.5).

In general links do not have to be defined for all composites. When they are, as in our two examples above, we call them *global links*. More generally, links may be defined only on tensors belonging to a specific exact relation. For example, in two-dimensional elasticity there is a link discovered by Lurie *et al* [14] that says that composite polycrystals whose constituents have a square symmetry and one and the same bulk modulus κ_0 must always have square symmetry (even for highly anisotropic microstructures) and that same bulk modulus κ_0. All two-dimensional elasticity tensors belonging to this exact relation have a common eigenvalue $2\kappa_0$, corresponding to the common eigenvector $\boldsymbol{I}_2$, and two other eigenvalues $2\mu_1$, $2\mu_2$. There is a link [8] defined only on elastic tensors belonging to the Lurie–Cherkaev–Fedorov exact relation that maps C to C' with the same eigenvectors and eigenvalues $2\kappa_0$, $2\mu_1'$, $2\mu_2'$, determined by the equations

$$\left(\frac{1}{\kappa_0} + \frac{1}{\mu_1}\right)\left(\frac{1}{\kappa_0} + \frac{1}{\mu_2'}\right) = \left(\frac{1}{\kappa_0} + \frac{1}{\mu_2}\right)\left(\frac{1}{\kappa_0} + \frac{1}{\mu_1'}\right) = c_0, \tag{7.6}$$

where $c_0 > \kappa_0^{-2}$ is an arbitrary constant. Those familiar with the Lurie–Cherkaev exact relation [13] will immediately recognize that tensors belonging to it are exactly the ones left invariant by the link (7.6). A final remark is that links need not be *functions* mapping L into L', they can relate (link) only parts of L and L', leaving remaining parts unrelated. For example, in the setting of fibrous conducting composites, the effective conductivity in the plane transversal to the fibers depends only on the local transversal conductivity. That means that two composites, whose local transversal conductivity tensors are the same, while other components of the full conductivity tensor are unrelated, will have effective conductivity tensors with the same transversal component. In the case studies in this part of the book we will encounter links of all of the described types, that range from simple and natural to unexpected and strange.

We find it convenient (and elegant) to think of exact relations geometrically, as submanifolds $\mathbb{M}$ (of positive codimension) in the space of material tensors. In our example of two-dimensional conductivity, the set of all material tensors is an open subset $\mathrm{Sym}^+(\mathbb{R}^2)$ of all positive definite matrices in the three-dimensional space of all 2×2 symmetric matrices. The two-dimensional submanifold $\mathbb{M} = \{\sigma \in \mathrm{Sym}^+(\mathbb{R}^2) : \det\sigma = d_0\}$, corresponding to the Keller–Dykhne exact relation (7.1), can be visualized as an entire sheet of a two-sheeted hyperboloid shown on the cover of the book. Very often it will be convenient to define the exact relation submanifolds $\mathbb{M}$

by parametric equations, instead of implicit equations, such as (7.1). An example of such a description is a well-know Hill's exact relation [10, 11], that says that if a composite is made of any number of isotropic materials with the same shear modulus μ_0 then, regardless of the microstructure, its effective tensor will always be isotropic and have the same shear modulus μ_0. Geometrically, this exact relation is a one-dimensional straight line in the 21-dimensional space $\mathrm{Sym}^+(\mathrm{Sym}(\mathbb{R}^3))$ of elastic tensors, which are symmetric positive definite operators on the six-dimensional space $\mathrm{Sym}(\mathbb{R}^3)$ of symmetric 3×3 matrices. This straight line is given by parametric equations

$$\mathbb{M}_{\mu_0} = \left\{\mathsf{C} \in \mathrm{Sym}^+\left(\mathrm{Sym}\left(\mathbb{R}^3\right)\right) : \mathsf{C} = 2\mu_0\mathsf{I} + \lambda \boldsymbol{I}_3 \otimes \boldsymbol{I}_3,\ \lambda \in \mathbb{R}\right\}. \tag{7.7}$$

The Lamé modulus λ plays the role of a variable parameter, determining the point on $\mathbb{M}_{\mu_0}$, while the constant shear modulus μ_0 labels a specific exact relation in an infinite family $\mathbb{M}_{\mu_0}$. Hill's theorem that each submanifold $\mathbb{M}_{\mu_0}$, $\mu_0 > 0$ represents an exact relation can be stated as a formula $\mathsf{C}^* = 2\mu_0\mathsf{I} + \lambda^*\boldsymbol{I}_3 \otimes \boldsymbol{I}_3$ for the effective elastic tensor of a composite $\mathsf{C}(\boldsymbol{x}) = 2\mu_0\mathsf{I} + \lambda(\boldsymbol{x})\boldsymbol{I}_3 \otimes \boldsymbol{I}_3$, where λ^* depends on the microstructure. As it turns out, the dependence of λ^* on the microstructure is fairly simple, given by

$$\frac{1}{2\mu_0 + \lambda^*} = \left\langle \frac{1}{2\mu_0 + \lambda} \right\rangle, \tag{7.8}$$

where $\langle\cdot\rangle$ denotes an average, understood in the context of the composite class under consideration. For periodic composites, it is the average over a period cell, for random composites it is an ensemble average, or an average over a representative volume element. For more mathematically inclined readers $\langle\cdot\rangle$ stands for the L^∞ weak-* limit of a sequence of oscillating fields, in which case (7.8) represents an equality between objects obtained via the G- or H-limit (denoted by stars) and objects obtained via a much simpler L^∞ weak-* limit. In our list of exact relations, Hill's exact relation (7.7), (7.8) will be listed as

$$\mathsf{C} = 2\mu_0\mathsf{I} + \lambda \boldsymbol{I}_3 \otimes \boldsymbol{I}_3, \quad \lambda \in \mathbb{R}, \qquad \left\langle \frac{1}{2\mu_0 + \lambda} \right\rangle. \tag{7.9}$$

An essential part of our system on notation is the consistent use of the zero subscript to denote constants, which do not vary throughout the composite and parameters without the zero subscript that are allowed to vary from point to point in a composite *arbitrarily*, while the explicit dependence on $\boldsymbol{x}$ is not shown in our notation.

The theory described in part 2 of the book makes it possible to compute *complete* lists of exact relations and links for conducting, elastic, piezoelectric, thermoelastic and thermoelectric composites. In cases where anisotropic constituents are used in a composite, it is often practically impossible to control the orientation in which each anisotropic crystallite occurs within a composite. We therefore focus almost exclusively on *polycrystalline* exact relations and links, whereby all anisotropic constituents are allowed to be used in arbitrary orientations. A partial exception to this rule occurs in fibrous composites, where anisotropic cylindrical fibers (of arbitrary cross-section) can be injected into an isotropic matrix. In this case, it is natural to assume that we can keep the orientation of the fiber axis fixed, while rotations of fibers around the fiber axis cannot be controlled. In our theory such composites are also regarded as polycrystalline, since in the idealized setting of perfectly parallel fibers the microstructure is essentially two-dimensional, determined by any cross-section perpendicular to the fibers. Hence, mathematically, such composites can be treated as two-dimensional polycrystalline. Another exception to the polycrystalline rule occurs in the study of conducting composites with the Hall effect, where the anisotropy is induced by a fixed orientation of the ambient magnetic field, which also breaks the symmetry of conductivity tensors. In this specific case all exact relations and links, without restriction, are computed.

The actual derivation of complete lists of exact relations and links, while feasible, is often very labor-intensive. In many cases it required the combined efforts of many students, both graduate and undergraduate, working over several academic years with the author to accomplish the task. Hence, at present, complete lists of exact relations and links are not available in some contexts. In other contexts, where complete results are available, the number of different families of exact relations and links may number into the hundreds. In order to make the information contained in these results presentable, it was expedient to eliminate logically redundant facts. For example, we could have chosen to omit the Keller–Dykhne exact relation (7.1) from the list of formulas for two-dimensional conducting composites because, as we have shown above, it is a logical corollary of Mendelson's link (7.4) (and the universal homogeneity link $\mathsf{L}' = \lambda_0 \mathsf{L}$). After elimination of logical redundancies the remaining several dozens essential results could then be effectively presented. Thus, one has to keep in mind that completeness of the lists in this part of the book is understood not in the sense that every single exact relation and link is listed, but in the sense that every exact relation and link can be obtained, if desired, as a logical consequence of listed results. As a rule, no effort has been made to make all of the lists maximally logically reduced, as long as they can be conveniently and concisely presented.

It is natural to ask if there is a simple way to understand or explain exact relation formulas in terms of standard physical principles. The answer is 'no' in the sense that the mathematical theory developed in this book delivering all exact relations and links in a unified way cannot be interpreted in terms of equally unified general physical principles. However, there are classes of exact relations and links that can be obtained by particular methods, described in [17]. One of the most well-known examples of such a class of exact relations is called, following Milton [18], *uniform field relations* (UFRs). These relations are consequences of the observation that if all constituents in a composite respond identically to a particular applied field, then any composite, made with these materials, will also respond in exactly the same way to that applied field. Physically, this is obvious, since, as far as that applied field is concerned, all constituent materials are indistinguishable from one another. The idea originated in [9, 11], in fact, Hill's exact relation (7.7) is a UFR. It was subsequently formalized in [3, 14], among others. This simple idea becomes more and more effective as the number of coupled physical properties grow. It results in a very large class of exact relations in the context of piezothermoelectroelastic fibrous composites exhibiting pyroelectricity [1]. All UFRs are also easily identifiable in the framework of our general theory. In each list the UFRs are identified, by the 'UFR' label.

Bibliography

[1] Benveniste Y 1993 Exact results in the micromechanics of fibrous piezoelectric composites exhibiting pyroelectricity *Proc. R. Soc. Lond.* A **441** 59–81

[2] Dunn M L 1993 Exact relations between the thermoelectroelastic moduli of heterogeneous materials *Proc. R. Soc. Lond.* A **441** 549–57

[3] Dvorak G J 1990 On uniform fields in heterogeneous media *Proc. R. Soc. Lond.* A **431** 89–110

[4] Dykhne A M 1971 Conductivity of a two-dimensional two-phase system *Sov. Phys.–JETP* **32** 63–5

Dykhne A M 1970 Conductivity of a two-dimensional two-phase system *Zh. Eksp. Teor. Fiz* **59** 110–5

[5] Grabovsky Y 2009 An application of the general theory of exact relations to fiber-reinforced conducting composites with Hall effect *Mech. Mater.* **41** 456–62

[6] Hashin Z and Rosen B W 1964 The elastic moduli of fiber-reinforced materials *J. Appl. Mech.* **31** 223–32

[7] Hegg M 2013 Links between effective tensors for fiber-reinforced elastic composites *Compt. Rend. Méc.* **341** 520–32

[8] Helsing J, Milton G W and Movchan A B 1997 Duality relations, correspondences and numerical results for planar elastic composites *J. Mech. Phys. Solids* **45** 565–90

[9] Hill R 1952 The elastic behavior of a crystalline aggregate *Proc. Phys. Soc. Lond.* A **65** 349–54

[10] Hill R 1963 Elastic properties of reinforced solids: some theoretical principles *J. Mech. Phys. Solids* **11** 357–72

[11] Hill R 1964 Theory of mechanical properties of fibre-strengthened materials: I. Elastic behaviour *J. Mech. Phys. Solids* **12** 199–212

[12] Keller J B 1964 A theorem on the conductivity of a composite medium *J. Math. Phys.* **5** 548–9

[13] Lurie K A and Cherkaev A V 1984 G-closure of some particular sets of admissible material characteristics for the problem of bending of thin plates *J. Optim. Theory Appl.* **42** 305–16

[14] Lurie K A, Cherkaev A V and Fedorov A V 1984 On the existence of solutions to some problems of optimal design for bars and plates *J. Optim. Theory Appl.* **42** 247–81

[15] Mendelson K S 1975 Effective conductivities of two-phase material with cylindrical phase boundaries *J. Appl. Phys.* **46** 917–8

[16] Mendelson K S 1975 A theorem on the conductivity of two-dimensional heterogeneous medium *J. Appl. Phys.* **46** 4740–1

[17] Milton G W 1997 Composites: a myriad of microstructure independent relations *Proc. 19th Int. Congress Theoretical and Applied Mechanics (Kyoto, 1996)* ed T Tatsumi, E Watanabe and T Kambe (Amsterdam: Elsevier) pp 443–59

[18] Milton G W 2002 *The theory of composites* (*Cambridge Monographs on Applied and Computational Mathematics* vol 6) (Cambridge: Cambridge University Press)

[19] Schulgasser K 1992 Relationships between the effective properties of transversely isotropic piezoelectric composites *J. Mech. Phys. Solids* **40** 473–9

Chapter 8

Conductivity with the Hall effect

Consider composites made with conductors exhibiting the Hall effect. That means that the electric field $\boldsymbol{e}$ and the current field $\boldsymbol{j}$ are related by a linear transformation $\boldsymbol{L}$, whose symmetric part $\boldsymbol{\sigma}$ represents the conductivity tensor of the material, and whose antisymmetric part depends (linearly) on the applied magnetic field. The system combining Maxwell equations with the linear constitutive law

$$\nabla \times \boldsymbol{e} = 0, \qquad \nabla \cdot \boldsymbol{j} = 0, \qquad \boldsymbol{j} = \boldsymbol{L}\boldsymbol{e}, \tag{8.1}$$

augmented with appropriate boundary conditions determines the electric and the current field inside the material uniquely.

8.1 Two-dimensional conductivity with the Hall effect

Physically, two-dimensional conductivity results would apply to fibrous conducting composites placed in a uniform magnetic field directed along the fibers, while both the electric field and the current field are perpendicular to the fibers. The arrangement and cross-section of fibers can be arbitrary, including the case where different fibers have different cross-sections. All exact relations and links in this section can be found elsewhere in the literature (see e.g. [4, 5, 11, 14]). The calculations using the theory of exact relations can be found in [6, 8]. The Hall conductivity tensors in this context will be represented as $\boldsymbol{L} = \boldsymbol{\sigma} + r\boldsymbol{R}_{\perp}$, where $\boldsymbol{\sigma}$ denotes the symmetric part of the conductivity tensor, and

$$\boldsymbol{R}_{\perp} = \begin{bmatrix} 0 & -1 \\ 1 & 0 \end{bmatrix}.$$

The Hall coefficient r is a multiple of the magnitude of the applied magnetic field. However, since we will not be studying the dependence of Hall conductivity on the magnetic field, the dependence of r on the magnetic field will not be made explicit.

doi:10.1088/978-0-7503-1048-2ch8

8.1.1 Exact relations and links

Exact relations

1. (Uniform field relation (UFR)) If $\boldsymbol{L}\boldsymbol{a}_0 = \boldsymbol{b}_0$
2. $r = r_0$
3. $(r - r_0)^2 + \det \boldsymbol{\sigma} = c_0$.

Links

1. $\boldsymbol{L}' = \boldsymbol{L}^T$
2. $\boldsymbol{L}' = \boldsymbol{L} + r_0 \boldsymbol{R}_\perp$.
3. $\boldsymbol{L}' = \boldsymbol{\sigma}' + r' \boldsymbol{R}_\perp$, where

$$\boldsymbol{\sigma}' = \frac{c_0 \boldsymbol{\sigma}}{(r_0 - r)^2 + \det \boldsymbol{\sigma}}, \qquad r' = \frac{c_0(r_0 - r)}{(r_0 - r)^2 + \det \boldsymbol{\sigma}} + r'_0.$$

Technically speaking, there are other results, but all of them are consequences of the ones listed above. Our theory also guarantees that there are no other genuinely different microstructure-independent equalities.

8.1.2 An application

As an example of the application of these formulas let us consider a polycrystalline composite made of a single crystal with conductivity tensor $\boldsymbol{L}_0 = \boldsymbol{\sigma}_0 + r_0 \boldsymbol{R}_\perp$. Then the effective tensor of such a polycrystal must have the form $\boldsymbol{L}^* = \boldsymbol{\sigma}^* + r_0 \boldsymbol{R}_\perp$, where $\det \boldsymbol{\sigma}^* = \det \boldsymbol{\sigma}_0$. In particular, if the polycrystalline texture of the composite is statistically isotropic, then $\boldsymbol{L}^* = \sqrt{\det \boldsymbol{\sigma}_0}\, \boldsymbol{I}_2 + r_0 \boldsymbol{R}_\perp$. A more powerful example is provided by composites made of two isotropic constituents

$$\boldsymbol{L}_i = \sigma_i \boldsymbol{I}_2 + r_i \boldsymbol{R}_\perp, \qquad i = 1, 2. \tag{8.2}$$

Then all components of the effective tensor $\boldsymbol{L}^*$ can be expressed in terms of a single microstructure-dependent function $\mathfrak{S}^*(\xi)$, which is the effective conductivity of the original composite in which materials $\boldsymbol{L}_1$ and $\boldsymbol{L}_2$ are replaced with materials $\boldsymbol{I}_2$ and $\xi \boldsymbol{I}_2$, respectively. Assuming $r_1 \neq r_2$, we first use the link to 'eliminate' the Hall effect. In other words, we need to choose the constants r_0, $\hat{r}_0$ and c_0, such that $\hat{r}_1 = \hat{r}_2 = 0$ and $\hat{\sigma}_1 = 1$. Thus, we need to solve the quadratic equation

$$\frac{r_0 - r_1}{(r_0 - r_1)^2 + \sigma_1^2} = \frac{r_0 - r_2}{(r_0 - r_2)^2 + \sigma_2^2} \tag{8.3}$$

for r_0, and then set

$$c_0 = \frac{(r_0 - r_1)^2 + \sigma_1^2}{\sigma_1}, \qquad \hat{r}_0 = -\frac{c_0(r_0 - r_1)}{(r_0 - r_1)^2 + \sigma_1^2}.$$

It is easy to see that equation (8.3) is equivalent to a quadratic equation with two distinct real roots (assuming $\boldsymbol{L}_1 \neq \boldsymbol{L}_2$). Let r_0 denote one of the roots. (Choosing either root for r_0 does not affect the final answer.) Then, $\boldsymbol{L}_2$ gets mapped to $\xi \boldsymbol{I}$, where

$$\xi = \frac{\sigma_2}{\sigma_1} \cdot \frac{(r_0 - r_1)^2 + \sigma_1^2}{(r_0 - r_2)^2 + \sigma_2^2} > 0.$$

We note that equation (8.3) stays the same if we interchange indices 1 and 2. However, ξ gets transformed to $1/\xi$. Hence, we may assume, without loss of generality, that $\xi > 1$. In fact, we may derive the equation for ξ in terms of σ_i, r_i, $i = 1, 2$:

$$\sigma_1\sigma_2\xi^2 - \left(\sigma_1^2 + \sigma_2^2 + (r_1 - r_2)^2\right)\xi + \sigma_1\sigma_2 = 0, \quad \xi > 1.$$

Then we compute

$$\hat{r}_0 = \frac{\sigma_2\xi - \sigma_1}{r_1 - r_2}, \qquad r_0 = r_1 - \sigma_1\hat{r}_0, \qquad c_0 = \sigma_2\xi + \sigma_1 + \hat{r}_0\frac{\sigma_2^2 - \sigma_1^2}{r_1 - r_2}.$$

Finally, solving the equations

$$\frac{c_0\boldsymbol{\sigma}^*}{(r_0 - r^*)^2 + \det\boldsymbol{\sigma}^*} = \mathfrak{S}^*(\xi\boldsymbol{I}_2), \qquad \frac{c_0(r_0 - r^*)}{(r_0 - r^*)^2 + \det\boldsymbol{\sigma}^*} + \hat{r}_0 = 0 \tag{8.4}$$

for $\boldsymbol{\sigma}^*$ and r^*, we obtain the effective Hall conductivity of the composite in terms of the properties of the constituents and the microstructure-dependent symmetric 2×2 matrix $\mathfrak{S}^* = \mathfrak{S}^*(\xi\boldsymbol{I}_2)$:

$$\boldsymbol{\sigma}^* = \frac{c_0\mathfrak{S}^*}{\hat{r}_0^2 + \det\mathfrak{S}^*}, \qquad r^* = r_0 - \frac{c_0\hat{r}_0}{\hat{r}_0^2 + \det\mathfrak{S}^*}. \tag{8.5}$$

Since the G- and G_θ-closure of two isotropic conductors $\boldsymbol{I}_2$ and $\xi\boldsymbol{I}_2$ is known, formulas (8.5) give an exact description of the G- and G_θ-closures of two isotropic conductors (8.2). In the three-dimensional space $(\lambda_1, \lambda_2, r)$, where λ_1, λ_2 are the eigenvalues of $\boldsymbol{\sigma}$, these G-closures lie on a two-dimensional surface (a hyperboloid, whose equation is $(8.4)_2$).

8.2 Three-dimensional conductivity with the Hall effect

Here we are assuming that a composite is placed in a uniform magnetic field, whose direction is arbitrary. In three space dimensions the constitutive relation $\boldsymbol{j} = \boldsymbol{L}\boldsymbol{e}$ can be written as follows

$$\boldsymbol{L}\boldsymbol{e} = \boldsymbol{\sigma}\boldsymbol{e} + \boldsymbol{r} \times \boldsymbol{e},$$

where $\boldsymbol{\sigma}$ is the symmetric part of $\boldsymbol{L}$ and the map $\boldsymbol{e} \mapsto \boldsymbol{r} \times \boldsymbol{e}$ constitutes its antisymmetric part. There are very few exact relations and links in the context of three-dimensional conductivity with the Hall effect. In fact the only link is

$$\boldsymbol{L}' = \boldsymbol{L}^T. \tag{8.6}$$

All of the exact relations can be obtained by means of intersections and applications of the transposition link (8.6) from a family of UFRs

$$Le_0 = j_0,$$

and a single non-UFR exact relation $L^T = L$ (the set of fixed points of (8.6)), saying that the Hall effect cannot be obtained from non-Hall conductors by means of making anisotropic composites. This statement was proved in the PhD dissertation of Hansun To [15].

8.3 Fibrous conducting composites with the Hall effect

By 'fibrous composites' we mean that there exists a direction $\hat{f}$ (fiber direction) so that all cross-sections of a composite by planes orthogonal to $\hat{f}$ have the same microstructure. That means that we consider a class of composites that include all fibrous ones, but is broader in the sense that we do not place any geometric connectivity constraints on each phase. No other restrictions are placed on the microstructure. We would like to emphasize here that the direction of the magnetic field is not assumed to be aligned in any way either with the crystallographic axes of the constituent materials nor with the fiber direction. By contrast with two and three dimensional conductivity there is a very large number of exact relations and links, some of which are highly non-trivial. This happens because we are working in a significantly larger space of three-dimensional material tensors, while the microstructure is two-dimensional. The results of this section were the eventual outcome of three consecutive Summer REU programs, involving a total of 14 undergraduate students who are named in the Preface. A complete report is published in [8].

The presence of the special fiber direction suggests that we should split the electric and the current fields into their transversal and fiber components:

$$e = e_{\mathsf{T}} + e_{\|}\hat{f}, \qquad j = j_{\mathsf{T}} + j_{\|}\hat{f},$$

where $\hat{f}$ is a unit vector directed along the fibers. The conductivity tensors will then be represented as 2×2 block-matrices

$$L = \begin{bmatrix} \Lambda_{\mathsf{T}} & p_{\mathsf{T}\|} \\ q_{\|\mathsf{T}} & \sigma_{\|} \end{bmatrix},$$

effecting the constitutive relation

$$j_{\mathsf{T}} = \Lambda_{\mathsf{T}} e_{\mathsf{T}} + e_{\|} p_{\mathsf{T}\|}, \quad j_{\|} = q_{\|\mathsf{T}} \cdot e_{\mathsf{T}} + \sigma_{\|} e_{\|}.$$

We formulate all exact relations and links in terms of the block-components Λ_{T}, $p_{\mathsf{T}\|}$, $q_{\|\mathsf{T}}$, $\sigma_{\|}$.

Exact relations

1. $p_{\mathsf{T}\|} = p_0$, $\langle \sigma_{\|} \rangle$ (see e.g. [9])
2. $\Lambda_{\mathsf{T}} = \Lambda_0$, $\langle p_{\mathsf{T}\|} \rangle$, $\langle q_{\|\mathsf{T}} \rangle$.

Links

1. $\boldsymbol{L}' = \begin{bmatrix} a_0\boldsymbol{I}_2 & 0 \\ \boldsymbol{q}_0 & \alpha_0 \end{bmatrix} \boldsymbol{L} \begin{bmatrix} \boldsymbol{I}_2 & \boldsymbol{p}_0 \\ 0 & \beta_0 \end{bmatrix} + \begin{bmatrix} r_0\boldsymbol{R}_\perp & \boldsymbol{p}_0' \\ \boldsymbol{q}_0' & \sigma_0 \end{bmatrix}.$
2. $\boldsymbol{L}' = \begin{bmatrix} \dfrac{\boldsymbol{\Lambda}_\mathrm{T}^T}{\det \boldsymbol{\Lambda}_\mathrm{T}} & \dfrac{\boldsymbol{\Lambda}_\mathrm{T}^T \boldsymbol{p}_{\mathrm{T}\|}^\perp}{\det \boldsymbol{\Lambda}_\mathrm{T}} \\ \dfrac{\boldsymbol{\Lambda}_\mathrm{T} \boldsymbol{q}_{\|\mathrm{T}}^\perp}{\det \boldsymbol{\Lambda}_\mathrm{T}} & \boldsymbol{\Lambda}^{-1}\boldsymbol{p}_{\mathrm{T}\|} \cdot \boldsymbol{q}_{\|\mathrm{T}} - \sigma_\| + \sigma_\|^0 \end{bmatrix}.$
3. $\boldsymbol{L}' = \boldsymbol{L}^T$.
4. $\boldsymbol{\Lambda}_\mathrm{T}^*$ depends only on $\boldsymbol{\Lambda}_\mathrm{T}$. In particular, restricting $\boldsymbol{\Lambda}_\mathrm{T}$ to an exact relation for two-dimensional Hall effect conductivity, described in section 8.1.1, while keeping $\boldsymbol{p}_{\mathrm{T}\|}$, $\boldsymbol{q}_{\|\mathrm{T}}$ and $\sigma_\|$ arbitrary produces an exact relation. An observation to this effect was made in [10].
5. $\boldsymbol{p}_{\mathrm{T}\|}^*$ depends only on $\boldsymbol{\Lambda}_\mathrm{T}$ and $\boldsymbol{p}_{\mathrm{T}\|}$.
6.

$$\boldsymbol{L} = \begin{bmatrix} \boldsymbol{\sigma} & \boldsymbol{p}_{\mathrm{T}\|} \\ \boldsymbol{q}_{\|\mathrm{T}} & \sigma_\| \end{bmatrix}, \qquad \boldsymbol{L}' = \begin{bmatrix} \boldsymbol{\sigma} & \boldsymbol{p}_{\mathrm{T}\|} + \boldsymbol{q}_{\|\mathrm{T}} \\ \boldsymbol{q}_{\|\mathrm{T}} & \sigma_\| \end{bmatrix}, \tag{8.7}$$

where $\boldsymbol{\sigma}^T = \boldsymbol{\sigma}$. This link is an example of a link defined only on a specific exact relation, and not on the entire space of conductivity tensors.

Combining the results in this section one can obtain many interesting corollaries. We refer the reader to a long series of papers by Bergman and Strelniker [1–3, 12, 13] for a taste of what is possible. Another application of these results is given in [7].

Bibliography

[1] Bergman D J, Li X and Strelniker Y M 2005 Macroscopic conductivity tensor of a three-dimensional composite with a one- or two-dimensional microstructure *Phys. Rev.* B **71** 035120

[2] Bergman D J and Strelniker Y M 1998 Duality transformation in a three dimensional conducting medium with two dimensional heterogeneity and an in-plane magnetic field *Phys. Rev. Lett.* **80** 3356–9

[3] Bergman D J and Strelniker Y M 1999 Magnetotransport in conducting composite films with a disordered columnar microstructure and an in-plane magnetic field *Phys. Rev.* B **60** 13016–27

[4] Dykhne A M 1971 Anomalous plasma resistance in a strong magnetic field *Sov. Phys.–JETP* **32** 348–51
Dykhne A M 1970 Anomalous plasma resistance in a strong magnetic field *Zh. Eksp. Teor. Fiz.* pp **59** 641–7 (in Russian)

[5] Dykhne A M and Ruzin I M 1994 On the theory of the fractional quantum Hall effect: the two-phase model J *Phys. Rev.* B **50** 2369–79

[6] Grabovsky Y 2004 Algebra, geometry and computations of exact relations for effective moduli of composites *Advances in Multifield Theories of Continua with Substructure* ed G Capriz and P M Mariano (*Modeling and Simulation in Science, Engineering and Technology*) (Boston, MA: Birkhäuser) pp 167–97
[7] Grabovsky Y 2009 An application of the general theory of exact relations to fiber-reinforced conducting composites with Hall effect *Mech. Mater.* **41** 456–62
[8] Grabovsky Y 2009 Exact relations for effective conductivity of fiber-reinforced conducting composites with the Hall effect via a general theory *SIAM J. Math. Anal.* **41** 973–1024
[9] Hashin Z 1968 Assessment of the self consistent scheme approximation: conductivity of particulate composites *J. Composite Mater* **2** 284–300
[10] Hashin Z 1983 Analysis of composite materials—a survey *J. Appl. Mech.* **50** 481–505
[11] Milton G W 1988 Classical Hall effect in two-dimensional composites: a characterization of the set of realizable effective conductivity tensors *Phys. Rev.* B **38** 11296–303
[12] Strelniker Y M and Bergman D J 2000 Exact relations between magnetoresistivity tensor components of conducting composites with a columnar microstructure *Phys. Rev.* B **61** 6288–97
[13] Strelniker Y M and Bergman D J 2003 Exact relations between macroscopic moduli of composite media in three dimensions: application to magnetoconductivity and magneto-optics of three-dimensional composites with related columnar microstructures *Phys. Rev.* B **67** 184416
[14] Stroud D and Bergman D J 1984 New exact results for the Hall-coefficient and magneto-resistance of inhomogeneous two-dimensional metals *Phys. Rev.* B **30** 447–9
[15] To H 2004 Homogenization of dynamic materials *PhD Thesis* (Philadelphia, PA: Temple University)

IOP Publishing

Composite Materials
Mathematical theory and exact relations
Yury Grabovsky

Chapter 9

Elasticity

Elastic properties of materials, in the context of linear elasticity, are described by the elasticity tensor $\mathsf{C} \in \mathrm{Sym}(\mathrm{Sym}(\mathbb{R}^d))$, $d = 2, 3$, relating the strain field $\boldsymbol{\varepsilon}(\boldsymbol{x})$ to the stress field $\boldsymbol{\sigma}(\boldsymbol{x})$

$$\boldsymbol{\sigma}(\boldsymbol{x}) = \mathsf{C}\boldsymbol{\varepsilon}(\boldsymbol{x}). \tag{9.1}$$

Both fields take values in the space of symmetric $d \times d$ matrices. The strain field is a symmetric part of the gradient of the displacement vector $\boldsymbol{u}$:

$$\boldsymbol{\varepsilon} = e(\boldsymbol{u}) = \frac{1}{2}\left(\nabla \boldsymbol{u} + (\nabla \boldsymbol{u})^T\right), \tag{9.2}$$

while the stress field satisfies the equation of equilibrium.

$$\nabla \cdot \boldsymbol{\sigma} = 0. \tag{9.3}$$

We equip the stress–strain space $\mathrm{Sym}(\mathbb{R}^d)$, with the Frobenius inner product

$$(\boldsymbol{A}, \boldsymbol{B})_F = \mathrm{Tr}\,(\boldsymbol{AB}), \qquad \{\boldsymbol{A}, \boldsymbol{B}\} \subset \mathrm{Sym}\left(\mathbb{R}^d\right), \; d = 2, 3,$$

so that the tensor product $\boldsymbol{C} \otimes \boldsymbol{C}$, $\boldsymbol{C} \in \mathrm{Sym}(\mathbb{R}^d)$ can be regarded as an operator on $\mathrm{Sym}(\mathbb{R}^d)$, acting by the rule

$$(\boldsymbol{C} \otimes \boldsymbol{C})\boldsymbol{\varepsilon} = (\boldsymbol{C}, \boldsymbol{\varepsilon})_F \boldsymbol{C}.$$

For example, an isotropic elastic tensor with shear modulus μ and Lamé modulus λ can be written as

$$\mathsf{C} = 2\mu \mathsf{I} + \lambda \boldsymbol{I}_d \otimes \boldsymbol{I}_d,$$

where I is the identity operator on $\mathrm{Sym}(\mathbb{R}^d)$.

We note an unfortunate circumstance of the absence of a canonical basis in $\mathrm{Sym}(\mathbb{R}^d)$. Any choice of basis permits one to represent an elastic tensor as a 6×6

doi:10.1088/978-0-7503-1048-2ch9

matrix. However, a different basis choice will lead to a different representation of the same elastic tensor, making comparisons of results from different papers difficult. Our results are written in coordinate-free form in the sense that they only use invariant (or covariant) objects, such as the trace, the determinant, the identity matrix, the identity operator and the tensor product $\otimes$. In this coordinate-free form the results acquire the most symmetric and beautiful form, underscoring their physical significance.

9.1 Two-dimensional elasticity

In this section and the next it will be convenient to list the few links that exist in the context of elastic composites together with the exact relations on which they are defined.

1. Hill's exact relation [10, 11]

$$\mathsf{C} = 2\mu_0 \mathsf{I} + \lambda \boldsymbol{I}_2 \otimes \boldsymbol{I}_2, \qquad \left\langle \frac{1}{2\mu_0 + \lambda} \right\rangle. \tag{9.4}$$

   Even though the exact relation itself is a uniform field relation (UFR), the additional volume average formula in (9.4) is not derivable from uniform field ideas.

2. We recall that an elastic tensor C has a square symmetry in two space dimensions if it is unchanged by 90° rotations. Such a tensor is characterized by the condition $\mathsf{C}\boldsymbol{I}_2 = 2\kappa \boldsymbol{I}_2$, where κ is called the bulk modulus. The remaining two eigenvalues of C are denoted $2\mu_1$ and $2\mu_2$, where μ_1 and μ_2 are called the shear moduli. Suppose that all constituents of a composite have square symmetry and one and the same bulk modulus κ_0. Then the effective elastic tensor of the composite will always possess square symmetry and that same bulk modulus κ_0. This is a UFR, since it can be described by the equation

$$\mathsf{C}\boldsymbol{I}_2 = 2\kappa_0 \boldsymbol{I}_2, \tag{9.5}$$

   and was first described in [14]. There is also a link associated with this exact relation. The link maps the set of square symmetric elastic tensors C, with the same bulk modulus κ_0 to the set of square symmetric elastic tensors C', that have another constant bulk modulus κ_0', such that C' has the same eigenstrains as C and shear moduli μ_1' and μ_2', given by the equations

$$\left(\frac{1}{\kappa_0} + \frac{1}{\mu_1}\right)\left(\frac{1}{\kappa_0'} + \frac{1}{\mu_2'}\right) = \left(\frac{1}{\kappa_0} + \frac{1}{\mu_2}\right)\left(\frac{1}{\kappa_0'} + \frac{1}{\mu_1'}\right) = c_0, \tag{9.6}$$

   where $c_0 > 1/\kappa_0\kappa_0'$ is a given constant. This link was obtained in [3] for incompressible materials (i.e. $\kappa_0 = +\infty$) and subsequently generalized to (9.6) in [9].

3. $\mathsf{C}\boldsymbol{I}_2 = 2\kappa_0$ and

$$\left(\frac{1}{\kappa_0} + \frac{1}{\mu_1}\right)\left(\frac{1}{\kappa_0} + \frac{1}{\mu_2}\right) = c_0. \tag{9.7}$$

   This beautiful exact relation was discovered by Lurie and Cherkaev in [12].

4. This exact relation first manifested itself in [1], where upper and lower bounds on the effective shear modulus of a two-dimensional isotropic elastic polycrystal collapsed under a somewhat strange condition on the elastic tensor of a crystallite. Graeme Milton had the insight that this strange condition is equivalent to a particular structure of the elastic tensor. The most convenient description is afforded by its quadratic form

$$(\mathsf{C}\boldsymbol{\varepsilon}, \boldsymbol{\varepsilon})_F = (\boldsymbol{C}, \boldsymbol{\varepsilon})_F^2 - 4\mu \det \boldsymbol{\varepsilon},$$

where μ is a scalar and $\boldsymbol{C}$ is a 2×2 symmetric matrix. We notice that $-2 \det \boldsymbol{\varepsilon}$ is an isotropic quadratic form $(\mathsf{T}\boldsymbol{\varepsilon}, \boldsymbol{\varepsilon})_F$, where the isotropic elastic tensor T has bulk modulus $-1/2$ and shear modulus $1/2$. In [6] Milton's insight was confirmed. It was shown that

$$\mathsf{C} = 2\mu_0 \mathsf{T} + \boldsymbol{C} \otimes \boldsymbol{C}, \quad \boldsymbol{C} \in \mathrm{Sym}(\mathbb{R}^2) \tag{9.8}$$

is an exact relation. Virtually the same result was obtained in [5], where it was also shown to apply to polycrystals made with materials that are rigid in certain directions, specifically, materials whose compliance tensor is rank-one.

There is also a link associated with this exact relation.

$$\mathsf{C} = 2\mu_0 \mathsf{T} + \boldsymbol{C} \otimes \boldsymbol{C}, \quad \mathsf{C}' = 2\mu_0 \mathsf{T} + \lambda' \boldsymbol{C} \otimes \boldsymbol{C}. \tag{9.9}$$

This link has a new feature that the new composite $\mathsf{C}'(\boldsymbol{x})$ is not completely determined by the original composite $\mathsf{C}(\boldsymbol{x})$. The scalar multiple $\lambda'(\boldsymbol{x})$ has nothing to do with the original composite. In other words, we only know that matrices $\boldsymbol{C}'(\boldsymbol{x})$ and $\boldsymbol{C}(\boldsymbol{x})$ are scalar multiples of one another. The link guarantees that matrices $\boldsymbol{C}^*$ and $\boldsymbol{C}'_*$ entering the effective tensors C^* and C'_*, respectively, are scalar multiples of one another.

5. Finally, there is also a two-parameter family of global links discovered in [4].

$$\mathsf{C}' = \alpha_0 \left(\mathsf{I} - \beta_0 \mathsf{C}\mathsf{T}\right)^{-1} \mathsf{C}, \tag{9.10}$$

where the tensor T is the same as in (9.8). In [4] this link was presented in a simpler and more elegant form $\mathsf{S}' = \alpha_0(\mathsf{S} - \beta_0 \mathsf{T})$, relating the compliance tensors $\mathsf{S} = \mathsf{C}^{-1}$. In fact, if we always keep in mind the homogeneity link $\mathsf{S}' = \alpha_0 \mathsf{S}$, then the link (9.10) can essentially be recorded as $\mathsf{S}' = \mathsf{S} - \beta_0 \mathsf{T}$. The reason for this simple form is that in two space dimensions the stress tensor can be given by the Airy stress potential and translating compliances by T does not change the underlying equations, because T is an *isotropic null-Lagrangian* [15, 13]. For this reason we called exact relation (9.8) the 'rank-one plus a null-Lagrangian' (RPN) exact relation.

9.2 Three-dimensional elasticity

In three dimensions there are many fewer exact relations and links. Specifically, exact relation (9.7) and links (9.6) and (9.10) are specific to two space dimensions and do not generalize. The remaining exact relations and links hold in any number of space dimensions. We list them explicitly again for reference purposes.

1. Hill's exact relation [10, 11]

$$\mathsf{C} = 2\mu_0\mathsf{I} + \lambda \boldsymbol{I}_3 \otimes \boldsymbol{I}_3, \qquad \left\langle \frac{1}{2\mu_0 + \lambda} \right\rangle. \tag{9.11}$$

   Even though the exact relation itself is a UFR, the additional volume average formula in (9.11) is not derivable from uniform field ideas.

2. The UFR [14] (see also [16])

$$\mathsf{C}\boldsymbol{I}_3 = 3\kappa_0 \boldsymbol{I}_3, \tag{9.12}$$

   is of course still valid in three dimensions. However, the associated link (9.6) is no longer valid in three dimensions, since the two-dimensional duality (due to Berdichevsky [3] for elasticity) is behind (9.6).

3. Both the RPN exact relation (9.8) and related link (9.9) have exactly the same form in three dimensions as in two dimensions. It is described by a one-parameter family of manifolds

$$\mathsf{C} = 2\mu_0\mathsf{T} + \boldsymbol{C} \otimes \boldsymbol{C} : \boldsymbol{C} \in \mathrm{Sym}\left(\mathbb{R}^3\right), \tag{9.13}$$

   where T is a unique (up to a multiple) isotropic null-Lagrangian, given in all space dimensions by

$$\mathsf{T}\boldsymbol{\varepsilon} = \boldsymbol{\varepsilon} - (\mathrm{Tr}\,\boldsymbol{\varepsilon})\boldsymbol{I}_d. \tag{9.14}$$

   The associated link

$$\mathsf{C} = 2\mu_0\mathsf{T} + \boldsymbol{C} \otimes \boldsymbol{C}, \quad \mathsf{C}' = 2\mu_0\mathsf{T} + \lambda'\boldsymbol{C} \otimes \boldsymbol{C},$$

   has the same form in all space dimensions. It is remarkable that the only available proof of (9.13) and (9.14) is via the general theory of exact relations described in this book.

   We note that all elastic tensors of the form (9.13) are orthotropic, since T is isotropic and all symmetric 3×3 matrices possess orthotropic symmetry (180° rotations around eigenvectors leave symmetric matrices invariant). However, not every orthotropic tensor has form (9.13). Indeed, in a basis in which $\mathsf{C}\boldsymbol{I}_3$ is a diagonal matrix, tensors of the form (9.13) will be described by diagonal matrices $\boldsymbol{C}$, which means that $\mathsf{C}\boldsymbol{\xi} = 2\mu_0\boldsymbol{\xi}$ for every matrix $\boldsymbol{\xi}$ with all zeros on the diagonal. Conversely, if C has this property, then the space of diagonal matrices must be an invariant subspace of C. In that case C will have the form (9.13) if and only if $\mathsf{C} - 2\mu_0\mathsf{T}$ is a non-negative semidefinite rank-one transformation on the three-dimensional space of diagonal matrices. In particular, elastic tensors of the form (9.13) respond isotropically to strains lying in the four-dimensional subspace of shear strains orthogonal to $\mathrm{dev}(\boldsymbol{C})$, the trace-free part of $\boldsymbol{C}$, where

$$\mathrm{dev}(\boldsymbol{C}) = \boldsymbol{C} - \frac{1}{d}(\mathrm{Tr}\,\boldsymbol{C})\boldsymbol{I}_d, \qquad \boldsymbol{C} \in \mathrm{Sym}\left(\mathbb{R}^d\right). \tag{9.15}$$

We emphasize that the list above is *complete*. It lists all microstructure-independent relations with none missing or not yet discovered.

9.3 Fibrous elastic composites

In the context of fibrous elastic composites, the microstructure is the same in any two-dimensional cross-section by a plane perpendicular to the fibers. To fix notation we designate the x_3-direction as the direction of the fibers. The existence of this special direction forces us to represent physical space $\mathbb{R}^3$ as a direct sum of the one-dimensional subspace $\| = \{x_3\boldsymbol{e}_3 : x_3 \in \mathbb{R}\}$—parallel to the fibers and the two-dimensional subspace $\top = \{\boldsymbol{x} \in \mathbb{R}^3 : \boldsymbol{x} \cdot \boldsymbol{e}_3 = 0\}$, transversal (orthogonal) to the fibers.

One of the essential features of polycrystalline exact relations and links is the existence of coordinate-free representations. In the fibrous setting the space direction along the fibers is singled out. The transversal plane, however, will remain coordinate-free. Accordingly, the strain space $\mathrm{Sym}(\mathbb{R}^3)$ acquires the following block-structure

$$\boldsymbol{\varepsilon} = \begin{bmatrix} \boldsymbol{\varepsilon}_\top & \boldsymbol{\varepsilon}_\times \\ \boldsymbol{\varepsilon}_\times & \varepsilon_\| \end{bmatrix}. \tag{9.16}$$

The block $\boldsymbol{\varepsilon}_\top : \mathbb{R}^2 \to \mathbb{R}^2$ is the strain in the transversal plane, $\varepsilon_\| \in \mathbb{R}$ is the strain along the fibers, describing their stretching or shrinking, while $\boldsymbol{\varepsilon}_\times \in \mathbb{R}^2$ describes the 'tilt' of the transversal plane. The direction of $\boldsymbol{\varepsilon}_\times$ is along the projection of the normal to the transversal plane after the deformation onto the transversal plane before the deformation, while the magnitude of $\boldsymbol{\varepsilon}_\times$ indicates the angle of the tilt. The action of operator $\boldsymbol{\varepsilon}$ on $\boldsymbol{x} = (x_1, x_2, x_3) = (\boldsymbol{x}_\top, x_3)$, where $\boldsymbol{x}_\top, \in \mathbb{R}^2$ is

$$\boldsymbol{\varepsilon}\boldsymbol{x} = \begin{bmatrix} \boldsymbol{\varepsilon}_\top & \boldsymbol{\varepsilon}_\times \\ \boldsymbol{\varepsilon}_\times & \varepsilon_\| \end{bmatrix} \begin{bmatrix} \boldsymbol{x}_\top \\ x_3 \end{bmatrix} = \begin{bmatrix} \boldsymbol{\varepsilon}_\top \boldsymbol{x}_\top + x_3\boldsymbol{\varepsilon}_\times \\ \boldsymbol{\varepsilon}_\times \cdot \boldsymbol{x}_\top + x_3\varepsilon_\| \end{bmatrix} \in \mathbb{R}^3.$$

In elasticity the strain space is the underlying vector space on which elasticity tensors act. We will therefore want to regard strains not as 2×2 block-matrices but as block-vectors with 'coordinates' $\boldsymbol{\varepsilon}_\top, \boldsymbol{\varepsilon}_\times, \varepsilon_\|$. Keeping in mind that there are standard inner products on $\mathrm{Sym}(\mathbb{R}^2)$ (Frobenius), $\mathbb{R}^2$ (dot product) and $\mathbb{R}$ (product of numbers), we quickly realize that identifying $\mathrm{Sym}(\mathbb{R}^3)$ with $\mathrm{Sym}(\mathbb{R}^2) \oplus \mathbb{R}^2 \oplus \mathbb{R}$ via the 'natural' coordinatization $\boldsymbol{\varepsilon} \mapsto [\boldsymbol{\varepsilon}_\top, \boldsymbol{\varepsilon}_\times, \varepsilon_\|]$ will lead to a representation of elastic tensors as non-symmetric block-matrices. In order to preserve the symmetry in the block-matrix representation of elastic tensors we write the Frobenius inner product on $\mathrm{Sym}(\mathbb{R}^3)$ in terms of the standard inner products on $\mathrm{Sym}(\mathbb{R}^2)$, $\mathbb{R}^2$ and $\mathbb{R}$:

$$(\boldsymbol{\sigma}, \boldsymbol{\varepsilon})_F = (\boldsymbol{\sigma}_\top, \boldsymbol{\varepsilon}_\top)_F + 2\boldsymbol{\sigma}_\times \cdot \boldsymbol{\varepsilon}_\times + \sigma_\|\varepsilon_\|.$$

This formula shows that if we keep a standard inner product on $\mathbb{R}^2$, then the inner products on $\mathrm{Sym}(\mathbb{R}^2)$ and $\mathbb{R}$ in which the elastic tensor will be represented as a symmetric block-matrix, must be

$$(\boldsymbol{\sigma}_{\mathsf{T}}, \boldsymbol{\varepsilon}_{\mathsf{T}})_{\mathsf{T}} = \frac{1}{2}(\boldsymbol{\sigma}_{\mathsf{T}}, \boldsymbol{\varepsilon}_{\mathsf{T}})_F, \qquad (\sigma_{\|}, \varepsilon_{\|})_{\|} = \frac{1}{2}\sigma_{\|}\varepsilon_{\|}.$$

The inner product $(\cdot, \cdot)_{\mathsf{T}}$ is very convenient, since $(\boldsymbol{I}_2, \boldsymbol{I}_2)_{\mathsf{T}} = 1$. However, the inner product $(\cdot, \cdot)_{\|}$ on $\mathbb{R}$ is neither natural, nor convenient for doing algebra. Thus, in order to simplify the process of computation of exact relations and links we decided to identify 3×3 symmetric matrices $\boldsymbol{\varepsilon}$ with block-vectors $[\boldsymbol{\varepsilon}_{\mathsf{T}}, \boldsymbol{\varepsilon}_{\measuredangle}, \varepsilon_{\|}/\sqrt{2}]$, in which case the elastic tensors will be represented by a symmetric 3×3 block-matrices

$$\mathsf{C} = \begin{bmatrix} \mathsf{C}_{\mathsf{T}} & \mathsf{C}_{\mathsf{T}\measuredangle} & \mathsf{C}_{\mathsf{T}\|} \\ \mathsf{C}_{\mathsf{T}\measuredangle}^T & \mathsf{C}_{\measuredangle} & \mathsf{C}_{\measuredangle\|} \\ \mathsf{C}_{\mathsf{T}\|}^T & \mathsf{C}_{\measuredangle\|}^T & \mathsf{C}_{\|} \end{bmatrix},$$

where $\mathsf{C}_{\mathsf{T}} : \mathrm{Sym}(\mathbb{R}^2) \to \mathrm{Sym}(\mathbb{R}^2)$, $\mathsf{C}_{\mathsf{T}\measuredangle} : \mathbb{R}^2 \to \mathrm{Sym}(\mathbb{R}^2)$, $\mathsf{C}_{\mathsf{T}\|} : \mathbb{R} \to \mathrm{Sym}(\mathbb{R}^2)$, $\mathsf{C}_{\measuredangle} : \mathbb{R}^2 \to \mathbb{R}^2$, $\mathsf{C}_{\measuredangle\|} : \mathbb{R} \to \mathbb{R}^2$, $\mathsf{C}_{\mathsf{T}\measuredangle}^T : \mathrm{Sym}(\mathbb{R}^2) \to \mathbb{R}^2$, $\mathsf{C}_{\mathsf{T}\|}^T : \mathrm{Sym}(\mathbb{R}^2) \to \mathbb{R}$, $\mathsf{C}_{\measuredangle\|}^T : \mathbb{R}^2 \to \mathbb{R}$, $\mathsf{C}_{\|} : \mathbb{R} \to \mathbb{R}$. We first observe that maps $\mathsf{C}_{\mathsf{T}\|} : \mathbb{R} \to \mathrm{Sym}(\mathbb{R}^2)$, $\mathsf{C}_{\measuredangle\|} : \mathbb{R} \to \mathbb{R}^2$ and $\mathsf{C}_{\|} : \mathbb{R} \to \mathbb{R}$ can be uniquely determined by the image of $1 \in \mathbb{R}$. In this way we identify the map $\mathsf{C}_{\mathsf{T}\|}$ with the matrix $\boldsymbol{C}_{\mathsf{T}\|} \in \mathrm{Sym}(\mathbb{R}^2)$, $\mathsf{C}_{\measuredangle\|}$ with the vector $\boldsymbol{c}_{\measuredangle\|} \in \mathbb{R}^2$, and $\mathsf{C}_{\|}$ with the number $c_{\|} \in \mathbb{R}$. The symmetric operator $\mathsf{C}_{\measuredangle} : \mathbb{R}^2 \to \mathbb{R}^2$, is naturally identified with a 2×2 symmetric matrix $\boldsymbol{C}_{\measuredangle}$. The operators $\mathsf{C}_{\mathsf{T}\measuredangle} : \mathbb{R}^2 \to \mathrm{Sym}(\mathbb{R}^2)$ will be denoted by $\mathcal{C}_{\mathsf{T}\measuredangle}$ to distinguish their special nature as third order tensors. The symmetry of the full elastic tensor C as an operator on $\mathrm{Sym}(\mathbb{R}^3)$ implies that the remaining operators are identified by duality with respect to standard inner products on $\mathbb{R}$, $\mathbb{R}^2$ and $\mathrm{Sym}(\mathbb{R}^2)$:

$$\mathsf{C}_{\mathsf{T}\measuredangle}^T\boldsymbol{\varepsilon}_{\mathsf{T}} \cdot \boldsymbol{\varepsilon}_{\measuredangle} = (\mathcal{C}_{\mathsf{T}\measuredangle}\boldsymbol{\varepsilon}_{\measuredangle}, \boldsymbol{\varepsilon}_{\mathsf{T}})_{\mathsf{T}}, \qquad \left(\mathsf{C}_{\mathsf{T}\|}^T\boldsymbol{\varepsilon}_{\mathsf{T}}\right)\varepsilon_{\|} = \varepsilon_{\|}\left(\boldsymbol{C}_{\mathsf{T}\|}, \boldsymbol{\varepsilon}_{\mathsf{T}}\right)_{\mathsf{T}}, \qquad \left(\mathsf{C}_{\measuredangle\|}^T\boldsymbol{\varepsilon}_{\measuredangle}\right)\varepsilon_{\|} = \varepsilon_{\|}(\boldsymbol{c}_{\measuredangle\|} \cdot \boldsymbol{\varepsilon}_{\measuredangle}).$$

These formulas show that we can encode the dual operator $\mathsf{C}_{\measuredangle\|}^T$ by the same vector $\boldsymbol{c}_{\measuredangle\|}$ that is identified with $\mathsf{C}_{\measuredangle\|}$. Similarly, we encode $\mathsf{C}_{\mathsf{T}\|}^T$ by the same symmetric 2×2 matrix $\boldsymbol{C}_{\mathsf{T}\|}$ that is identified with $\mathsf{C}_{\mathsf{T}\|}$. Even though the operator $\mathsf{C}_{\mathsf{T}\measuredangle}^T$ is identified with the third order tensor $\mathcal{C}_{\mathsf{T}\measuredangle}$ we will denote it as $\mathcal{C}_{\measuredangle\mathsf{T}}$ for improved readability. In summary, the full elastic tensor C will be written in the block form as

$$\mathsf{C} = \begin{bmatrix} \mathsf{C}_{\mathsf{T}} & \mathcal{C}_{\mathsf{T}\measuredangle} & \boldsymbol{C}_{\mathsf{T}\|} \\ \mathcal{C}_{\measuredangle\mathsf{T}} & \boldsymbol{C}_{\measuredangle} & \boldsymbol{c}_{\measuredangle\|} \\ \boldsymbol{C}_{\mathsf{T}\|} & \boldsymbol{c}_{\measuredangle\|} & c_{\|} \end{bmatrix}, \tag{9.17}$$

and understood to describe the constitutive relation

$$\begin{cases} \boldsymbol{\sigma}_{\mathsf{T}} = \mathsf{C}_{\mathsf{T}}\boldsymbol{\varepsilon}_{\mathsf{T}} + \mathcal{C}_{\mathsf{T}\angle}\boldsymbol{\varepsilon}_{\angle} + \varepsilon_{\|}\dfrac{\mathcal{C}_{\mathsf{T}\|}}{\sqrt{2}}, \\ \boldsymbol{\sigma}_{\angle} = \mathcal{C}_{\angle\mathsf{T}}\boldsymbol{\varepsilon}_{\mathsf{T}} + \mathbf{C}_{\angle}\boldsymbol{\varepsilon}_{\angle} + \varepsilon_{\|}\dfrac{\boldsymbol{c}_{\angle\|}}{\sqrt{2}}, \\ \sigma_{\|} = \left(\dfrac{\mathcal{C}_{\mathsf{T}\|}}{\sqrt{2}}, \boldsymbol{\varepsilon}_{\mathsf{T}}\right)_F + \sqrt{2}\,\boldsymbol{c}_{\angle\|}\cdot\boldsymbol{\varepsilon}_{\angle} + c_{\|}\varepsilon_{\|}, \end{cases}$$

where $(\mathcal{C}_{\mathsf{T}\angle}\boldsymbol{\varepsilon}_{\angle}, \boldsymbol{\varepsilon}_{\mathsf{T}})_{\mathsf{T}} = (\mathcal{C}_{\angle\mathsf{T}}\boldsymbol{\varepsilon}_{\mathsf{T}})\cdot\boldsymbol{\varepsilon}_{\angle}$, and

$$\boldsymbol{\sigma} = \begin{bmatrix} \boldsymbol{\sigma}_{\mathsf{T}} & \boldsymbol{\sigma}_{\angle} \\ \boldsymbol{\sigma}_{\angle} & \sigma_{\|} \end{bmatrix}, \qquad \boldsymbol{\varepsilon} = \begin{bmatrix} \boldsymbol{\varepsilon}_{\mathsf{T}} & \boldsymbol{\varepsilon}_{\angle} \\ \boldsymbol{\varepsilon}_{\angle} & \varepsilon_{\|} \end{bmatrix}$$

is the block-matrix representation (9.16) of the stress and strain tensors. In our notation, scalars are denoted by lower case letters, $\mathbb{R}^2$ vectors by bold lower case letters, matrices by bold upper case letters, third order tensors by script upper case letters and fourth order tensors, i.e. operators on Sym($\mathbb{R}^2$), are denoted by sans serif upper case letters. The subscripts T, $\angle$, $\|$ will help the reader understand the physical meaning of the corresponding objects, denoting transversal, tilt and longitudinal components, respectively.

9.3.1 Exact relations

In the context of fibrous elastic composites there is a veritable plethora of exact relations and links. This fact made it imperative to eliminate those exact relations and links that could be derived from the remaining set. One very effective way to do this was to derive all global links first. This was done by Meredith Hegg in her PhD dissertation [7] (see also [8]). The fact that a global link would map an exact relation into an exact relation means that all exact relations split into equivalence classes of exact relations related by a global link. Only one exact relation from each equivalence class need to be listed. In general, exact relations come in infinite families, parametrized by one or more real parameters. Remarkably, exact relations in each such infinite family could be mapped into one another by global links, leaving only finitely many equivalence classes. While the number of such equivalence classes were into several hundreds, a great many of them could be obtained as intersections of other exact relations. The finiteness of the number of equivalence classes meant that the search for intersections could be automated. This was done during the summer of 2013 by a graduate student Tatyana Nuzhnaya, reducing the number of essential exact relations to a few dozen. During the 2013–14 academic year two undergraduate students Mark Mikida and Andrew Schneider computed the final form of essential exact relations. In the list below each exact relation consists of all positive definite elastic tensors of the form (9.17), where the block components have either specified form or satisfy specified constraints. For example, the first exact relation in our list is specified as $\boldsymbol{C}_{\mathsf{T}\|} = 0$, $\boldsymbol{c}_{\measuredangle\|} = 0$, $\langle c_{\|}\rangle$. That means that it consists of tensors C of the form

$$\mathsf{C} = \begin{bmatrix} \mathsf{C}_{\mathsf{T}} & \mathcal{C}_{\mathsf{T}\measuredangle} & 0 \\ \mathcal{C}_{\measuredangle\mathsf{T}} & \boldsymbol{C}_{\measuredangle} & 0 \\ 0 & 0 & c_{\|} \end{bmatrix}.$$

So that any fibrous composite made of materials of this form will have an effective tensor of the same form, where the blocks $\mathsf{C}_{\mathsf{T}}^{*}$, $\mathcal{C}_{\mathsf{T}\measuredangle}^{*}$, $\boldsymbol{C}_{\measuredangle}^{*}$, $c_{\|}^{*}$ depend on the microstructure, but in such a way that $c_{\|}^{*} = \langle c_{\|}\rangle$. We list exact relations in the order of increasing complexity.

1. $\boldsymbol{C}_{\mathsf{T}\|} = 0$, $\boldsymbol{c}_{\measuredangle\|} = 0$, $\langle c_{\|}\rangle$.
2. $\mathcal{C}_{\mathsf{T}\measuredangle} = 0$.
3. $\mathcal{C}_{\mathsf{T}\measuredangle} = 0$, $\boldsymbol{c}_{\measuredangle\|} = 0$. This exact relation says that if a fibrous composite is made of materials with monoclinic symmetry, whose two-fold axis coincides with fiber direction, then the composite will always have this monoclinic symmetry. This is the only elasticity symmetry class that is stable under homogenization in the context of fibrous composites.
4. $\mathcal{C}_{\mathsf{T}\measuredangle} = 0$, $\boldsymbol{C}_{\mathsf{T}\|} = 0$.
5. $\mathcal{C}_{\mathsf{T}\measuredangle} = 0$, $\det \boldsymbol{C}_{\measuredangle} = d_0$.
6. $\mathcal{C}_{\mathsf{T}\measuredangle} = 0$, $\boldsymbol{C}_{\measuredangle} = \sigma_0 \boldsymbol{I}_2$, $\langle \boldsymbol{c}_{\measuredangle\|}\rangle$.
7. $\mathsf{C}_{\mathsf{T}}\boldsymbol{I}_2 = 2\kappa_0 \boldsymbol{I}_2$, $\mathcal{C}_{\measuredangle\mathsf{T}}\boldsymbol{I}_2 = 0$, $\langle \mathrm{Tr}\, \boldsymbol{C}_{\mathsf{T}\|}\rangle$.
8. $\mathcal{C}_{\mathsf{T}\measuredangle} = 0$, $\mathsf{C}_{\mathsf{T}}\boldsymbol{I}_2 = 2\kappa_0 \boldsymbol{I}_2$, $\langle \mathrm{Tr}\, \boldsymbol{C}_{\mathsf{T}\|}\rangle$. Let $2\mu_1$ and $2\mu_2$ be the other two eigenvalues of C_{T} (κ_0, μ_1 and μ_2 have the physical meaning of the bulk and two shear moduli of a square symmetric two-dimensional elastic material). This exact relation is then described by the constraints above and

$$\left(\frac{1}{\kappa_0} + \frac{1}{\mu_1}\right)\left(\frac{1}{\kappa_0} + \frac{1}{\mu_2}\right) = c_0.$$

9. $\mathsf{C}_{\mathsf{T}} = 2\mu_0 \mathsf{T} + \boldsymbol{B} \otimes \boldsymbol{B}$, $\mathcal{C}_{\mathsf{T}\measuredangle} = \boldsymbol{B} \otimes \boldsymbol{b}$, $\boldsymbol{B} \in \mathrm{Sym}(\mathbb{R}^2)$, $\boldsymbol{b} \in \mathbb{R}^2$, where T is a two-dimensional isotropic elasticity tensor with bulk modulus $-1/2$ and shear modulus $1/2$, given by (9.14). The expression $\mathcal{C}_{\mathsf{T}\measuredangle} = \boldsymbol{B} \otimes \boldsymbol{b}$ means

$$\mathcal{C}_{\mathsf{T}\measuredangle}\boldsymbol{\varepsilon}_{\measuredangle} = (\boldsymbol{b} \cdot \boldsymbol{\varepsilon}_{\measuredangle})\boldsymbol{B}, \qquad \mathcal{C}_{\measuredangle\mathsf{T}}\boldsymbol{\varepsilon}_{\mathsf{T}} = (\boldsymbol{\varepsilon}_{\mathsf{T}}, \boldsymbol{B})_{\mathsf{T}}\boldsymbol{b}.$$

10. $\mathsf{C}_{\mathsf{T}} = 2\mu_0 \mathsf{T} + \boldsymbol{B} \otimes \boldsymbol{B}$, $\mathcal{C}_{\mathsf{T}\measuredangle} = \boldsymbol{B} \otimes \boldsymbol{b}$, $\boldsymbol{C}_{\mathsf{T}\|} = \beta \boldsymbol{B}$.
11. $\mathsf{C}_{\mathsf{T}} = 2\mu_0 \mathsf{T} + \boldsymbol{B} \otimes \boldsymbol{B}$, $\mathcal{C}_{\mathsf{T}\measuredangle} = \boldsymbol{B} \otimes \boldsymbol{b}$, $\boldsymbol{C}_{\mathsf{T}\|} = \beta \boldsymbol{B}$, $\boldsymbol{c}_{\measuredangle\|} = \beta \boldsymbol{b}$, $\langle c_{\|} - \beta^2\rangle$.
12. $\mathsf{C}_{\mathsf{T}} = 2\mu_0 \mathsf{T} + \boldsymbol{B} \otimes \boldsymbol{B}$, $\mathcal{C}_{\mathsf{T}\measuredangle} = \boldsymbol{B} \otimes \boldsymbol{b}$, $\boldsymbol{C}_{\measuredangle} = \boldsymbol{\sigma} + \boldsymbol{b} \otimes \boldsymbol{b}$, $\det \boldsymbol{\sigma} = d_0$.
13. $\mathsf{C}_{\mathsf{T}} = 2\mu_0 \mathsf{T} + \boldsymbol{B} \otimes \boldsymbol{B}$, $\mathcal{C}_{\mathsf{T}\measuredangle} = \boldsymbol{B} \otimes \boldsymbol{b}$, $\boldsymbol{C}_{\measuredangle} = \sigma_0 \boldsymbol{I}_2 + \boldsymbol{b} \otimes \boldsymbol{b}$.
14. $\mathsf{C}_{\mathsf{T}} = 2\mu_0 \mathsf{T} + \boldsymbol{B} \otimes \boldsymbol{B}$, $\mathcal{C}_{\mathsf{T}\measuredangle} = \boldsymbol{B} \otimes \boldsymbol{b}$, $\boldsymbol{C}_{\mathsf{T}\|} = \beta \boldsymbol{B}$, $\boldsymbol{C}_{\measuredangle} = \sigma_0 \boldsymbol{I}_2 + \boldsymbol{b} \otimes \boldsymbol{b}$, $\langle \boldsymbol{c}_{\measuredangle\|} - \beta \boldsymbol{b}\rangle$.

15. $\mathsf{C}_{\mathsf{T}} = 2\mu_0 \mathsf{I} + 2\lambda \boldsymbol{I}_2 \otimes \boldsymbol{I}_2$, $\mathcal{C}_{\mathsf{T}\measuredangle} = \boldsymbol{I}_2 \otimes \boldsymbol{b}$, $\langle \mathrm{dev}(\boldsymbol{C}_{\mathsf{T}\|}) \rangle$, where $\mathrm{dev}(\cdot)$ denotes the trance-free part of a matrix, and was defined in (9.15).
16. $\mathsf{C}_{\mathsf{T}} = 2\mu_0 \mathsf{I} + 2\lambda \boldsymbol{I}_2 \otimes \boldsymbol{I}_2$, $\mathcal{C}_{\mathsf{T}\measuredangle} = 0$, $\langle \frac{1}{2\mu_0 + \lambda} \rangle$, $\langle \mathrm{dev}(\boldsymbol{C}_{\mathsf{T}\|}) \rangle$.
17. $\mathsf{C}_{\mathsf{T}} = 2\mu_0 \mathsf{I} + 2\lambda \boldsymbol{I}_2 \otimes \boldsymbol{I}_2$, $\mathcal{C}_{\mathsf{T}\measuredangle} = 0$, $\boldsymbol{C}_{\mathsf{T}\|} = \beta \boldsymbol{I}_2$, $\langle \frac{1}{2\mu_0 + \lambda} \rangle$, $\langle \frac{\beta}{2\mu_0 + \lambda} \rangle$.
18. $\mathsf{C}_{\mathsf{T}} = 2\mu_0 \mathsf{I} + 2\lambda \boldsymbol{I}_2 \otimes \boldsymbol{I}_2$, $\mathcal{C}_{\mathsf{T}\measuredangle} = \boldsymbol{I}_2 \otimes \boldsymbol{b}$, $\boldsymbol{C}_{\measuredangle} = \sigma_0 \boldsymbol{I}_2 + \frac{\boldsymbol{b} \otimes \boldsymbol{b}}{2(\lambda + 2\mu_0)}$, $\langle \frac{\boldsymbol{b}}{\lambda + 2\mu_0} \rangle$, $\langle \mathrm{dev}(\boldsymbol{C}_{\mathsf{T}\|}) \rangle$.
19. $\mathsf{C}_{\mathsf{T}} = 2\mu_0 \mathsf{I} + 2\lambda \boldsymbol{I}_2 \otimes \boldsymbol{I}_2$, $\mathcal{C}_{\mathsf{T}\measuredangle} = \boldsymbol{I}_2 \otimes \boldsymbol{b}$, $\boldsymbol{C}_{\mathsf{T}\|} = \beta \boldsymbol{I}_2$, $\boldsymbol{c}_{\measuredangle\|} = \frac{\beta \boldsymbol{b}}{2(\lambda + 2\mu_0)}$, $\langle \frac{\beta}{\lambda + 2\mu_0} \rangle$, $\langle c_{\|} - \frac{\beta^2}{2(\lambda + 2\mu_0)} \rangle$.
20. $\mathsf{C}_{\mathsf{T}} = 2\mu_0 \mathsf{I} + 2\lambda \boldsymbol{I}_2 \otimes \boldsymbol{I}_2$, $\mathcal{C}_{\mathsf{T}\measuredangle} = 0$, $\boldsymbol{C}_{\mathsf{T}\|} = \beta \boldsymbol{I}_2$, $\boldsymbol{c}_{\measuredangle\|} = 0$, $c_{\|} = c_0 + \frac{\beta^2 + \alpha_0 \beta + \beta_0}{2(\lambda + 2\mu_0)}$ $\langle \frac{1}{2\mu_0 + \lambda} \rangle$, $\langle \frac{\beta}{\lambda + 2\mu_0} \rangle$.
21. $\mathsf{C}_{\mathsf{T}} = 2\mu_0 \mathsf{I} + 2\lambda \boldsymbol{I}_2 \otimes \boldsymbol{I}_2$, $\mathcal{C}_{\mathsf{T}\measuredangle} = \boldsymbol{I}_2 \otimes \boldsymbol{b}$, $\boldsymbol{C}_{\mathsf{T}\|} = \beta \boldsymbol{I}_2$, $\boldsymbol{C}_{\measuredangle} = \sigma_0 \boldsymbol{I}_2 + \frac{\boldsymbol{b} \otimes \boldsymbol{b}}{2(\lambda + 2\mu_0)}$, $\boldsymbol{c}_{\measuredangle\|} = \frac{\beta_0 \boldsymbol{b}^{\perp} + \beta \boldsymbol{b}}{2(\lambda + 2\mu_0)}$, $\langle \frac{\boldsymbol{b}}{\lambda + 2\mu_0} \rangle$, $\langle c_{\|} - \frac{\beta_0^2 + \beta^2}{2(\lambda + 2\mu_0)} - \frac{4\mu_0^2 \, |\boldsymbol{b}|^2}{\sigma_0 (\lambda + 2\mu_0)^2} \rangle$, where $\boldsymbol{b}^{\perp} = \boldsymbol{R}_{\perp} \boldsymbol{b}$.
22. $\mathsf{C}_{\mathsf{T}} = 2\mu_0 \mathsf{I} + 2\lambda \boldsymbol{I}_2 \otimes \boldsymbol{I}_2$, $\mathcal{C}_{\mathsf{T}\measuredangle} = \boldsymbol{I}_2 \otimes \boldsymbol{b}$, $\boldsymbol{C}_{\mathsf{T}\|} = \beta \boldsymbol{I}_2$, $\boldsymbol{c}_{\measuredangle\|} = \frac{\beta \boldsymbol{b}}{2(\lambda + 2\mu_0)}$, $c_{\|} = c_0 + \frac{\beta(\beta + \beta_0)}{2(\lambda + 2\mu_0)}$, $\langle \frac{\beta}{\lambda + 2\mu_0} \rangle$.
23. An elasticity tensor C belongs to this exact relations if and only if its blocks satisfy the following constraints.
    (a) $\mathsf{C}_{\mathsf{T}} \boldsymbol{I}_2 = 2\kappa_0 \boldsymbol{I}_2$, $\mathcal{C}_{\measuredangle\mathsf{T}} \boldsymbol{I}_2 = 0$, where $\kappa_0 > 0$ is an arbitrary constant.
    (b) Choose arbitrary constants $\sigma_0 > 0$ and $0 < \tau_0 < \kappa_0$, and define Schur complements S_{T}, $\boldsymbol{S}_{\measuredangle}$ and tensor $\mathcal{S}_{\mathsf{T}\measuredangle}$ by the formula

$$\begin{bmatrix} \mathsf{C}_{\mathsf{T}} + \tau_0 \mathsf{I} & \mathcal{C}_{\mathsf{T}\measuredangle} \\ \mathcal{C}_{\measuredangle\mathsf{T}} & \boldsymbol{C}_{\measuredangle} + \sigma_0 \boldsymbol{I}_2 \end{bmatrix}^{-1} = \begin{bmatrix} (\mathsf{S}_{\mathsf{T}} + \tau_0 \mathsf{I})^{-1} & \mathcal{S}_{\mathsf{T}\measuredangle} \\ \mathcal{S}_{\measuredangle\mathsf{T}} & (\boldsymbol{S}_{\measuredangle} + \sigma_0 \boldsymbol{I}_2)^{-1} \end{bmatrix}. \tag{9.18}$$

    (c) $\mathcal{S}_{\mathsf{T}\measuredangle} : \mathbb{R}^2 \to \mathrm{Sym}(\mathbb{R}^2)$ is given by its action on an arbitrary vector $\boldsymbol{u} \in \mathbb{R}^2$:

$$\mathcal{S}_{\mathsf{T}\measuredangle} \boldsymbol{u} = \psi(cu), \quad \text{or} \quad \mathcal{S}_{\mathsf{T}\measuredangle} \boldsymbol{u} = \psi(c\bar{u}), \tag{9.19}$$

    where $u \in \mathbb{C}$ is a complex number represented by the vector $\boldsymbol{u}$, $c \in \mathbb{C}$, and the map $\psi : \mathbb{C} \to \mathrm{Sym}(\mathbb{R}^2)$ is defined by

$$\psi(a_1 + \mathrm{i}a_2) = \begin{bmatrix} a_1 & a_2 \\ a_2 & -a_1 \end{bmatrix}. \tag{9.20}$$

    (d) We require that $\det(\boldsymbol{S}_{\measuredangle}) = \sigma_0^2$.
    (e) Let μ_1 and μ_2 be the shear moduli of S_{T} (its bulk modulus is κ_0). We require that

$$\left(\frac{1}{\kappa_0}+\frac{1}{\mu_1}\right)\left(\frac{1}{\kappa_0}+\frac{1}{\mu_2}\right)=\frac{1}{\rho_0^2}, \qquad \tau_0=\left(\frac{\kappa_0^{-1}+\rho_0^{-1}}{2}\right)^{-1}.$$

(f) $\mathrm{Tr}\, \boldsymbol{C}^*_{\mathsf{T}\|} = \mathrm{Tr}\, \langle \boldsymbol{C}_{\mathsf{T}\|} \rangle$.

In order to explain this exact relation we describe how to construct a tensor that satisfies all the requirements. First, we choose any positive definite two-dimensional elastic tensor S_T with square symmetry, i.e. $\mathsf{S}_\mathsf{T}\boldsymbol{I}_2 = 2\kappa_0\boldsymbol{I}_2$. Then, we choose a complex number c and construct an operator $\mathcal{S}_{\mathsf{T}\times}$: $\mathbb{R}^2 \to \mathrm{Sym}(\mathbb{R}^2)$ given by one of the two forms in (9.19). Finally, we choose two operators on $\mathbb{R}^2$, denoted as $\boldsymbol{S}_\times > 0$ and $\boldsymbol{C}_{\mathsf{T}\|}$, a vector $\boldsymbol{c}_{\times\|} \in \mathbb{R}^2$, and a scalar $c_\| > 0$ arbitrarily. Once these choices have been made, we compute

$$\sigma_0=\sqrt{\det \boldsymbol{S}_\times}, \qquad \rho_0=\frac{1}{\sqrt{\left(\kappa_0^{-1}+\mu_1^{-1}\right)\left(\kappa_0^{-1}+\mu_2^{-1}\right)}}, \qquad \tau_0=\left(\frac{\kappa_0^{-1}+\rho_0^{-1}}{2}\right)^{-1},$$

where $2\mu_1$ and $2\mu_2$ are the other two eigenvalues of the operator S_T. We define operators C_T, $\mathcal{C}_{\mathsf{T}\times}$ and $\boldsymbol{C}_\times$ by (9.18). Explicitly,

$$\tilde{\mathsf{C}}_\mathsf{T}=\left(\tilde{\mathsf{S}}_\mathsf{T}^{-1}-\mathcal{S}_{\mathsf{T}\times}\tilde{\boldsymbol{S}}_\times\mathcal{S}_{\times\mathsf{T}}\right)^{-1}, \quad \boldsymbol{C}_\times=\left(\tilde{\boldsymbol{S}}_\times^{-1}-\mathcal{S}_{\times\mathsf{T}}\tilde{\mathsf{S}}_\mathsf{T}\mathcal{S}_{\mathsf{T}\times}\right)^{-1}, \quad \mathcal{C}_{\mathsf{T}\times}=-\tilde{\mathsf{S}}_\mathsf{T}\mathcal{S}_{\mathsf{T}\times}\boldsymbol{C}_\times,$$

where $\tilde{\mathsf{C}}_\mathsf{T} = \mathsf{C}_\mathsf{T} + \tau_0\mathsf{I}$, $\tilde{\mathsf{S}}_\mathsf{T} = \mathsf{S}_\mathsf{T} + \tau_0\mathsf{I}$, and $\tilde{\boldsymbol{S}}_\times = \boldsymbol{S}_\times + \sigma_0\boldsymbol{I}_2$. If we use such a material to make a transversely isotropic polycrystal

$$\mathsf{C}^*=\begin{bmatrix} 2\mu^*(\mathsf{I}-\boldsymbol{I}_2\otimes\boldsymbol{I}_2)+2\kappa^*\boldsymbol{I}_2\otimes\boldsymbol{I}_2 & 0 & m^*\boldsymbol{I}_2 \\ 0 & \sigma^*\boldsymbol{I}_2 & 0 \\ m^*\boldsymbol{I}_2 & 0 & c_\|^* \end{bmatrix},$$

four out of its five parameters are texture-independent:

$$\kappa^*=\kappa_0, \quad \sigma^*=\sqrt{\det \boldsymbol{S}_\times}, \quad 2m^*=\mathrm{Tr}\left\langle \boldsymbol{C}_{\mathsf{T}\|}\right\rangle,$$

$$\frac{1}{\mu^*}=\sqrt{\left(\frac{1}{\kappa_0}+\frac{1}{\mu_1}\right)\left(\frac{1}{\kappa_0}+\frac{1}{\mu_2}\right)}-\frac{1}{\kappa_0}.$$

The fifth parameter $c_\|^*$ is texture-dependent.

9.3.2 Links

All links (or rather families of links) in the context of fibrous elastic composites could be identified using Nuzhnaya's MATLAB® code. Their number was staggering: more than 500. However, eliminating logically redundant families of links was a far more difficult task. It became possible only due to the joint work of graduate student Adam Jacoby and the author during the summer of 2014, whose Maple™ code reduced the number of non-redundant links to several dozen. The computation of links in its present form was started by the undergraduate student Patrick Wynne, working with the author over the 2014–15 academic year, followed by the joint efforts of three undergraduates Abraham Lyle, Hansen Pei and Patrick Wynne, over the summer of 2015, followed by the work of Hansen Pei during the fall of 2015.

The links in the list below are grouped into categories determined by the exact relations on which they are defined. We list them in the order of increasing complexity.

1. Global links
   (a) $\mathsf{C}_{\mathsf{T}}^{*}$, $\mathcal{C}_{\mathsf{T}\measuredangle}^{*}$, $\boldsymbol{C}_{\measuredangle}^{*}$, are independent of the remaining blocks $\boldsymbol{C}_{\mathsf{T}\|}$, $\boldsymbol{c}_{\measuredangle\|}$, $c_{\|}$.
   (b) $\mathsf{C}_{\mathsf{T}}^{*}$, $\mathcal{C}_{\mathsf{T}\measuredangle}^{*}$, $\boldsymbol{C}_{\measuredangle}^{*}$, $\boldsymbol{C}_{\mathsf{T}\|}^{*}$, $\boldsymbol{c}_{\measuredangle\|}^{*}$, is independent of $c_{\|}$.
   (c) The derivation of this family of global links constitutes an essential part of Meredith Hegg's PhD dissertation [7]. It can be represented as a composition of a six-parameter family of affine global links,

$$\mathsf{C}' = \mathfrak{F}_{\mathsf{Q},\mathsf{F}}(\mathsf{C}) = \mathsf{QCQ}^{T} + \mathsf{F}, \tag{9.21}$$

where

$$\mathsf{Q} = \begin{bmatrix} q_1\mathsf{I} & 0 & 0 \\ 0 & q_2\boldsymbol{I}_2 & 0 \\ q_4\boldsymbol{I}_2 & 0 & q_3 \end{bmatrix}, \qquad \mathsf{F} = \begin{bmatrix} 0 & 0 & f_1\boldsymbol{I}_2 \\ 0 & 0 & 0 \\ f_1\boldsymbol{I}_2 & 0 & f_2 \end{bmatrix},$$

and a one-parameter family of nonlinear global links

$$\mathsf{C}' = \Lambda_a(\mathsf{C}) = \mathsf{C} - \mathsf{C}\begin{bmatrix} \Theta_a(\mathsf{C}_{\mathsf{T}}) & 0 & 0 \\ 0 & 0 & 0 \\ 0 & 0 & 0 \end{bmatrix}\mathsf{C}, \tag{9.22}$$

where

$$\Theta_a(\mathsf{C}_{\mathsf{T}}) = \left(\frac{1}{a}\mathsf{T} + \mathsf{C}_{\mathsf{T}}\right)^{-1}.$$

Transformations $\mathfrak{F}_{\mathsf{Q},\mathsf{F}}$ and Λ_a generate a group of global links and have the following composition properties:

$$\Lambda_a \circ \Lambda_b = \Lambda_{a+b}, \qquad \lim_{a\to 0}\Lambda_a(\mathsf{C}) = \mathsf{C}, \qquad \Lambda_a \circ \mathfrak{F}_{\mathsf{Q},\mathsf{F}} = \mathfrak{F}_{\mathsf{Q}',\mathsf{F}'} \circ \Lambda_{a'}$$

where $a' = aq_1^2$ and

$$\mathsf{Q}' = \begin{bmatrix} q_1 & 0 & 0 \\ 0 & q_2 & 0 \\ af_1q_1 + q_4 & 0 & q_3 \end{bmatrix}, \quad \mathsf{F}' = \begin{bmatrix} 0 & 0 & f_1\boldsymbol{I}_2 \\ 0 & 0 & 0 \\ f_1\boldsymbol{I}_2 & 0 & f_2 + af_1^2. \end{bmatrix}$$

Explicitly we have

$$\mathfrak{F}_{\mathsf{Q},\,\mathsf{F}}(\mathsf{C}) = \begin{bmatrix} q_1^2\mathsf{C}_{\mathsf{T}} & q_1q_2\mathcal{C}_{\mathsf{T}\curlywedge} & q_1q_3\boldsymbol{C}_{\mathsf{T}\|} + q_1q_4\mathsf{C}_{\mathsf{T}}\boldsymbol{I}_2 + f_1\boldsymbol{I}_2 \\ * & q_2^2\boldsymbol{C}_{\curlywedge} & q_2q_3\boldsymbol{c}_{\curlywedge\|} + q_2q_4\mathcal{C}_{\curlywedge\mathsf{T}}\boldsymbol{I}_2 \\ * & * & q_3^2c_{\|} + 2q_3q_4(\boldsymbol{C}_{\mathsf{T}\|}, \boldsymbol{I}_2)_{\mathsf{T}} + q_4^2(\mathsf{C}_{\mathsf{T}}\boldsymbol{I}_2, \boldsymbol{I}_2)_{\mathsf{T}} + f_2 \end{bmatrix}$$

and

$$\Lambda_a(\mathsf{C}) = \begin{bmatrix} \frac{1}{a}\mathsf{T}\Theta_a(\mathsf{C}_{\mathsf{T}})\mathsf{C}_{\mathsf{T}} & \frac{1}{a}\mathsf{T}\Theta_a(\mathsf{C}_{\mathsf{T}})\mathcal{C}_{\mathsf{T}\curlywedge} & \frac{1}{a}\mathsf{T}\Theta_a(\mathsf{C}_{\mathsf{T}})\boldsymbol{C}_{\mathsf{T}\|} \\ * & \boldsymbol{C}_{\curlywedge} - \mathcal{C}_{\curlywedge\mathsf{T}}\Theta_a(\mathsf{C}_{\mathsf{T}})\mathcal{C}_{\mathsf{T}\curlywedge} & \boldsymbol{c}_{\curlywedge\|} - \mathcal{C}_{\curlywedge\mathsf{T}}\Theta_a(\mathsf{C}_{\mathsf{T}})\boldsymbol{C}_{\mathsf{T}\|} \\ * & * & c_{\|} - (\Theta_a(\mathsf{C}_{\mathsf{T}})\boldsymbol{C}_{\mathsf{T}\|}, \boldsymbol{C}_{\mathsf{T}\|})_{\mathsf{T}} \end{bmatrix}.$$

A particular instance of this link (but in a more general context of fibrous piezoelectric composites) was discovered by Benveniste [2].

2. $\mathsf{C} = \begin{bmatrix} \mathsf{C}_{\mathsf{T}} & 0 & \mathcal{C}_{\mathsf{T}\|} \\ 0 & \boldsymbol{C}_{\curlywedge} & \boldsymbol{c}_{\curlywedge\|} \\ \mathcal{C}_{\mathsf{T}\|} & \boldsymbol{c}_{\curlywedge\|} & c_{\|} \end{bmatrix}.$

   (a) $\mathsf{C}_{\mathsf{T}}^*$ depends only on C_{T}.

   (b) $\boldsymbol{C}_{\curlywedge}^*$ depends only on $\boldsymbol{C}_{\curlywedge}$.

   (c) $\boldsymbol{c}_{\curlywedge\|}^*$ depends only on $\boldsymbol{C}_{\mathsf{T}\|}$ and $\boldsymbol{c}_{\curlywedge\|}$.

   (d) $\boldsymbol{C}_{\mathsf{T}\|}^*$ depends only on $\boldsymbol{C}_{\mathsf{T}\|}$ and C_{T}.

   (e) $\mathsf{C}_{\mathsf{T}}^*$, $\boldsymbol{C}_{\curlywedge}^*$ and $\boldsymbol{C}_{\mathsf{T}\|}^*$ are independent of $\boldsymbol{c}_{\curlywedge\|}$ and $c_{\|}$.

3. $\mathsf{C} = \begin{bmatrix} \mathsf{C}_{\mathsf{T}} & 0 & 0 \\ 0 & \boldsymbol{C}_{\curlywedge} & \boldsymbol{c}_{\curlywedge\|} \\ 0 & \boldsymbol{c}_{\curlywedge\|} & c_{\|} \end{bmatrix}.$

   (a) $c_{\|}^*$ is independent of C_{T}.

   (b) $\mathsf{C}' = \begin{bmatrix} \mathsf{C}_{\mathsf{T}} & 0 & 0 \\ 0 & \dfrac{\boldsymbol{C}_{\curlywedge}}{\det \boldsymbol{C}_{\curlywedge}} & \dfrac{\boldsymbol{C}_{\curlywedge}\boldsymbol{c}_{\curlywedge\|}^{\perp}}{\det \boldsymbol{C}_{\curlywedge}} \\ 0 & \dfrac{\boldsymbol{C}_{\curlywedge}\boldsymbol{c}_{\curlywedge\|}^{\perp}}{\det \boldsymbol{C}_{\curlywedge}} & \boldsymbol{C}_{\curlywedge}^{-1}\boldsymbol{c}_{\curlywedge\|} \cdot \boldsymbol{c}_{\curlywedge\|} - c_{\|} + c_0. \end{bmatrix}.$

4. $\mathsf{C} = \begin{bmatrix} \mathsf{C}_{\mathsf{T}} & 0 & C_{\mathsf{T}\parallel} \\ 0 & C_{\times} & 0 \\ C_{\mathsf{T}\parallel} & 0 & c_{\parallel} \end{bmatrix}$.

(a) $c_{\parallel}^{*}$ is independent of $C_{\times}$.

(b) $\mathsf{C}' = \begin{bmatrix} \mathsf{C}_{\mathsf{T}} & 0 & C_{\mathsf{T}\parallel} \\ 0 & \dfrac{C_{\times}}{\det C_{\times}} & 0 \\ C_{\mathsf{T}\parallel} & 0 & c_{\parallel} \end{bmatrix}$.

5. $\mathsf{C} = \begin{bmatrix} \mathsf{C}_{\mathsf{T}} & 0 & C_{\mathsf{T}\parallel} \\ 0 & C_{\times} & c_{\times\parallel} \\ C_{\mathsf{T}\parallel} & c_{\times\parallel} & c_{\parallel} \end{bmatrix}$, $\quad \mathsf{C}_{\mathsf{T}} I_2 = 2\kappa_0 I_2, \quad \mathrm{Tr}\, C_{\mathsf{T}\parallel} = t_0.$

The blocks in C' are constructed according to the following rules. $\mathsf{C}'_{\mathsf{T}} I_2 = 2\kappa_0 I_2$. Let $\mu_0 \in (0, \kappa_0)$ be a constant. Then the two shear moduli μ_1', μ_2' of C'_{T} are defined by

$$\left(\frac{1}{\mu_1'} + \frac{1}{\kappa_0}\right)\left(\frac{1}{\mu_2} + \frac{1}{\kappa_0}\right) = \left(\frac{1}{\mu_2'} + \frac{1}{\kappa_0}\right)\left(\frac{1}{\mu_1} + \frac{1}{\kappa_0}\right) = \frac{1}{\kappa_0}\left(\frac{1}{\kappa_0} + \frac{1}{\mu_0}\right) = \frac{1}{a_0^2}.$$

We also have $\mathrm{Tr}\, C'_{\mathsf{T}\parallel} = t_0$ and

$$\mathrm{dev}\left(C'_{\mathsf{T}\parallel}\right) = \frac{2\kappa_0\mu_0}{a_0} R_{\perp}(\mathsf{C}_{\mathsf{T}} - 2\mu_0 \mathsf{I})^{-1}\mathrm{dev}\left(C_{\mathsf{T}\parallel}\right),$$

where $\mathrm{dev}(\cdot)$ was defined in (9.15),

$$C'_{\times} = \sigma_0^2 \frac{C_{\times}}{\det C_{\times}}, \qquad c'_{\times\parallel} = \sigma_0 \frac{C_{\times}}{\det C_{\times}} c^{\perp}_{\times\parallel},$$

$$c'_{\parallel} = b_0 - c_{\parallel} + C_{\times}^{-1} c_{\times\parallel} \cdot c_{\times\parallel} + \left((\mathsf{C}_{\mathsf{T}} - 2a_0\mathsf{I})^{-1} C_{\mathsf{T}\parallel}, C_{\mathsf{T}\parallel}\right)_{\mathsf{T}}.$$

6. $\mathsf{C} = \begin{bmatrix} 2\mu_0\mathsf{T} + B \otimes B & B \otimes b & C_{\mathsf{T}\parallel} \\ b \otimes B & C_{\times} & c_{\times\parallel} \\ C_{\mathsf{T}\parallel} & c_{\times\parallel} & c_{\parallel} \end{bmatrix}$.

(a) $C'_{\times} - b' \otimes b' = C_{\times} - b \otimes b$.

(b) $B' = \Lambda B$, where $\Lambda = \Lambda(x)$ is a scalar.

(c) $B' = \Lambda B$, $C'_{\times} - b' \otimes b' = C_{\times} - b \otimes b$, $C'_{\mathsf{T}\parallel} = C_{\mathsf{T}\parallel} + \nu B$, where $\nu = \nu(x)$ is a scalar.

(d) $B' = \Lambda B$, $C'_{\times} - b' \otimes b' = C_{\times} - b \otimes b$, $C'_{\mathsf{T}\parallel} = C_{\mathsf{T}\parallel} + \nu B$, $c'_{\times\parallel} = c_{\times\parallel} + \nu b$.

7. $\mathsf{C} = \begin{bmatrix} 2\mu_0 \mathsf{T} + \boldsymbol{B} \otimes \boldsymbol{B} & \boldsymbol{B} \otimes \boldsymbol{b} & \beta \boldsymbol{B} \\ \boldsymbol{b} \otimes \boldsymbol{B} & \boldsymbol{C}_{\measuredangle} & \boldsymbol{c}_{\measuredangle\|} \\ \beta \boldsymbol{B} & \boldsymbol{c}_{\measuredangle\|} & c_{\|}. \end{bmatrix}.$

(a) $\boldsymbol{C}'_{\measuredangle} - \boldsymbol{b}' \otimes \boldsymbol{b}' = \boldsymbol{C}_{\measuredangle} - \boldsymbol{b} \otimes \boldsymbol{b}$, $\boldsymbol{c}'_{\measuredangle\|} - \beta' \boldsymbol{b}' = \boldsymbol{c}_{\measuredangle\|} - \beta \boldsymbol{b}$, $\boldsymbol{B}' = \Lambda \boldsymbol{B}$, $\beta' = \Lambda \beta$.

(b) $\boldsymbol{C}'_{\measuredangle} - \boldsymbol{b}' \otimes \boldsymbol{b}' = \boldsymbol{C}_{\measuredangle} - \boldsymbol{b} \otimes \boldsymbol{b}$, $\boldsymbol{c}'_{\measuredangle\|} - \beta' \boldsymbol{b}' = \boldsymbol{c}_{\measuredangle\|} - \beta \boldsymbol{b}$, $c'_{\|} - (\beta')^2 = c_{\|} - \beta^2$.

8. $\mathsf{C} = \begin{bmatrix} 2\mu_0 \mathsf{T} + \boldsymbol{B} \otimes \boldsymbol{B} & \boldsymbol{B} \otimes \boldsymbol{b} & \beta \boldsymbol{B} \\ \boldsymbol{b} \otimes \boldsymbol{B} & \boldsymbol{C}_{\measuredangle} & \beta \boldsymbol{b} \\ \beta \boldsymbol{B} & \beta \boldsymbol{b} & c_{\|} \end{bmatrix}.$

(a) $\mathsf{C}' = \begin{bmatrix} 2\mu_0 \mathsf{T} + \boldsymbol{B} \otimes \boldsymbol{B} & \boldsymbol{B} \otimes \boldsymbol{b} & \beta \boldsymbol{B} \\ \boldsymbol{b} \otimes \boldsymbol{B} & \boldsymbol{C}_{\measuredangle} & \beta \boldsymbol{b} \\ \beta \boldsymbol{B} & \beta \boldsymbol{b} & a_0 c_{\|} + (1 - a_0)\beta^2 \end{bmatrix}.$

(b) $\mathsf{C}' = \begin{bmatrix} 2\mu_0 \mathsf{T} + \Lambda \boldsymbol{B} \otimes \boldsymbol{B} & \boldsymbol{B} \otimes \boldsymbol{b}' & \Lambda \beta \boldsymbol{B} \\ \boldsymbol{b}' \otimes \boldsymbol{B} & \boldsymbol{C}'_{\measuredangle} & \beta \boldsymbol{b}' \\ \Lambda \beta \boldsymbol{B} & \beta \boldsymbol{b}' & c'_{\|} \end{bmatrix}.$

9. $\mathsf{C} = \begin{bmatrix} 2\mu_0 \mathsf{T} + \boldsymbol{B} \otimes \boldsymbol{B} & 0 & \beta \boldsymbol{B} \\ 0 & \boldsymbol{C}_{\measuredangle} & \boldsymbol{c}_{\measuredangle\|} \\ \beta \boldsymbol{B} & \boldsymbol{c}_{\measuredangle\|} & c_{\|} \end{bmatrix}.$

(a) $\mathsf{C}' = \begin{bmatrix} 2\mu_0 \mathsf{T} + \boldsymbol{B} \otimes \boldsymbol{B} & 0 & c_0 \beta \boldsymbol{B} \\ 0 & \boldsymbol{C}_{\measuredangle} & \boldsymbol{c}_{\measuredangle\|} \\ c_0 \beta \boldsymbol{B} & \boldsymbol{c}_{\measuredangle\|} & c_{\|} + (c_0^2 - 1)\beta^2 \end{bmatrix}.$

(b) $\mathsf{C}' = \begin{bmatrix} 2\mu_0 \mathsf{T} + \boldsymbol{B} \otimes \boldsymbol{B} & 0 & \beta \boldsymbol{B} \\ 0 & \dfrac{\boldsymbol{C}_{\measuredangle}}{\det \boldsymbol{C}_{\measuredangle}} & \dfrac{\boldsymbol{C}_{\measuredangle} \boldsymbol{c}^{\perp}_{\measuredangle\|}}{\det \boldsymbol{C}_{\measuredangle}} \\ \beta \boldsymbol{B} & \dfrac{\boldsymbol{C}_{\measuredangle} \boldsymbol{c}^{\perp}_{\measuredangle\|}}{\det \boldsymbol{C}_{\measuredangle}} & \boldsymbol{C}^{-1}_{\measuredangle} \boldsymbol{c}_{\measuredangle\|} \cdot \boldsymbol{c}_{\measuredangle\|} + 2\beta^2 - c_{\|} + c_0 \end{bmatrix}.$

10. $\mathsf{C} = \begin{bmatrix} 2\lambda \boldsymbol{I}_2 \otimes \boldsymbol{I}_2 + 2\mu_0 \mathsf{I} & 0 & \beta \boldsymbol{I}_2 \\ 0 & \boldsymbol{C}_{\measuredangle} & \boldsymbol{c}_{\measuredangle\|} \\ \beta \boldsymbol{I}_2 & \boldsymbol{c}_{\measuredangle\|} & c_{\|} \end{bmatrix},$

$$\frac{\lambda'}{\lambda' + 2\mu_0} = a_0 \frac{\lambda}{\lambda + 2\mu_0} + b_0 \frac{\beta}{\lambda + 2\mu_0}, \quad \frac{\beta'}{\lambda' + 2\mu_0} = \frac{\beta}{\lambda + 2\mu_0},$$

$$c_{\|}' - \frac{(\beta')^2}{2(\lambda' + 2\mu_0)} = c_{\|} - \frac{\beta^2}{2(\lambda + 2\mu_0)} + \frac{c_0}{\lambda + 2\mu_0}.$$

11. $$\mathsf{C} = \begin{bmatrix} 2\lambda \boldsymbol{I}_2 \otimes \boldsymbol{I}_2 + 2\mu_0 \mathsf{I} & 0 & \boldsymbol{C}_{\top\|} \\ 0 & \boldsymbol{C}_{\angle} & \boldsymbol{c}_{\angle\|} \\ \boldsymbol{C}_{\top\|} & \boldsymbol{c}_{\angle\|} & c_{\|} \end{bmatrix}, \qquad c_{\|}' = c_{\|} + \frac{a_0}{\lambda + 2\mu_0}.$$

12. $$\mathsf{C} = \begin{bmatrix} 2\lambda \boldsymbol{I}_2 \otimes \boldsymbol{I}_2 + 2\mu_0 \mathsf{I} & \boldsymbol{I}_2 \otimes \boldsymbol{b} & \beta \boldsymbol{I}_2 \\ \boldsymbol{b} \otimes \boldsymbol{I}_2 & \boldsymbol{C}_{\angle} & \boldsymbol{c}_{\angle\|} \\ \beta \boldsymbol{I}_2 & \boldsymbol{c}_{\angle\|} & c_{\|} \end{bmatrix}.$$

(a) $$\frac{1}{\lambda' + 2\mu_0} = \frac{1}{\lambda + 2\mu_0} + b_0, \qquad \frac{\boldsymbol{b}'}{\lambda' + 2\mu_0} = \frac{\boldsymbol{b}}{\lambda + 2\mu_0},$$

$$\frac{\beta'}{\lambda' + 2\mu_0} = \frac{\beta}{\lambda + 2\mu_0}, \qquad \boldsymbol{C}_{\angle}' - \frac{\boldsymbol{b}' \otimes \boldsymbol{b}'}{2(\lambda' + 2\mu_0)} = \boldsymbol{C}_{\angle} - \frac{\boldsymbol{b} \otimes \boldsymbol{b}}{2(\lambda + 2\mu_0)},$$

$$\boldsymbol{c}_{\angle\|}' - \frac{\beta' \boldsymbol{b}'}{2(\lambda' + 2\mu_0)} = \boldsymbol{c}_{\angle\|} - \frac{\beta \boldsymbol{b}}{2(\lambda + 2\mu_0)}, \qquad c_{\|}' - \frac{(\beta')^2}{2(\lambda' + 2\mu_0)} = c_{\|} - \frac{\beta^2}{2(\lambda + 2\mu_0)} + c_0.$$

(b) $$\frac{2\mu_0^2}{\lambda' + 2\mu_0} = \frac{\beta^2}{2(\lambda + 2\mu_0)} - c_{\|} + b_0, \qquad \frac{\mu_0 \boldsymbol{b}'}{\lambda' + 2\mu_0} = \boldsymbol{c}_{\angle\|} - \frac{\beta \boldsymbol{b}}{2(\lambda + 2\mu_0)},$$

$$\boldsymbol{c}_{\angle\|}' - \frac{\beta' \boldsymbol{b}'}{2(\lambda' + 2\mu_0)} = a_0 \left(\boldsymbol{c}_{\angle\|} - \frac{(\beta - \beta_0) \boldsymbol{b}}{2(\lambda + 2\mu_0)} \right), \qquad \frac{\mu_0 \beta'}{\lambda' + 2\mu_0} = a_0 \left(c_{\|} - \frac{\beta(\beta - \beta_0)}{2(\lambda + 2\mu_0)} \right),$$

$$\boldsymbol{C}_{\angle}' - \frac{\boldsymbol{b}' \otimes \boldsymbol{b}'}{2(\lambda' + 2\mu_0)} = \boldsymbol{C}_{\angle} - \frac{\boldsymbol{b} \otimes \boldsymbol{b}}{2(\lambda + 2\mu_0)},$$

$$c_{\|}' - \frac{(\beta')^2}{2(\lambda' + 2\mu_0)} = a_0^2 \left(c_{\|} - \frac{(\beta - \beta_0)^2}{2(\lambda + 2\mu_0)} \right) + c_0.$$

(c) This exact relation looks better if we use the bulk modulus $\kappa = \lambda + \mu_0$ instead of the Lamé modulus λ:

$$\frac{\mu_0^2}{\kappa'} = \kappa - \frac{1}{2} \boldsymbol{C}_{\angle}^{-1} \boldsymbol{b} \cdot \boldsymbol{b} \qquad \frac{\boldsymbol{b}'}{2\kappa'} = \frac{\boldsymbol{C}_{\angle} \boldsymbol{b}^{\perp}}{\det \boldsymbol{C}_{\angle}},$$

$$\boldsymbol{C}_{\angle}' - \frac{\boldsymbol{b}' \otimes \boldsymbol{b}'}{2\kappa'} = \frac{4\mu_0^2 \boldsymbol{C}_{\angle}}{\det \boldsymbol{C}_{\angle}}, \qquad \frac{\mu_0 \beta'}{\kappa'} = \boldsymbol{C}_{\angle}^{-1} \boldsymbol{b} \cdot \boldsymbol{c}_{\angle\|} - \beta,$$

$$c'_{\measuredangle\|} - \frac{\beta' \boldsymbol{b}'}{2\kappa'} = \frac{2\mu_0 \boldsymbol{C}_{\measuredangle} \boldsymbol{c}^{\perp}_{\measuredangle\|}}{\det \boldsymbol{C}_{\measuredangle}}, \qquad c_{\|}' - \frac{(\beta')^2}{2\kappa'} = c_0 - c_{\|} + \boldsymbol{C}_{\measuredangle}^{-1} \boldsymbol{c}_{\measuredangle\|} \cdot \boldsymbol{c}_{\measuredangle\|}.$$

13. $$\mathsf{C} = \begin{bmatrix} 2\mu_0 \mathsf{I} + 2\lambda \boldsymbol{I}_2 \otimes \boldsymbol{I}_2 & \boldsymbol{I}_2 \otimes \boldsymbol{b} & \beta \boldsymbol{I}_2 \\ \boldsymbol{b} \otimes \boldsymbol{I}_2 & \boldsymbol{\sigma} + \dfrac{\boldsymbol{b} \otimes \boldsymbol{b}}{2(\lambda + 2\mu_0)} & \boldsymbol{c}_{\measuredangle\|} \\ \beta \boldsymbol{I}_2 & \boldsymbol{c}_{\measuredangle\|} & c_{\|} \end{bmatrix}, \qquad \det \boldsymbol{\sigma} = d_0^2.$$

(a) In C' the parameters λ' and β' are arbitrary and independent of C. Then, for arbitrary constant angles ϕ_0, β_0 we have

$$\frac{2\mu_0 \boldsymbol{b}'}{\lambda' + 2\mu_0} = \frac{1}{\cos\phi_0}\left(\cos\beta_0 \boldsymbol{w}_1 - \frac{\sin\beta_0}{d_0} \boldsymbol{\sigma} \boldsymbol{w}_2^{\perp}\right),$$

$$\boldsymbol{c}'_{\measuredangle\|} - \frac{\beta' \boldsymbol{b}'}{2(\lambda' + 2\mu_0)} = \frac{1}{\cos\phi_0}\left(-\cos\beta_0 \boldsymbol{w}_2 + \frac{\sin\beta_0}{d_0} \boldsymbol{\sigma} \boldsymbol{w}_1^{\perp}\right),$$

where

$$\boldsymbol{w}_1 = \frac{2\mu_0 \boldsymbol{b}}{\lambda + 2\mu_0} \mp \left(\boldsymbol{c}_{\measuredangle\|} - \frac{\beta \boldsymbol{b}}{2(\lambda + 2\mu_0)}\right)\sin\phi_0,$$

$$\boldsymbol{w}_2 = \frac{2\boldsymbol{b}\mu_0 \sin\phi_0}{\lambda + 2\mu_0} \mp \left(\boldsymbol{c}_{\measuredangle\|} - \frac{\beta \boldsymbol{b}}{2(\lambda + 2\mu_0)}\right).$$

$$\begin{aligned} c'_{\measuredangle\|} - \frac{(\beta')^2}{2(\lambda' + 2\mu_0)} - \frac{4\mu_0 \lambda'}{\lambda' + 2\mu_0} &= c_{\|} - \frac{\beta^2}{2(\lambda + 2\mu_0)} - \frac{4\mu_0 \lambda}{\lambda + 2\mu_0} \\ &+ \sin^2\beta_0 \left(\frac{(\beta^2 - 16\mu_0^2)\boldsymbol{\sigma}^{-1}\boldsymbol{b} \cdot \boldsymbol{b}}{4(\lambda + 2\mu_0)^2} + \boldsymbol{\sigma}^{-1} \boldsymbol{c}_{\measuredangle\|} \cdot \boldsymbol{c}_{\measuredangle\|} - \frac{\beta \boldsymbol{\sigma}^{-1}\boldsymbol{b} \cdot \boldsymbol{c}_{\measuredangle\|}}{\lambda + 2\mu_0}\right) \pm \frac{2\sin 2\beta_0}{d_0} \boldsymbol{b}^{\perp} \cdot \boldsymbol{c}_{\measuredangle\|}. \end{aligned}$$

(b) Moduli λ' and β' are arbitrary and independent of C, while

$$\frac{\boldsymbol{b}'}{\lambda' + 2\mu_0} = \frac{\boldsymbol{b}}{\lambda + 2\mu_0}, \qquad \boldsymbol{c}'_{\measuredangle\|} - \frac{\beta' \boldsymbol{b}'}{\lambda' + 2\mu_0} = \boldsymbol{\sigma}\left(\boldsymbol{c}_{\measuredangle\|} - \frac{\beta \boldsymbol{b}}{\lambda + 2\mu_0}\right)^{\perp}.$$

$$c_{\|}' - \frac{(\beta')^2}{2(\lambda' + 2\mu_0)} - \frac{4\mu_0\lambda'}{\lambda' + 2\mu_0} = c_0 - \left(c_{\|} - \frac{\beta^2}{2(\lambda + 2\mu_0)}\right) - \frac{4\mu_0\lambda}{\lambda + 2\mu_0}$$
$$+\boldsymbol{\sigma}\left(\boldsymbol{c}_{\measuredangle\|} - \frac{\beta\boldsymbol{b}}{\lambda + 2\mu_0}\right)^{\perp} \cdot \left(\boldsymbol{c}_{\measuredangle\|} - \frac{\beta\boldsymbol{b}}{\lambda + 2\mu_0}\right)^{\perp}.$$

(c) Lamé modulus λ' is independent of C, $\boldsymbol{\sigma}' = \boldsymbol{\sigma}$, and

$$\frac{\boldsymbol{b}'}{2(\lambda' + 2\mu_0)} = a_0\left(\boldsymbol{c}_{\measuredangle\|} - \frac{\beta\boldsymbol{b}}{2(\lambda + 2\mu_0)}\right) + b_0\boldsymbol{\sigma}\left(\boldsymbol{c}_{\measuredangle\|}^{\perp} - \frac{(\beta \pm 8\mu_0)\boldsymbol{b}^{\perp}}{2(\lambda + 2\mu_0)}\right),$$

$$\frac{2a_0\beta' + 1}{2(\lambda' + 2\mu_0)} = a_0^2\left(c_{\|} - \frac{\beta^2}{2(\lambda + 2\mu_0)}\right)$$
$$+d_0^2b_0^2\left(c_{\|} - \frac{(\beta \pm 8\mu_0)^2}{2(\lambda + 2\mu_0)} + \boldsymbol{\sigma}^{-1}\left(\boldsymbol{c}_{\measuredangle\|} - \frac{(\beta \pm 8\mu_0)\boldsymbol{b}}{2(\lambda + 2\mu_0)}\right) \cdot \left(\boldsymbol{c}_{\measuredangle\|} - \frac{(\beta \pm 8\mu_0)\boldsymbol{b}}{2(\lambda + 2\mu_0)}\right)\right).$$

$$\boldsymbol{c}_{\measuredangle\|}' - \frac{\beta'\boldsymbol{b}'}{2(\lambda' + 2\mu_0)} = \boldsymbol{c}_{\measuredangle\|} - \frac{\beta\boldsymbol{b}}{2(\lambda + 2\mu_0)},$$
$$c_{\|}' - \frac{\beta'^2}{2(\lambda' + 2\mu_0)} = c_{\|} - \frac{\beta^2}{2(\lambda + 2\mu_0)} + c_0.$$

14. $$\mathsf{C} = \begin{bmatrix} 2\mu_0\mathsf{I} + 2\lambda\boldsymbol{I}_2 \otimes \boldsymbol{I}_2 & \boldsymbol{I}_2 \otimes \boldsymbol{b} & \beta\boldsymbol{I}_2 \\ \boldsymbol{b} \otimes \boldsymbol{I}_2 & \sigma_0\boldsymbol{I}_2 + \dfrac{\boldsymbol{b} \otimes \boldsymbol{b}}{2(\lambda + 2\mu_0)} & \dfrac{\beta_0\boldsymbol{b}^{\perp} + \beta\boldsymbol{b}}{2(\lambda + 2\mu_0)} \\ \beta\boldsymbol{I}_2 & \dfrac{\beta_0\boldsymbol{b}^{\perp} + \beta\boldsymbol{b}}{2(\lambda + 2\mu_0)} & c_{\|} \end{bmatrix}.$$

Denoting

$$\boldsymbol{u}(\lambda, \boldsymbol{b}) = \frac{\boldsymbol{b}}{\lambda + 2\mu_0}, \quad \theta(\lambda, \boldsymbol{b}) = \frac{\beta_0^2}{2(\lambda + 2\mu_0)} + \frac{\beta_0^2}{4\sigma_0}\frac{|\boldsymbol{b}|^2}{(\lambda + 2\mu_0)^2}$$

$$\alpha(\lambda, \beta) = \frac{\beta}{\lambda + 2\mu_0}, \quad \gamma(\lambda, \beta, c_{\|}) = c_{\|} - \frac{\beta^2}{2(\lambda + 2\mu_0)},$$

the link is given by the equations

$$\boldsymbol{u}(\lambda', \boldsymbol{b}') = \boldsymbol{u}(\lambda, \boldsymbol{b}),$$

$$\theta(\lambda', \boldsymbol{b}') = a_0\theta(\lambda, \boldsymbol{b}) + (1 - a_0)\gamma\big(\lambda, \beta, c_{\parallel}\big),$$

$$\alpha(\lambda', \beta') = \alpha(\lambda, \beta) + b_0\Big(\gamma\big(\lambda, \beta, c_{\parallel}\big) - \theta(\lambda, \boldsymbol{b})\Big),$$

$$\gamma\big(\lambda', \beta', c_{\parallel}'\big) = c_0\gamma\big(\lambda, \beta, c_{\parallel}\big) + (1 - c_0)\theta(\lambda, \boldsymbol{b}),$$

where in view of the first equation, the second can be rewritten as a formula for λ':

$$\frac{\beta_0^2}{2\big(\lambda' + 2\mu_0\big)} = \frac{a_0\beta_0^2}{2\big(\lambda + 2\mu_0\big)} + (1 - a_0)\left(c_{\parallel} - \frac{\beta^2}{2\big(\lambda + 2\mu_0\big)} - \frac{\beta_0^2}{4\sigma_0}\frac{|\boldsymbol{b}|^2}{\big(\lambda + 2\mu_0\big)^2}\right).$$

15. Suppose C belongs to the exact relation in item 23 in section 9.3.1. Then there is a link described by the formulas
    - $\mathsf{C}_{\mathsf{T}}' = \mathsf{C}_{\mathsf{T}}$, $\mathcal{C}_{\mathsf{T}\measuredangle}' = \mathcal{C}_{\mathsf{T}\measuredangle}$, $\boldsymbol{C}_{\measuredangle}' = \boldsymbol{C}_{\measuredangle}$.
    - Let $\hat{\mathsf{S}} = (\mathsf{C} + \mathsf{C}_0)^{-1}$ and $\hat{\mathsf{S}}' = (\mathsf{C}' + \mathsf{C}_0)^{-1}$, where $\mathsf{C}_0 = \begin{bmatrix} \tau_0\mathsf{I} & 0 & 0 \\ 0 & \sigma_0\boldsymbol{I}_2 & 0 \\ 0 & 0 & 0 \end{bmatrix}$.

    Then

$$\frac{1}{\hat{s}_{\parallel}} + \frac{1}{\hat{s}_{\parallel}'} = \frac{1}{\zeta_0}, \qquad \frac{\hat{\boldsymbol{s}}_{\measuredangle\parallel}'}{\hat{s}_{\parallel}'} = \pm\frac{\hat{\boldsymbol{s}}_{\measuredangle\parallel}^{\perp}}{\hat{s}_{\parallel}}, \qquad \frac{\mathrm{dev}\Big(\hat{\boldsymbol{S}}_{\mathsf{T}\parallel}'\Big)}{\hat{s}_{\parallel}'} = \pm\frac{\boldsymbol{R}_{\perp}\mathrm{dev}\Big(\hat{\boldsymbol{S}}_{\mathsf{T}\parallel}\Big)}{\hat{s}_{\parallel}},$$

    where the '+' sign corresponds to $\mathcal{S}_{\mathsf{T}\measuredangle}\boldsymbol{u} = \psi(c\bar{u})$ and the '–' sign to $\mathcal{S}_{\mathsf{T}\measuredangle}\boldsymbol{u} = \psi(cu)$.

Bibliography

[1] Avellaneda M, Cherkaev A V, Gibiansky L V, Milton G W and Rudelson M 1996 A complete characterization of the possible bulk and shear moduli of planar polycrystals *J. Mech. Phys. Solids* **44** 1179–218

[2] Benveniste Y 1995 Correspondence relations among equivalent classes of heterogeneous piezoelectric solids under anti-plane mechanical and in-plane electrical fields *J. Mech. Phys. Solids* **43** 553–71

[3] Berdichevsky V L 1983 *Variational Principles of Continuum Mechanics* (Moscow: Nauka) (in Russian)

[4] Cherkaev A V, Lurie K A and Milton G W 1992 Invariant properties of the stress in plane elasticity and equivalence classes of composites *Proc. R. Soc. Lond.* A **438** 519–29

[5] Gibiansky L V 1998 Effective moduli of plane polycrystals with a rank-one local compliance tensor *Proc. R. Soc. Edin.* A **128** 1325–54

[6] Grabovsky Y and Milton G W 1998 Rank one plus a null-Lagrangian is an inherited property of two-dimensional compliance tensors under homogenization *Proc. R. Soc. Edin.* A **128** 283–99

[7] Hegg M 2012 Exact relations and links for fiber-reinforced elastic composites *PhD Thesis* (Philadelphia, PA: Temple University)

[8] Hegg M 2013 Links between effective tensors for fiber-reinforced elastic composites *Compt. Rend. Méc.* **341** 520–32

[9] Helsing J, Milton G W and Movchan A B 1997 Duality relations, correspondences and numerical results for planar elastic composites *J. Mech. Phys. Solids* **45** 565–90

[10] Hill R 1963 Elastic properties of reinforced solids: some theoretical principles *J. Mech. Phys. Solids* **11** 357–72

[11] Hill R 1964 Theory of mechanical properties of fibre-strengthened materials: I. Elastic behaviour *J. Mech. Phys. Solids* **12** 199–212

[12] Lurie K A and Cherkaev A V 1984 G-closure of some particular sets of admissible material characteristics for the problem of bending of thin plates *J. Optim. Theory Appl.* **42** 305–16

[13] Lurie K A and Cherkaev A V 1986 Effective characteristics of composite materials and the optimal design of structural elements *Usp. Mekh.* **9** 3–81 (in Russian)
Lurie K A and Cherkaev A V 1997 Effective characteristics of composite materials and the optimal design of structural elements*Topics in the Mathematical Modeling of Composite Materials* (*Progress in Nonlinear Differential equations and their Applications* vol. 31) ed A Cherkaev and R Kohn (Boston, MA: Birkhäuser) pp 175–258

[14] Lurie K A, Cherkaev A V and Fedorov A V 1984 On the existence of solutions to some problems of optimal design for bars and plates *J. Optim. Theory Appl.* **42** 247–81

[15] Tartar L 1983 Estimation fines des coefficients homogénéisés *E De Giorgi Colloquium* (*Paris, 1983*) ed P Kree (London: Pitman) pp 168–87

[16] Walpole L 1985 Evaluation of the elastic moduli of a transversely isotropic aggregate of cubic crystals *J. Mech. Phys. Solids* **33** 623–36

IOP Publishing

Composite Materials
Mathematical theory and exact relations
Yury Grabovsky

Chapter 10

Piezoelectricity

A piezoelectric material is described by the constitutive relation

$$\begin{bmatrix}\boldsymbol{\sigma}(\boldsymbol{x})\\ \boldsymbol{e}(\boldsymbol{x})\end{bmatrix}=\mathsf{L}(\boldsymbol{x})\begin{bmatrix}\boldsymbol{\varepsilon}(\boldsymbol{x})\\ \boldsymbol{j}(\boldsymbol{x})\end{bmatrix},\qquad \mathsf{L}(\boldsymbol{x})=\begin{bmatrix}\mathsf{C}(\boldsymbol{x}) & \mathcal{P}(\boldsymbol{x})\\ \mathcal{P}^T(\boldsymbol{x}) & \rho(\boldsymbol{x})\end{bmatrix}, \tag{10.1}$$

where symmetric matrix fields $\boldsymbol{\varepsilon}(\boldsymbol{x})$ and $\boldsymbol{\sigma}(\boldsymbol{x})$ denote the elastic strain and stress fields, satisfying differential constraints (9.2) and (9.3), respectively, while vector fields $\boldsymbol{e}(\boldsymbol{x})$ and $\boldsymbol{j}(\boldsymbol{x})$, satisfying (8.1), denote the electric field and the dielectric displacement, respectively. Thus $\rho(\boldsymbol{x})^{-1}$ has the meaning of a dielectric tensor, while $\mathcal{P}(\boldsymbol{x})$ is a piezoelectric coupling tensor.

Most exact relations and links are the same in two and three dimensions, with a few two-dimensional specific results that we list in a separate section. We therefore, use d to denote dimension $d = 2$ or 3.

10.1 Exact relations

The complete list of exact relations for three-dimensional piezoelectricity was first published in [4]. There just five exact relations. Two of them are uniform field relations (UFR), one is related to Hill's exact relation (9.11) and two to the 'rank-one plus a null-Lagrangian' exact relation (9.13).

1. (UFR) $\mathsf{C}\boldsymbol{I}_d = d\kappa_0\boldsymbol{I}_d$, $\mathcal{P}^T\boldsymbol{I}_d = 0$.
2. $\mathsf{C} = 2\mu_0\mathsf{T} + \boldsymbol{B}\otimes\boldsymbol{B}$, $\mathcal{P} = \boldsymbol{B}\otimes\boldsymbol{b}$.
3. (UFR) $\mathsf{C} = 2\mu_0\mathsf{I} + \lambda\boldsymbol{I}_d\otimes\boldsymbol{I}_d$, $\mathcal{P} = \boldsymbol{I}_d\otimes\boldsymbol{b}$.
4. $\mathsf{C} = 2\mu_0\mathsf{T} + \boldsymbol{B}\otimes\boldsymbol{B}$, $\mathcal{P} = \boldsymbol{B}\otimes\boldsymbol{b}$, $\rho = \rho_0\boldsymbol{I}_d + \boldsymbol{b}\otimes\boldsymbol{b}$.
5. $\mathsf{C} = 2\mu_0\mathsf{I} + \lambda\boldsymbol{I}_d\otimes\boldsymbol{I}_d$, $\mathcal{P} = \boldsymbol{I}_d\otimes\boldsymbol{b}$, $\rho = \rho_0\boldsymbol{I}_d + \frac{\boldsymbol{b}\otimes\boldsymbol{b}}{\lambda+2\mu_0}$, $\left\langle\frac{\boldsymbol{b}}{\lambda+2\mu_0}\right\rangle$.

10.2 Links

Links have not been presented anywhere to the best of the author's knowledge. Their computation (in three space dimensions) became possible due to the work of undergraduate student Daniel Lapsley during the summer of 2016, who wrote Maple™ code for the multiplication of SO(3) irreducible representations, based on the theory presented in appendix C.

1. Global link. $\mathsf{L}' = \begin{bmatrix} \mathsf{C} & -\mathcal{P} \\ -\mathcal{P}^T & \rho \end{bmatrix}$
2. $\mathsf{L} = \begin{bmatrix} 2\mu_0\mathsf{T} + \boldsymbol{B} \otimes \boldsymbol{B} & \boldsymbol{B} \otimes \boldsymbol{b} \\ \boldsymbol{b} \otimes \boldsymbol{B} & \rho \end{bmatrix}$.
   (a) $\boldsymbol{B}'$ and $\boldsymbol{b}'$ are unrelated to L, while $\rho' - \boldsymbol{b}' \otimes \boldsymbol{b}' = \rho - \boldsymbol{b} \otimes \boldsymbol{b}$.
   (b) $\boldsymbol{b}'$ and ρ' are unrelated to L, $\boldsymbol{B}' = \lambda'\boldsymbol{B}$, where λ' is a scalar, unrelated to L.
   (c) $\boldsymbol{b}'$ is unrelated to L, $\boldsymbol{B}' = \lambda'\boldsymbol{B}$, $\rho' - \boldsymbol{b}' \otimes \boldsymbol{b}' = \rho - \boldsymbol{b} \otimes \boldsymbol{b}$, where λ' is a scalar, unrelated to L.
   (d) $\boldsymbol{b}' = \Lambda\boldsymbol{b}$, $\boldsymbol{B}' = a_0\Lambda(\boldsymbol{B} + b_0(\mathrm{Tr}\,\boldsymbol{B})\boldsymbol{I}_d)$, $\rho' - \boldsymbol{b}' \otimes \boldsymbol{b}' = \rho - \boldsymbol{b} \otimes \boldsymbol{b}$.
3. $\mathsf{L} = \begin{bmatrix} 2\mu_0\mathsf{I} + \lambda\boldsymbol{I}_d \otimes \boldsymbol{I}_d & \boldsymbol{I}_d \otimes \boldsymbol{b} \\ \boldsymbol{b} \otimes \boldsymbol{I}_d & \rho \end{bmatrix}$,

$$\frac{1}{\lambda' + 2\mu_0'} = \frac{a_0^2}{\lambda + 2\mu_0} + b_0, \qquad \frac{\boldsymbol{b}'}{\lambda' + 2\mu_0'} = a_0\frac{\boldsymbol{b}}{\lambda + 2\mu_0},$$

$$\rho' - \frac{\boldsymbol{b}' \otimes \boldsymbol{b}'}{\lambda' + 2\mu_0'} = \rho - \frac{\boldsymbol{b} \otimes \boldsymbol{b}}{\lambda + 2\mu_0}.$$

10.3 Two-dimension-specific relations and links

We note that the effective behavior of a two-dimensional piezoelectric composite can be determined from the three-dimensional effective elasticity tensor of a fibrous composite with the same microstructure [2, 3, 6]. However, we must note that such relations between the effective moduli of composites in different physical contexts have long been known and used (see e.g. [5]).

In order to match equations of elasticity for a fibrous composite and of two-dimensional piezoelectricity we need to 'interchange' $\boldsymbol{e}$ and $\boldsymbol{j}$ in (10.1). In general this is impossible without destroying either the symmetry of L or its positive definiteness [7], section 2.4. However, in two space dimensions there is another possibility:

$$\boldsymbol{\varepsilon}_{\times} = \boldsymbol{j}^{\perp}, \qquad \boldsymbol{\sigma}_{\times} = \boldsymbol{e}^{\perp}. \tag{10.2}$$

The vector fields $\boldsymbol{\varepsilon}_{\times}$ and $\boldsymbol{\sigma}_{\times}$ are then curl-free and divergence-free, respectively. In this case

$$\mathsf{C} = \mathsf{C}_{\mathsf{T}}, \qquad \mathcal{P} = \mathcal{C}_{\mathsf{T}\angle}\boldsymbol{R}_{\perp}^{T}, \qquad \boldsymbol{\rho} = \boldsymbol{R}_{\perp}\boldsymbol{C}_{\angle}\boldsymbol{R}_{\perp}^{T}. \tag{10.3}$$

Using the conversion rule (10.3) we can read-off all the exact relations and links for two-dimensional piezoelectric composites from the complete lists in section 9.3. As we have mentioned before, most exact relations and links hold for both two- and three-dimensional piezoelectric polycrystals. There is a three-parameter family of global links [1, 7], given by

$$\mathsf{L}' = \mathfrak{F}_{a,q_1,q_2}(\mathsf{L}) = \begin{bmatrix} q_1^2\mathsf{T}\Phi_a(\mathsf{C})\mathsf{C} & q_1q_2\mathsf{T}\Phi_a(\mathsf{C})\mathcal{P} \\ q_1q_2\mathcal{P}^T\Phi_a(\mathsf{C})\mathsf{T} & q_2^2\left(\boldsymbol{\rho} - a\mathcal{P}^T\Phi_a(\mathsf{C})\mathcal{P}\right) \end{bmatrix}, \tag{10.4}$$

where $\Phi_a(\mathsf{C}) = (\mathsf{T} + a\mathsf{C})^{-1}$.

There are also two special exact relations

1. $\mathsf{L} = \begin{bmatrix} 2\mu_0\mathsf{T} + \boldsymbol{B}\otimes\boldsymbol{B} & \boldsymbol{B}\otimes\boldsymbol{b} \\ \boldsymbol{b}\otimes\boldsymbol{B} & \boldsymbol{\rho} + \boldsymbol{b}\otimes\boldsymbol{b} \end{bmatrix}, \qquad \det\boldsymbol{\rho} = \rho_0.$
2. A piezoelectric tensor L belongs to this exact relation if and only if its blocks C, $\mathcal{P}$ and $\boldsymbol{\rho}$ satisfy the following constraints.
   - $\mathsf{C}\boldsymbol{I}_2 = 2\kappa_0\boldsymbol{I}_2$, $\mathcal{P}^T\boldsymbol{I}_2 = 0$, where $\kappa_0 > 0$ is an arbitrary constant.
   - Choose arbitrary constants $\sigma_0 > 0$ and $0 < \tau_0 < \kappa_0$, and define tensors S, $\boldsymbol{\alpha}$ and $\mathcal{Q}$ by

$$\begin{bmatrix} \mathsf{C} + \tau_0\mathsf{I} & \mathcal{P} \\ \mathcal{P}^T & \boldsymbol{\rho} + \rho_0\boldsymbol{I}_2 \end{bmatrix}^{-1} = \begin{bmatrix} (\mathsf{S} + \tau_0\mathsf{I})^{-1} & \mathcal{Q} \\ \mathcal{Q} & \left(\boldsymbol{\alpha} + \rho_0\boldsymbol{I}_2\right)^{-1} \end{bmatrix}. \tag{10.5}$$

   - The range of the operator $\mathcal{Q} : \mathbb{R}^2 \to \mathrm{Sym}(\mathbb{R}^2)$ must be $\mathrm{Sym}_0(\mathbb{R}^2)$, consisting of all trace-free matrices in $\mathrm{Sym}(\mathbb{R}^2)$, and $\mathcal{Q}^T\mathcal{Q} : \mathbb{R}^2 \to \mathbb{R}^2$ must be a (possibly variable) multiple of $\boldsymbol{I}_2$,

$$\mathcal{Q}^T\mathcal{Q} = q\boldsymbol{I}_2. \tag{10.6}$$

   - We require that $\det\boldsymbol{\alpha} = \rho_0^2$.
   - Let μ_1 and μ_2 be the shear moduli of S (its bulk modulus is κ_0). We require that

$$\left(\frac{1}{\kappa_0} + \frac{1}{\mu_1}\right)\left(\frac{1}{\kappa_0} + \frac{1}{\mu_2}\right) = \frac{1}{\alpha_0^2}, \qquad \tau_0 = \left(\frac{\kappa_0^{-1} + \alpha_0^{-1}}{2}\right)^{-1}.$$

Of course, instead of the implicit definition (10.5), one can easily find explicit formulas for S, $\boldsymbol{\alpha}$ and $\mathcal{Q}$. However, (10.5) seems like the most elegant way to define these operators.

In order to explain this exact relation we describe how to construct a tensor that satisfies all the requirements. First, we choose any positive definite two-dimensional elastic tensor S with square symmetry, i.e. $\mathsf{S}\boldsymbol{I}_2 = 2\kappa_0\boldsymbol{I}_2$. Then, we choose a complex number c and construct[1] an operator $\mathcal{Q} : \mathbb{R}^2 \to \mathrm{Sym}_0(\mathbb{R}^2)$, satisfying (10.6). Finally, we choose an arbitrary operator $\boldsymbol{\alpha} > 0$ on $\mathbb{R}^2$. Once these choices have been made, we compute

$$\rho_0 = \sqrt{\det \boldsymbol{\alpha}}, \quad \alpha_0 = \frac{1}{\sqrt{\left(\kappa_0^{-1} + \mu_1^{-1}\right)\left(\kappa_0^{-1} + \mu_2^{-1}\right)}}, \quad \tau_0 = \left(\frac{\kappa_0^{-1} + \alpha_0^{-1}}{2}\right)^{-1},$$

where $2\mu_1$ and $2\mu_2$ are the other two eigenvalues of the operator S. We define operators C, $\mathcal{P}$ and $\boldsymbol{\rho}$ by (10.5). Explicitly,

$$\tilde{\mathsf{C}} = \left(\tilde{\mathsf{S}}^{-1} - \mathcal{Q}\tilde{\boldsymbol{\alpha}}\mathcal{Q}^T\right)^{-1}, \quad \boldsymbol{\rho} = (\tilde{\boldsymbol{\alpha}}^{-1} - \mathcal{Q}^T\tilde{\mathsf{S}}\mathcal{Q})^{-1}, \quad \mathcal{P} = -\tilde{\mathsf{S}}\mathcal{Q}\boldsymbol{\rho},$$

where $\tilde{\mathsf{C}} = \mathsf{C} + \tau_0\mathsf{I}$, $\tilde{\mathsf{S}} = \mathsf{S} + \tau_0\mathsf{I}$, and $\tilde{\boldsymbol{\alpha}} = \boldsymbol{\alpha} + \rho_0\boldsymbol{I}_2$. If we use such a material to make a transversely isotropic polycrystal

$$\mathsf{L}^* = \begin{bmatrix} 2\mu^*(\mathsf{I} - \boldsymbol{I}_2 \otimes \boldsymbol{I}_2) + 2\kappa^*\boldsymbol{I}_2 \otimes \boldsymbol{I}_2 & 0 \\ 0 & \rho^*\boldsymbol{I}_2 \end{bmatrix},$$

all three parameters are texture-independent:

$$\kappa^* = \kappa_0, \quad \rho^* = \sqrt{\det \boldsymbol{\alpha}}, \quad \frac{1}{\mu^*} = \sqrt{\left(\frac{1}{\kappa_0} + \frac{1}{\mu_1}\right)\left(\frac{1}{\kappa_0} + \frac{1}{\mu_2}\right)} - \frac{1}{\kappa_0}.$$

Bibliography

[1] Benveniste Y 1995 Correspondence relations among equivalent classes of heterogeneous piezoelectric solids under anti-plane mechanical and in-plane electrical fields *J. Mech. Phys. Solids* **43** 553–71

[2] Chen T 1998 Further correspondences between plane piezoelectricity and generalized plane strain in elasticity *Proc. R. Soc. Lond.* A **454** 873–84

[3] Chen T and Lai D 1997 An exact correspondence between plane piezoelectricity and generalized plane strain in elasticity *Proc. R. Soc. Lond.* A **453** 2689–713

[4] Grabovsky Y, Milton G W and Sage D S 2000 Exact relations for effective tensors of polycrystals: necessary conditions and sufficient conditions *Commun. Pure. Appl. Math.* **53** 300–53

[1] The set of admissible operators $\mathcal{Q}$ consists of two components, each parametrized by $c \in \mathbb{C}$. We describe $\mathcal{Q}$ by its action on an arbitrary vector $\boldsymbol{u} \in \mathbb{R}^2$. This is accomplished by identifying $\boldsymbol{u} = (u_1, u_2) \in \mathbb{R}^2$ with a complex number $u = u_1 + iu_2$, and vice versa. Then either $\mathcal{Q}\boldsymbol{u} = \psi(cu)$, or $\mathcal{Q}\boldsymbol{u} = \psi(c\bar{u})$, where $\psi : \mathbb{C} \to \mathrm{Sym}_0(\mathbb{R}^2)$ is defined in (9.20).

[5] Hashin Z 1983 Analysis of composite materials–a survey *J. Appl. Mech.* **50** 481–505
[6] Milton G W 1997 Composites: a myriad of microstructure independent relations *Proc. 19th Int. Congr. Theoretical and Applied Mechanics (Kyoto, 1996)* ed T Tatsumi, E Watanabe and T Kambe (Amsterdam: Elsevier) pp 443–59
[7] Milton G W 2002 *The Theory of Composites Cambridge Monographs on Applied and Computational Mathematics* vol 6 (Cambridge: Cambridge University Press)

Chapter 11

Thermoelasticity

The properties of thermoelastic materials are described by the elastic compliance tensor S, thermal expansion tensor $\boldsymbol{\alpha}$ and the specific heat c_σ per unit volume at constant stress (see [7], section 2.5), so that the constitutive relation has the form

$$\boldsymbol{\varepsilon} = \mathsf{S}\boldsymbol{\sigma} + \boldsymbol{\alpha}\theta, \qquad \varsigma = (\boldsymbol{\alpha}, \boldsymbol{\sigma})_F + c_\sigma \frac{\theta}{T_0}, \tag{11.1}$$

where $\boldsymbol{\varepsilon}$ and $\boldsymbol{\sigma}$ are the strain and the stress fields, satisfying (9.2) and (9.3), respectively. The scalar field ς and a constant scalar θ are the change in entropy and temperature, respectively, relative to the stress-free state and temperature T_0. In order to have an easy comparison of our results for thermoelasticity and elasticity it will be convenient to rewrite the constitutive relation (11.1) in the block-matrix form

$$\begin{bmatrix} \boldsymbol{\sigma} \\ -\varsigma \end{bmatrix} = \begin{bmatrix} \mathsf{C} & \boldsymbol{A} \\ \boldsymbol{A} & -c_\varepsilon/T_0 \end{bmatrix} \begin{bmatrix} \boldsymbol{\varepsilon} \\ \theta \end{bmatrix}, \tag{11.2}$$

where C is the elasticity tensor, $\boldsymbol{A}$ is the thermal stress tensor and c_ε is the specific heat per unit volume at constant strain. We have

$$\mathsf{C} = \mathsf{S}^{-1}, \qquad \boldsymbol{A} = -\mathsf{S}^{-1}\boldsymbol{\alpha}, \qquad \frac{c_\varepsilon}{T_0} = \frac{c_\sigma}{T_0} - \left(\mathsf{S}^{-1}\boldsymbol{\alpha}, \boldsymbol{\alpha}\right)_F. \tag{11.3}$$

We note that we cannot apply our theory to a non-positive definite material tensor in (11.2). Instead we modify the constitutive law (11.3)

$$\begin{bmatrix} \boldsymbol{\sigma} \\ -\varsigma \end{bmatrix} = \begin{bmatrix} \mathsf{C} & \boldsymbol{A} \\ \boldsymbol{A} & (c_0 - c_\varepsilon)/T_0 \end{bmatrix} \begin{bmatrix} \boldsymbol{\varepsilon} \\ \theta \end{bmatrix}, \tag{11.4}$$

doi:10.1088/978-0-7503-1048-2ch11

where the positive constant c_0 is chosen so that $c_0 - c_\varepsilon > 0$ everywhere in the composite. The general theory is now applicable. It is a matter of simple comparison of cell problems for (11.1) and (11.4) to see that

$$\mathsf{C}^* = (\mathsf{S}^*)^{-1}, \qquad A^* = -(\mathsf{S}^*)^{-1}\boldsymbol{\alpha}^*, \qquad \frac{c_\varepsilon^*}{T_0} = \frac{c_\sigma^*}{T_0} - \left((\mathsf{S}^*)^{-1}\boldsymbol{\alpha}^*, \boldsymbol{\alpha}^*\right)_F. \tag{11.5}$$

11.1 Two-dimensional thermoelasticity

We note that all the exact relations for two-dimensional thermoelasticity are already contained in the exact relations for fibrous elastic composites. Indeed, making the identification

$$\boldsymbol{\sigma}_\mathrm{T} = \boldsymbol{\sigma}, \quad \boldsymbol{\varepsilon}_\mathrm{T} = \boldsymbol{\varepsilon}, \quad \sigma_\| = -\varsigma, \quad \varepsilon_\| = \theta,$$

$$\mathsf{C}_\mathrm{T} = \mathsf{C}, \qquad C_{\mathrm{T}\|} = \sqrt{2}\boldsymbol{A}, \qquad c_\| = \frac{c_0 - c_\varepsilon}{T_0}.$$

We reduce (11.4) to the constitutive relation for fibrous elastic composites

$$\boldsymbol{\sigma}_\mathrm{T} = \mathsf{C}_\mathrm{T}\boldsymbol{\varepsilon}_\mathrm{T} + \varepsilon_\|\frac{C_{\mathrm{T}\|}}{\sqrt{2}}, \qquad \sigma_\| = \left(\frac{C_{\mathrm{T}\|}}{\sqrt{2}}, \boldsymbol{\varepsilon}_\mathrm{T}\right)_F + c_\|\varepsilon_\|, \tag{11.6}$$

restricted to the exact relation

$$\mathsf{C} = \begin{bmatrix} \mathsf{C}_\mathrm{T} & 0 & C_{\mathrm{T}\|} \\ 0 & c_\angle^0 \boldsymbol{I}_2 & 0 \\ C_{\mathrm{T}\|} & 0 & c_\| \end{bmatrix}.$$

This permits us to simply read off all the exact relations and links for two-dimensional thermoelasticity from the ones for fibrous elastic composites.

11.1.1 Exact relations

1. $\mathsf{C}\boldsymbol{I}_2 = 2\kappa_0\boldsymbol{I}_2$, $\langle \mathrm{Tr}\, \boldsymbol{A}\rangle$,
2. $\mathsf{C}\boldsymbol{I}_2 = 2\kappa_0\boldsymbol{I}_2$, $(\frac{1}{\kappa_0} + \frac{1}{\mu_1})(\frac{1}{\kappa_0} + \frac{1}{\mu_2}) = c_0$, $\langle \mathrm{Tr}\, \boldsymbol{A}\rangle$, where $2\mu_1$ and $2\mu_2$ are the other two eigenvalues of C
3. $\mathsf{C} = 2\mu_0\mathsf{T} + \boldsymbol{B}\otimes\boldsymbol{B}$
4. $\mathsf{C} = 2\mu_0\mathsf{T} + \boldsymbol{B}\otimes\boldsymbol{B}$, $\boldsymbol{A} = \beta\boldsymbol{B}$, $\langle c_\varepsilon + 2T_0\beta^2\rangle$
5. $\mathsf{C} = 2\mu_0\mathsf{I} + 2\lambda\boldsymbol{I}_2\otimes\boldsymbol{I}_2$, $\langle\frac{1}{2\mu_0 + \lambda}\rangle$, $\langle \mathrm{dev}(\boldsymbol{A})\rangle$
6. $\mathsf{C} = 2\mu_0\mathsf{I} + 2\lambda\boldsymbol{I}_2\otimes\boldsymbol{I}_2$, $\boldsymbol{A} = \beta\boldsymbol{I}_2$, $\langle\frac{1}{2\mu_0 + \lambda}\rangle$, $\langle\frac{\beta}{2\mu_0 + \lambda}\rangle$, $\langle c_\varepsilon + \frac{T_0\beta^2}{\lambda + 2\mu_0}\rangle$.

11.1.2 Links

1. Global links
   (a) The block C^* is independent of the blocks $\boldsymbol{A}$, c_ε.
   (b) Blocks C^*, $\boldsymbol{A}^*$ are independent of c_ε.
   (c)
$$\mathsf{L}' = \mathfrak{F}_{a,\mathsf{Q},\mathsf{F}}(\mathsf{L}) = \mathsf{Q}\Lambda_a(\mathsf{L})\mathsf{Q}^T + \mathsf{F}, \tag{11.7}$$
   where
$$\mathsf{Q} = \begin{bmatrix} q_1\mathsf{I} & 0 \\ q_4\boldsymbol{I}_2 & q_3 \end{bmatrix}, \qquad \mathsf{F} = \begin{bmatrix} 0 & f_1\boldsymbol{I}_2 \\ f_1\boldsymbol{I}_2 & f_2 \end{bmatrix},$$
   and
$$\Lambda_a(\mathsf{L}) = \mathsf{L} - \mathsf{L}\begin{bmatrix} \Theta_a(\mathsf{C}) & 0 \\ 0 & 0 \end{bmatrix}\mathsf{L}, \qquad \Theta_a(\mathsf{C}) = \left(\frac{1}{a}\mathsf{T} + \mathsf{C}\right)^{-1}. \tag{11.8}$$

2. $\mathsf{C}\boldsymbol{I}_2 = 2\kappa_0\boldsymbol{I}_2$, $\operatorname{Tr}\boldsymbol{A} = t_0$.
   The blocks in C' are constructed according to the following rules. $\mathsf{C}'\boldsymbol{I}_2 = 2\kappa_0\boldsymbol{I}_2$. Let $\mu_0 \in (0, \kappa_0)$ be a constant. Then the two shear moduli μ_1', μ_2' of C' are defined by
$$\left(\frac{1}{\mu_1'} + \frac{1}{\kappa_0}\right)\left(\frac{1}{\mu_2} + \frac{1}{\kappa_0}\right) = \left(\frac{1}{\mu_1'} + \frac{1}{\kappa_0}\right)\left(\frac{1}{\mu_1} + \frac{1}{\kappa_0}\right) = \frac{1}{\kappa_0}\left(\frac{1}{\kappa_0} + \frac{1}{\mu_0}\right) = \frac{1}{a_0^2}.$$
   We also have $\operatorname{Tr}\boldsymbol{A}' = t_0$ and
$$\operatorname{dev}(\boldsymbol{A}') = \frac{2\kappa_0\mu_0}{a_0}\boldsymbol{R}_\perp(\mathsf{C} - 2\mu_0\mathsf{I})^{-1}\operatorname{dev}(\boldsymbol{A}).$$
$$c_\varepsilon' = b_0 - c_\varepsilon - T_0\left((\mathsf{C} - 2a_0\mathsf{I})^{-1}\boldsymbol{A}, \boldsymbol{A}\right)_F,$$
   where the constant b_0 is chosen in such a way as to make C' positive definite.

3. $\mathsf{C}\boldsymbol{I}_2 = 2\kappa_0\boldsymbol{I}_2$, $\left(\frac{1}{\kappa_0} + \frac{1}{\mu_1}\right)\left(\frac{1}{\kappa_0} + \frac{1}{\mu_2}\right) = \frac{1}{\rho_0^2}$, $\operatorname{Tr}\boldsymbol{A} = t_0$. The link, or rather a family of links parametrized by a constant $\zeta_0 > 0$ can be described in the following way
   - $\mathsf{C}' = \mathsf{C}$, $\operatorname{Tr}\boldsymbol{A}' = t_0$.
   - Let $\begin{bmatrix} \mathsf{S} & \boldsymbol{\alpha} \\ \boldsymbol{\alpha} & \xi \end{bmatrix} = \begin{bmatrix} \mathsf{C} + \tau_0\mathsf{I} & \boldsymbol{A} \\ \boldsymbol{A} & -c_\varepsilon/T_0 \end{bmatrix}^{-1}$ and $\begin{bmatrix} \mathsf{S}' & \boldsymbol{\alpha}' \\ \boldsymbol{\alpha}' & \xi' \end{bmatrix} = \begin{bmatrix} \mathsf{C} + \tau_0\mathsf{I} & \boldsymbol{A}' \\ \boldsymbol{A}' & -c_\varepsilon'/T_0 \end{bmatrix}^{-1}$,
     where $\tau_0 = \left(\frac{\kappa_0^{-1} + \rho_0^{-1}}{2}\right)^{-1}$.

- Then

$$\frac{1}{\xi} + \frac{1}{\xi'} = \frac{1}{\zeta_0}, \qquad \frac{\operatorname{dev}(\boldsymbol{\alpha}')}{\xi'} = \pm \frac{\boldsymbol{R}_\perp \operatorname{dev}(\boldsymbol{\alpha})}{\xi}.$$

4. $\mathsf{C} = 2\mu_0 \mathsf{T} + \boldsymbol{B} \otimes \boldsymbol{B}$.
   (a) $\boldsymbol{B}' = \lambda' \boldsymbol{B}$, where λ' is independent of L.
   (b) $\boldsymbol{B}' = \lambda' \boldsymbol{B}$, $\boldsymbol{A}' = \boldsymbol{A} + \nu' \boldsymbol{B}$, where λ' and ν' are independent of L.
5. $\mathsf{C} = 2\mu_0 \mathsf{T} + \boldsymbol{B} \otimes \boldsymbol{B}$, $\boldsymbol{A} = \beta \boldsymbol{B}$.
   (a) $\boldsymbol{B}' = \lambda' \boldsymbol{B}$, $\beta' = \lambda' \beta$.
   (b) $c_\varepsilon' + 2T_0(\beta')^2 = c_\varepsilon + 2T_0\beta^2$.
   (c) $\boldsymbol{B}' = \boldsymbol{B}$, $\beta' = \beta$, $c_\varepsilon' = c_0 + a_0 c_\varepsilon - 2T_0(1 - a_0)\beta^2$.
   (d) $\boldsymbol{B}' = \boldsymbol{B}$, $\beta' = c_0\beta$, $c_\varepsilon' + 2T_0(\beta')^2 = c_\varepsilon + 2T_0\beta^2$.
6. $\mathsf{C} = 2\mu_0 \mathsf{I} + 2\lambda \boldsymbol{I}_2 \otimes \boldsymbol{I}_2$, $\lambda' = \lambda$, $c_\varepsilon' = c_\varepsilon + \frac{a_0}{\lambda + 2\mu_0}$.
7. $\mathsf{C} = 2\mu_0 \mathsf{I} + 2\lambda \boldsymbol{I}_2 \otimes \boldsymbol{I}_2$, $\boldsymbol{A} = \beta \boldsymbol{I}_2$.
   (a) $\dfrac{1}{\lambda' + 2\mu_0} = \dfrac{1}{\lambda + 2\mu_0} + b_0, \quad \dfrac{\beta'}{\lambda' + 2\mu_0} = \dfrac{\beta}{\lambda + 2\mu_0},$

$$c_\varepsilon' + \frac{T_0(\beta')^2}{\lambda' + 2\mu_0} = c_\varepsilon + \frac{T_0\beta^2}{\lambda + 2\mu_0} + c_0.$$

   (b) $\dfrac{2\mu_0^2 T_0}{\lambda' + 2\mu_0} = \dfrac{T_0\beta^2}{\lambda + 2\mu_0} + c_\varepsilon + b_0, \quad \dfrac{\mu_0 T_0 \beta'}{\lambda' + 2\mu_0} = a_0\left(c_\varepsilon + \dfrac{T_0\beta(\beta - \beta_0)}{\lambda + 2\mu_0}\right),$

$$c_\varepsilon' + \frac{T_0(\beta')^2}{\lambda' + 2\mu_0} = 2a_0^2\left(c_\varepsilon + \frac{T_0(\beta - \beta_0)^2}{\lambda + 2\mu_0}\right) + c_0.$$

   (c) This exact relation looks better if we use the bulk modulus $\kappa = \lambda + \mu_0$ instead of the Lamé modulus λ:

$$\frac{\mu_0^2}{\kappa'} = \kappa, \qquad \frac{\mu_0\beta'}{\kappa'} = -\beta, \qquad c_\varepsilon' + \frac{T_0(\beta')^2}{\kappa'} = c_0 - c_\varepsilon.$$

   (d) Here instead of giving rather cumbersome explicit formulas for the moduli λ', β' and c_ε' we determine them from a system of equations that can be written nicely as

$$\begin{cases} \dfrac{T_0\beta_0^2}{\lambda' + 2\mu_0} = a_0\left(c_\varepsilon + \dfrac{T_0\beta^2}{\lambda + 2\mu_0}\right) + (a_0 + b_0)\dfrac{T_0\beta_0^2}{\lambda + 2\mu_0}, \\[2ex] \dfrac{\beta'}{\lambda' + 2\mu_0} = \dfrac{b_0\beta}{\lambda + 2\mu_0} + c_0\left(c_\varepsilon + T_0\dfrac{\beta^2 + \beta_0^2}{\lambda + 2\mu_0}\right), \\[2ex] c_\varepsilon' + \dfrac{T_0(\beta')^2}{\lambda + 2\mu_0} = d_0\dfrac{T_0\beta_0^2}{\lambda + 2\mu_0} + (b_0 + d_0)\left(c_\varepsilon + \dfrac{T_0\beta^2}{\lambda + 2\mu_0}\right). \end{cases}$$

(e) Moduli λ' and β' are arbitrary and independent of L, while

$$c_\varepsilon' + \frac{T_0(\beta')^2}{\lambda' + 2\mu_0} = c_\varepsilon + \frac{T_0\beta^2}{\lambda + 2\mu_0}.$$

(f) Moduli λ' and β' are arbitrary, while

$$c_\varepsilon' + \frac{T_0(\beta')^2}{\lambda' + 2\mu_0} = c_0 - \left(c_\varepsilon + \frac{T_0\beta^2}{\lambda + 2\mu_0}\right).$$

(g) Lamé modulus λ' is independent of L, and

$$\frac{2a_0^2\beta_0(\beta' + \beta_0)}{\lambda' + 2\mu_0} = d_0 - a_0^2\left(c_\varepsilon + \frac{T_0\beta^2}{\lambda + 2\mu_0}\right) - b_0^2\left(c_\varepsilon + \frac{T_0(\beta \pm \mu_0)^2}{\lambda + 2\mu_0}\right).$$

$$c_\varepsilon' + \frac{T_0\beta'^2}{\lambda' + 2\mu_0} = c_\varepsilon + \frac{T_0\beta^2}{\lambda + 2\mu_0} + c_0.$$

11.2 Three-dimensional thermoelasticity

This complete list of exact relations for three-dimensional thermoelasticity is taken from [1].

1. $\mathsf{C} = 2\mu_0\mathsf{T} + \boldsymbol{B} \otimes \boldsymbol{B}$, $\boldsymbol{A} = \alpha_0\boldsymbol{I}_3 + \beta\boldsymbol{B}$.
2. Another exact relation is obtained if we replace β in item 1 by $\nu_0 \operatorname{Tr} \boldsymbol{B}$.
3. $\mathsf{C}\boldsymbol{I}_3 = \kappa_0\boldsymbol{I}_3 + a_0\boldsymbol{A}$. This exact relation was first discovered by Rosen and Hashin [8] and Laws [5] for the case of an isotropic composite of two isotropic phases. It was then generalized further by Hashin [2] to the case of an isotropic polycrystal, and to the general case by Schulgasser [9].
4. $\mathsf{C} = 2\mu_0\mathsf{I} + \lambda\boldsymbol{I}_3 \otimes \boldsymbol{I}_3$, $\boldsymbol{A} = a\boldsymbol{I}_3$. As we have mentioned at the beginning of this section, the effective elastic tensor C^* does not depend on the local thermoelastic coefficient $\boldsymbol{A}$. Therefore, according to Hill's exact relation for elasticity [3, 4], C^* will be isotropic with the same shear modulus μ_0 and the effective Lamé modulus λ^* given by (4.1).

   Our new thermoelastic exact relation says that the effective thermal stress coefficient $\boldsymbol{A}^*$ will always be isotropic and

$$\frac{a^*}{\lambda^* + 2\mu_0} = \left\langle \frac{a}{\lambda + 2\mu_0} \right\rangle. \tag{11.9}$$

5. If in the previous item we assume in addition that there are constants a_0 b_0 and c_0 such that $a_0\alpha + b_0\lambda = c_0$ then $a_0\alpha^* + b_0\lambda^* = c_0$. In particular for a two-phase composite this exact relation can be written as

$$\det \begin{vmatrix} a_1 & a_2 & a^* \\ \lambda_1 & \lambda_2 & \lambda^* \\ 1 & 1 & 1 \end{vmatrix} = 0.$$

This exact relation is due to Levin [6].

Bibliography

[1] Grabovsky Y, Milton G W and Sage D S 2000 Exact relations for effective tensors of polycrystals: necessary conditions and sufficient conditions *Commun. Pure. Appl. Math.* **53** 300–53

[2] Hashin Z 1984 Thermal expansion of polycrystalline aggregates: I. Exact analysis *J. Mech. Phys. Solids* **32** 149–57

[3] Hill R 1963 Elastic properties of reinforced solids: some theoretical principles *J. Mech. Phys. Solids* **11** 357–72

[4] Hill R 1964 Theory of mechanical properties of fibre-strengthened materials: I. Elastic behaviour *J. Mech. Phys. Solids* **12** 199–212

[5] Laws N 1973 On thermostatics of composite materials *J. Mech. Phys. Solids* **21** 9–17

[6] Levin V M 1967 Thermal expansion coefficients of heterogeneous materials *Mech. Solids* **2** 88–94

[7] Milton G W 2002 *The Theory of Composites* (*Cambridge Monographs on Applied and Computational Mathematics* vol. 6) (Cambridge: Cambridge University Press)

[8] Rosen B W and Hashin Z 1970 Effective thermal expansion coefficients and specific heats of composite materials *Int. J. Eng. Sci.* **8** 157–73

[9] Schulgasser K 1987 Thermal expansion of polycrystalline aggregates with texture *J. Mech. Phys. Solids* **35** 34–42

IOP Publishing

Composite Materials
Mathematical theory and exact relations
Yury Grabovsky

Chapter 12

Three-dimensional thermoelectricity

The constitutive equation for the linear thermoelectric effect can be formulated as a linear relation between a pair of curl-free intensity fields $e_1(x)$ and $e_2(x)$ and a pair of divergence-free flux fields $j_1(x)$ and $j_2(x)$. Such a formulation is mathematically convenient. The physical interpretation of these quantities is as follows (see [2, 6]). j_1 is the total energy flux density, while $-j_2$ is the electron current density. Conservation of energy and the number of electrons are responsible for the divergence-free constraints placed on these fields. $e_1 = \nabla(1/T)$, where T is the temperature field, while $e_2 = \nabla(\mu/T)$, where μ is an electrochemical potential (per particle). The constitutive law

$$\begin{bmatrix} j_1(x) \\ j_2(x) \end{bmatrix} = \mathsf{L}(x) \begin{bmatrix} e_1(x) \\ e_2(x) \end{bmatrix}, \qquad (12.1)$$

is effected by a *symmetric* positive definite tensor $\mathsf{L}(x)$, according to the Onsager principle (see [3, 6]). We can represent $\mathsf{L}(x)$ as a 2×2 symmetric block-matrix, where each block of $\mathsf{L}(x)$ is a 3×3 matrix. If the material is isotropic then all four blocks must be scalar multiples of I_3. We have the following three non-trivial exact relations.

1. $\mathsf{L} = \begin{bmatrix} \sigma & -\nu \\ \nu & \sigma \end{bmatrix}$,

   where σ is symmetric and ν is a skew-symmetric 3×3 matrix, such that L is positive definite. This exact relation can be conveniently described using complex arithmetic. The pairs of fields e_1, e_2, and j_1, j_2, can be combined into two complex-valued fields

   $$j = j_1 + ij_2 \in \mathbb{C}^3, \qquad e = e_1 + ie_2 \in \mathbb{C}^3.$$

   Then thermoelectric tensors L belonging to this exact relation are characterized by the constitutive relation $j = \mathsf{K}e$, where K is a complex-Hermitian

3×3 matrix. We refer to the review article [1] and references therein for a discussion of the physical interpretation of complex conductivity.

2. (Uniform field relation) Let $\boldsymbol{n}_0 = [n_0^1, n_0^2]$ be a fixed unit vector in $\mathbb{R}^2$ and let

$$\mathsf{L} = \begin{bmatrix} \sigma_0 \boldsymbol{I}_3 + (n_0^1)^2 \boldsymbol{\sigma} & \nu_0 \boldsymbol{I}_3 + n_0^1 n_0^2 \boldsymbol{\sigma} \\ \nu_0 \boldsymbol{I}_3 + n_0^1 n_0^2 \boldsymbol{\sigma} & \gamma_0 \boldsymbol{I}_3 + (n_0^2)^2 \boldsymbol{\sigma} \end{bmatrix},$$

or using a tensor product notation

$$\mathsf{L} = \begin{bmatrix} \sigma_0 & \nu_0 \\ \nu_0 & \gamma_0 \end{bmatrix} \otimes \boldsymbol{I}_3 + (\boldsymbol{n}_0 \otimes \boldsymbol{n}_0) \otimes \boldsymbol{\sigma},$$

where $\boldsymbol{\sigma}$ is a 3×3 symmetric matrix.

As an example, consider an isotropic composite made of two isotropic materials L_1 and L_2 such that $\det(\mathsf{L}_1 - \mathsf{L}_2) = 0$. Then L^* must satisfy $\det(\mathsf{L}^* - \mathsf{L}_1) = 0$ and $\det(\mathsf{L}^* - \mathsf{L}_2) = 0$, which is also equivalent to the formula

$$\mathsf{L}^* = \alpha^* \mathsf{L}_1 + (1 - \alpha^*) \mathsf{L}_2$$

for some microstructure-dependent scalar α^*.

3. This exact relation is described by the condition that the four blocks comprising L satisfy a given linear relation, i.e. there are four constants c_{ij}^0, $i, j = 1, 2$ such that

$$c_{11}^0 \boldsymbol{L}_{11} + c_{12}^0 \boldsymbol{L}_{12} + c_{21}^0 \boldsymbol{L}_{12}^T + c_{22}^0 \boldsymbol{L}_{22} = 0.$$

This exact relation is due to Milgrom and Shtrikman [5] (see also Milgrom [4]).

As an example consider a composite made with two isotropic thermoelectric materials

$$\mathsf{L}_i = \begin{bmatrix} \sigma_i \boldsymbol{I}_3 & \nu_i \boldsymbol{I}_3 \\ \nu_i \boldsymbol{I}_3 & \gamma_i \boldsymbol{I}_3 \end{bmatrix}, \qquad i = 1, 2.$$

We can easily find three numbers c_1, c_2 and c_3 such that the vector $\boldsymbol{c} = [c_1, c_2, c_3]$ is orthogonal to the two vectors $\boldsymbol{l}_1 = [\sigma_1, \gamma_1, \nu_1]$ and $\boldsymbol{l}_2 = [\sigma_2, \gamma_2, \nu_2]$. Then our exact relation tells us that the vector $\boldsymbol{l}^* = (\sigma^*, \gamma^*, \nu^*)$ made with components of the isotropic effective tensor L^* is also orthogonal to $\boldsymbol{c}$. In other words, the three vectors $\boldsymbol{l}_1$, $\boldsymbol{l}_2$ and $\boldsymbol{l}^*$ are linearly dependent, i.e.

$$\det \begin{vmatrix} \sigma^* & \gamma^* & \nu^* \\ \sigma_1 & \gamma_1 & \nu_1 \\ \sigma_2 & \gamma_2 & \nu_2 \end{vmatrix} = 0.$$

Bibliography

[1] Bergman D J and Stroud D 1992 Physical properties of macroscopically inhomogeneous media *Solid State Phys.* **46** 147–269

[2] Callen H B 1960 *Thermodynamics* (New York: Wiley)

[3] Lifshitz E M and Pitaevskii L P 1980 *Statistical Physics* 3rd ed (Oxford: Pergamon)

[4] Milgrom M 1997 Some more exact results concerning multifield moduli of two-phase composites *J. Mech. Phys. Solids* **45** 399–404

[5] Milgrom M and Shtrikman S 1989 Linear response of two-phase composites with cross moduli: exact universal relations *Phys. Rev.* A **40** 1568–75

[6] Milton G W 2002 *The Theory of Composites Cambridge Monographs on Applied and Computational Mathematics* vol 6 (Cambridge: Cambridge University Press)

Part IV

Appendices

IOP Publishing

Composite Materials
Mathematical theory and exact relations
Yury Grabovsky

Appendix A

$\mathcal{E}$- and $\mathcal{J}$-regularity for conductivity and elasticity

A.1 Conductivity

$$\mathcal{E} = \left\{ \boldsymbol{E} \in L^2\left(\mathbb{R}^d; \mathbb{R}^d\right): \hat{\boldsymbol{E}}(\boldsymbol{\xi}) = \frac{\left(\hat{\boldsymbol{E}}(\boldsymbol{\xi}) \cdot \boldsymbol{\xi}\right)\boldsymbol{\xi}}{|\boldsymbol{\xi}|^2} \right\}.$$

$$\mathcal{J} = \left\{ \boldsymbol{J} \in L^2\left(\mathbb{R}^d; \mathbb{R}^d\right): \hat{\boldsymbol{J}}(\boldsymbol{\xi}) \cdot \boldsymbol{\xi} = 0 \right\}.$$

Theorem A.1. *Let $\Omega \subset \mathbb{R}^d$ be an open, bounded, connected set with Lipschitz boundary. Then*

$$\mathcal{E}(\Omega) = \left\{ \nabla\phi: \phi \in H^1(\Omega) \right\}. \tag{A.1}$$

Moreover, $\mathcal{E}(\Omega)$ is a closed subspace of $L^2(\Omega; \mathbb{R}^d)$.

Proof. Let $\mathcal{X} = \{\nabla\phi : \phi \in H^1(\Omega)\}$. The closedness of $\mathcal{X}$ in $L^2(\Omega)$ follows from the Poincaré inequality

$$\|\phi - \langle\phi\rangle\|_{H^1(\Omega)} \leqslant C(\Omega)\|\nabla\phi\|_{L^2(\Omega)}, \tag{A.2}$$

where $\langle\phi\rangle$ is the average of ϕ over Ω. Indeed, if $\nabla\phi_n \to \boldsymbol{f}$ in $L^2(\Omega)$, then $\psi_n = \phi_n - \langle\phi_n\rangle$ is bounded in $H^1(\Omega)$ by the Poincaré inequality. Extracting a weakly convergent subsequence, we conclude that there exists $\phi_0 \in H^1(\Omega)$, such that $\boldsymbol{f} = \nabla\phi_0$. Hence, $\boldsymbol{f} \in \mathcal{X}$.

By the density theorem 2.12 every $\boldsymbol{E} \in \mathcal{E}$ is a limit in $L^2(\mathbb{R}^d; \mathbb{R}^d)$ of a sequence of functions $\boldsymbol{E}_n$ of the form

$$\boldsymbol{E}_n(\boldsymbol{x}) = \nabla\left(\nabla \cdot \boldsymbol{\phi}_n\right)$$

for some $\boldsymbol{\phi}_n \in C_0^\infty(\mathbb{R}^d; \mathbb{R}^d)$. Thus, the restriction of $\boldsymbol{E}_n$ to Ω is in $\mathcal{X}$. Since $\mathcal{X}$ is closed, we conclude that the restriction of any $\boldsymbol{E} \in \mathcal{E}$ to Ω is in $\mathcal{X}$. This proves that $\mathcal{E}(\Omega) \subset \mathcal{X}$. To prove the reverse inclusion we need to show that every $\boldsymbol{E} \in \mathcal{X}$ can be extended to $\tilde{\boldsymbol{E}} \in \mathcal{E}$. The existence of such an extension is guaranteed by a theorem of Calderón [1] (see also Stein [2]), that says that there exists a (bounded) extension operator from all of $H^1(\Omega)$ to $H^1(\mathbb{R}^d)$. Therefore, every function $\nabla\psi \in \mathcal{X}$ is a restriction of some function $\nabla\Psi$ to Ω, where $\Psi \in H^1(\mathbb{R}^d)$. But then $\nabla\Psi \in \mathcal{E}$. Hence, $\mathcal{X} \subset \mathcal{E}(\Omega)$, establishing the theorem. □

The proof that

$$\mathcal{J}(B_1) = \mathcal{Y}(B_1) = \left\{\boldsymbol{J} \in L^2\left(B_1; \mathbb{R}^d\right): \nabla \cdot \boldsymbol{J} = 0\right\}$$

is performed using a completely different idea. It is obvious that $\mathcal{J}(B_1) \subset \mathcal{Y}(B_1)$. It is also obvious that $\mathcal{Y}(B_1)$ is a closed subspace in $L^2(B_1; \mathbb{R}^d)$. The non-trivial part is the possibility of extension of $\boldsymbol{J} \in \mathcal{Y}(B_1)$ to $\tilde{\boldsymbol{J}} \in \mathcal{J}$. Such an extension can be constructed explicitly using the Kelvin transform.

Lemma A.2. *Let $\boldsymbol{J} \in \mathcal{Y}(B_1)$. Let $\alpha > 0$ and let*

$$\boldsymbol{K}_\alpha(\boldsymbol{x}) = \frac{\boldsymbol{x}}{|\boldsymbol{x}|^{1+\alpha}} \tag{A.3}$$

be the Kelvin transform. Let

$$\tilde{\boldsymbol{J}}(\boldsymbol{x}) = \begin{cases} \boldsymbol{J}(\boldsymbol{x}), & |\boldsymbol{x}| < 1, \\ \operatorname{cof}(\nabla \boldsymbol{K}_\alpha(\boldsymbol{x}))^T \boldsymbol{J}(\boldsymbol{K}_\alpha(\boldsymbol{x})), & |\boldsymbol{x}| > 1. \end{cases}$$

Then $\tilde{\boldsymbol{J}} \in \mathcal{J}$.

Proof. The Kelvin transform maps interior of the unit ball onto the exterior and vice versa, while $\boldsymbol{K}_\alpha(\boldsymbol{x}) = \boldsymbol{x}$, when $|\boldsymbol{x}| = 1$. If $|\boldsymbol{x}| < 1$, then $\nabla \cdot \tilde{\boldsymbol{J}}(\boldsymbol{x}) = \nabla \cdot \boldsymbol{J} = 0$. If $|\boldsymbol{x}| > 1$, then

$$\nabla \cdot \tilde{\boldsymbol{J}}(\boldsymbol{x}) = \det(\nabla \boldsymbol{K}_\alpha)(\nabla \cdot \boldsymbol{J})(\boldsymbol{K}_\alpha(\boldsymbol{x})) = 0.$$

Finally, we compute $\nabla \boldsymbol{K}_\alpha(\boldsymbol{x}) = \boldsymbol{I}_d - (\alpha + 1)\boldsymbol{n} \otimes \boldsymbol{n}$ at $|\boldsymbol{x}| = 1$, where $\boldsymbol{n} = \boldsymbol{x}/|\boldsymbol{x}|$ is the outward unit normal to B_1. Therefore, on $|\boldsymbol{x}| = 1$

$$\operatorname{cof}(\boldsymbol{I}_d - (\alpha + 1)\boldsymbol{n} \otimes \boldsymbol{n})^T \boldsymbol{J}(\boldsymbol{K}_\alpha(\boldsymbol{x})) \cdot \boldsymbol{n} = \boldsymbol{J}(\boldsymbol{x}) \cdot \boldsymbol{n}.$$

Thus, $\nabla \cdot \tilde{\boldsymbol{J}} = 0$ on $\mathbb{R}^d$ in the sense of distributions. It remains to verify that $\tilde{\boldsymbol{J}} \in L^2(B_1^c; \mathbb{R}^d)$. We compute, making the substitution $\boldsymbol{y} = \boldsymbol{K}_\alpha(\boldsymbol{x})$,

$$\int_{B_1^c} \left|\tilde{\boldsymbol{J}}(\boldsymbol{x})\right|^2 \mathrm{d}\boldsymbol{x} =$$

$$\int_{B_1} |\boldsymbol{y}|^{(1+1/\alpha)(d-2)} |\mathrm{cof}(\boldsymbol{I}_d - (\alpha + 1)\hat{\boldsymbol{y}} \otimes \hat{\boldsymbol{y}})\boldsymbol{J}(\boldsymbol{y})|^2 \det(\boldsymbol{I}_d - (\alpha + 1)\hat{\boldsymbol{y}} \otimes \hat{\boldsymbol{y}})\mathrm{d}\boldsymbol{y},$$

where $\hat{\boldsymbol{y}} = \boldsymbol{y}/|\boldsymbol{y}|$. This calculation shows that $\tilde{\boldsymbol{J}} \in L^2(B_1^c; \mathbb{R}^d)$, when $d \geqslant 2$. □

Lemma A.2 is sufficient to establish the $\mathcal{J}$-regularity for conductivity. The same method can be used to prove that $\mathcal{J}(\Omega) = \mathcal{Y}(\Omega)$ for domains Ω diffeomorphic to a unit ball (where the diffeomorphism extends up to the boundary).

A.2 Elasticity

The proof of $\mathcal{E}$- and $\mathcal{J}$-regularity is a fairly straightforward extension of the same proof for conductivity:

$$\mathcal{E} = \left\{ \boldsymbol{E} \in L^2\left(\mathbb{R}^d; \mathrm{Sym}(\mathbb{R}^d)\right) : \hat{\boldsymbol{E}}(\boldsymbol{\xi}) = \frac{2\hat{\boldsymbol{E}}(\boldsymbol{\xi})\boldsymbol{\xi} \odot \boldsymbol{\xi}}{|\boldsymbol{\xi}|^2} - \frac{\left(\hat{\boldsymbol{E}}(\boldsymbol{\xi})\boldsymbol{\xi} \cdot \boldsymbol{\xi}\right)\boldsymbol{\xi} \otimes \boldsymbol{\xi}}{|\boldsymbol{\xi}|^4} \right\},$$

$$\mathcal{J} = \left\{ \boldsymbol{\sigma} \in L^2\left(\mathbb{R}^d; \mathrm{Sym}(\mathbb{R}^d)\right) : \hat{\boldsymbol{\sigma}}(\boldsymbol{\xi})\boldsymbol{\xi} = \boldsymbol{0} \right\},$$

where

$$\boldsymbol{a} \odot \boldsymbol{b} = \frac{1}{2}(\boldsymbol{a} \otimes \boldsymbol{b} + \boldsymbol{b} \otimes \boldsymbol{a}).$$

Theorem A.3. *Let $\Omega \subset \mathbb{R}^d$ be an open, bounded, connected Lipschitz domain. Then*

$$\mathcal{E}(\Omega) = \left\{ e(\boldsymbol{u}) : \boldsymbol{u} \in H^1\left(\Omega; \mathbb{R}^d\right) \right\}, \tag{A.4}$$

where

$$e(\boldsymbol{u}) = \frac{1}{2}\left(\nabla \boldsymbol{u} + (\nabla \boldsymbol{u})^T\right).$$

Moreover, $\mathcal{E}(\Omega)$ is a closed subspace of $L^2(\Omega; \mathrm{Sym}(\mathbb{R}^d))$.

Proof. Let $\mathcal{X} = \{e(\boldsymbol{u}) : \boldsymbol{u} \in H^1(\Omega; \mathbb{R}^d)\}$. Let us first prove that $\mathcal{X}$ is closed in $L^2(\Omega)$. Suppose, that $e(\boldsymbol{u}_n) \to \boldsymbol{E}$ in $L^2(\Omega)$. By the Korn inequality there exists $C = C(\Omega)$ and skew-symmetric matrices $\boldsymbol{S}_n \in \mathrm{Skew}(\mathbb{R}^d)$, such that

$$\|\nabla \boldsymbol{u}_n - \boldsymbol{S}_n\|_{L^2(\Omega)} \leqslant C(\Omega)\|e(\boldsymbol{u}_n)\|_{L^2(\Omega)}.$$

Thus, $\|\nabla \boldsymbol{v}_n\|_{L^2(\Omega)}$ is bounded, where $\boldsymbol{v}_n = \boldsymbol{u}_n - \boldsymbol{S}_n\boldsymbol{x}$. By the Poincaré inequality (A.2), $\boldsymbol{w}_n = \boldsymbol{v}_n - \langle \boldsymbol{v}_n \rangle$ is bounded in $H^1(\Omega; \mathbb{R}^d)$. Extracting a weakly convergent subsequence, we conclude that there exists $\boldsymbol{w}_0 \in H^1(\Omega; \mathbb{R}^d)$, such that $\boldsymbol{E} = e(\boldsymbol{w}_0)$. Hence, $\boldsymbol{E} \in \mathcal{X}$.

By the density theorem 2.12 every $\boldsymbol{E} \in \mathcal{E}$ is a limit in $L^2(\mathbb{R}^d; \mathrm{Sym}(\mathbb{R}^d))$ of a sequence of functions $\boldsymbol{E}_n$ of the form $\boldsymbol{E}_n = e(\boldsymbol{\phi}_n)$, for some $\boldsymbol{\phi}_n \in C_0^\infty(\mathbb{R}^d; \mathbb{R}^d)$. Specifically, theorem 2.12 says that

$$\boldsymbol{\phi}_n = 2\nabla \cdot \boldsymbol{\Phi}_n - \nabla(\nabla \cdot (\nabla \cdot \boldsymbol{\Phi}_n))$$

for some $\boldsymbol{\Phi}_n \in C_0^\infty(\mathbb{R}^d; \mathrm{Sym}(\mathbb{R}^d))$, but this is not important. It follows that the restriction of $\boldsymbol{E}_n$ to Ω is in $\mathcal{X}$. Since $\mathcal{X}$ is closed, we conclude that, the restriction of any $\boldsymbol{E} \in \mathcal{E}$ to Ω is in $\mathcal{X}$. This proves that $\mathcal{E}(\Omega) \subset \mathcal{X}$. To prove the reverse inclusion we need to show that every $\boldsymbol{E} \in \mathcal{X}$ can be extended to $\tilde{\boldsymbol{E}} \in \mathcal{E}$. The existence of such an extension is guaranteed by a theorem of Calderón [1] (see also Stein [2]), that says that there exists a bounded extension operator from $H^1(\Omega; \mathbb{R}^d)$ to $H^1(\mathbb{R}^d; \mathbb{R}^d)$. Therefore, every function $e(\boldsymbol{u}) \in \mathcal{X}$ is a restriction of some function $e(\boldsymbol{U})$ for some $\boldsymbol{U} \in H^1(\mathbb{R}^d; \mathbb{R}^d)$. Observing that $e(\boldsymbol{U}) \in \mathcal{E}$ for every $\boldsymbol{U} \in H^1(\mathbb{R}^d; \mathbb{R}^d)$, implies $\mathcal{X} \subset \mathcal{E}(\Omega)$, establishing the theorem. □

To prove $\mathcal{J}$-regularity for elasticity we modify the Kelvin transform method that we used for conductivity. Suppose $\boldsymbol{\sigma} \in L^2(B_1; \mathrm{Sym}(\mathbb{R}^d)$ is such that $\nabla \cdot \boldsymbol{\sigma} = 0$. We first define operators

$$X_\alpha \boldsymbol{\sigma} = \mathrm{cof}(\nabla \boldsymbol{K}_\alpha(\boldsymbol{x}))^T \boldsymbol{\sigma}(\boldsymbol{K}_\alpha(\boldsymbol{x}))\mathrm{cof}(\nabla \boldsymbol{K}_\alpha(\boldsymbol{x})), \quad \alpha > 0,$$

where $\boldsymbol{K}_\alpha$ is given by (A.3). Then

$$\nabla \cdot (X_\alpha \boldsymbol{\sigma}) = \det(\nabla \boldsymbol{K}_\alpha)\mathrm{cof}(\nabla \boldsymbol{K}_\alpha(\boldsymbol{x}))^T (\nabla \cdot \boldsymbol{\sigma})(\boldsymbol{K}_\alpha(\boldsymbol{x})),$$

proving that $\nabla \cdot (X_\alpha \boldsymbol{\sigma}) = 0$ on B_1^c. We also compute

$$\mathrm{cof}(\nabla \boldsymbol{K}_\alpha) = \boldsymbol{I}_d + (\alpha + 1)(\boldsymbol{n} \otimes \boldsymbol{n} - \boldsymbol{I}_d), \quad |\boldsymbol{x}| = 1.$$

Therefore,

$$(X_\alpha \boldsymbol{\sigma})\boldsymbol{n} = \boldsymbol{\sigma n} + (\alpha + 1)(\boldsymbol{n} \otimes \boldsymbol{n} - \boldsymbol{I}_d)\boldsymbol{\sigma n}, \quad |\boldsymbol{x}| = 1.$$

This shows that no X_α can work as an extension operator. The idea is then to define the extension operator as

$$X_{\mathcal{J}}(\Omega) = \sum_{j=1}^{N} \gamma_j X_{\alpha_j}$$

where the real numbers γ_j are chosen in such a way as to satisfy

$$\sum_{j=1}^{N} \gamma_j = 1, \qquad \sum_{j=1}^{N} \gamma_j(\alpha_j + 1) = 0.$$

It is enough to choose $N = 2$, any $\alpha_1 \neq \alpha_2$ and

$$\gamma_1 = \frac{\alpha_2 + 1}{\alpha_2 - \alpha_1}, \qquad \gamma_2 = \frac{\alpha_1 + 1}{\alpha_1 - \alpha_2}.$$

Defining

$$\tilde{\boldsymbol{\sigma}} = \gamma_1 X_{\alpha_1}\boldsymbol{\sigma} + \gamma_2 X_{\alpha_2}\boldsymbol{\sigma}$$

we obtain a divergence-free extension of $\boldsymbol{\sigma}$ to all of $\mathbb{R}^d$. It remains to verify that $\tilde{\boldsymbol{\sigma}} \in L^2(B_1^c)$, which would follow from the estimate

$$\|X_\alpha \boldsymbol{\sigma}\|_{L^2(B_1^c)} \leqslant C\,\|\boldsymbol{\sigma}\|_{L^2(B_1)}. \tag{A.5}$$

We prove (A.5) in the same way as in the case of conductivity, by changing variables $\boldsymbol{y} = \boldsymbol{K}_\alpha(\boldsymbol{x})$ in the integral

$$\|X_\alpha \boldsymbol{\sigma}\|^2_{L^2(B_1^c)} = \int_{B_1^c} \left|\operatorname{cof}(\nabla \boldsymbol{K}_\alpha(\boldsymbol{x}))^T \boldsymbol{\sigma}(\boldsymbol{K}_\alpha(\boldsymbol{x}))\operatorname{cof}(\nabla \boldsymbol{K}_\alpha(\boldsymbol{x}))\right|^2 \mathrm{d}\boldsymbol{x}.$$

Using formulas

$$\nabla \boldsymbol{K}_\alpha(\boldsymbol{x}) = |\boldsymbol{y}|^{1+1/\alpha}(\boldsymbol{I}_d - (\alpha + 1)\hat{\boldsymbol{y}} \otimes \hat{\boldsymbol{y}}),$$

$$\mathrm{d}\boldsymbol{x} = \det\left(|\boldsymbol{y}|^{-(1+1/\alpha)}\left(\boldsymbol{I}_d - \left(1 + \frac{1}{\alpha}\right)\hat{\boldsymbol{y}} \otimes \hat{\boldsymbol{y}}\right)\right)\mathrm{d}\boldsymbol{y}.$$

We obtain the estimate

$$\|X_\alpha \boldsymbol{\sigma}\|^2_{L^2(B_1^c)} \leqslant C \int_{B_1} |\boldsymbol{y}|^{(3d-4)(1+1/\alpha)} |\boldsymbol{\sigma}(\boldsymbol{y})|^2 \mathrm{d}\boldsymbol{y} \leqslant C' \|\boldsymbol{\sigma}\|^2_{L^2(B_1)},$$

whenever $d \geqslant 2$.

Bibliography

[1] Calderón A 1961 Lebesgue spaces of differentiable functions *Proc. Symp. Pure Math.* **4** 33–49

[2] Stein E M 1970 *Singular Integrals and Differentiability Properties of Functions* (*Princeton Mathematical Series* vol 30) (Princeton, NJ: Princeton University Press)

Appendix B

A polycrystalline L-relation that is not exact

In this section we describe the construction of a polycrystalline L-relation that is not exact from [1]. This example has important implications in Calculus of variations. However, this discussion is not directly relevant to the subject of the book and we omit it here.

Let us recall that for two-dimensional composites theorem 4.12 states that the three-chain property (see definition 4.7) must be satisfied by all Jordan $\mathcal{A}$-multialgebra corresponding to exact relations. In this chapter we construct a rotationally invariant Jordan $\mathcal{A}$-multialgebra that fails the three-chain property, and hence leads to an explicit example of an L-relation that is not exact. The example is found in the context of two-dimensional multifield response composite materials [3–5], coupling four curl-free fields $\boldsymbol{E} = (\nabla\phi_1, \dots, \nabla\phi_4)$ and four divergence-free fields $\boldsymbol{J} = (j_1, \dots, j_4)$. In this case

$$\mathcal{T} = \underbrace{\mathbb{R}^2 \oplus \cdots \oplus \mathbb{R}^2}_{4} = \mathbb{R}^4 \otimes \mathbb{R}^2$$

can be interpreted in two very different ways. On the one hand we can identify $\mathbb{R}^2$ with the set of complex numbers, so that $\mathcal{T} \cong \mathbb{C}^4$. This representation of the space $\mathcal{T}$ makes it convenient to describe the action of rotations. If $\boldsymbol{R}_\theta$ denotes the counter-clockwise rotation of the plane $\mathbb{R}^2$ through the angle θ, then

$$\boldsymbol{R}_\theta \cdot \boldsymbol{u} = e^{i\theta}\boldsymbol{u}, \qquad \boldsymbol{u} \in \mathcal{T} \cong \mathbb{C}^4.$$

On the other hand we can identify $\mathbb{R}^4$ with the set of quaternions $\mathbb{H}$, so that $\mathcal{T} \cong \mathbb{H}^2$. In the first interpretation every $\mathsf{K} \in \mathrm{Sym}(\mathcal{T})$ is uniquely determined by a complex Hermitian 4×4 matrix X and a complex symmetric 4×4 matrix Y by the rule

$$\mathsf{K}\boldsymbol{u} = X\boldsymbol{u} + Y\overline{\boldsymbol{u}}, \qquad \boldsymbol{u} \in \mathbb{C}^4.$$

Henceforth we will write $\mathsf{K}(X, Y)$ to indicate this parametrization of $\mathrm{Sym}(\mathcal{T})$. We easily compute the action of rotations $\boldsymbol{R}_\theta$ on $\mathsf{K}(X, Y)$:

$$\boldsymbol{R}_\theta \cdot \mathsf{K}(X, Y) = \mathsf{K}\left(X, e^{2i\theta} Y\right). \tag{B.1}$$

doi:10.1088/978-0-7503-1048-2ch14

Therefore, if Π is an SO(2)-invariant subspace of $\mathrm{Sym}(\mathcal{T})$ then

$$\Pi = \Pi_{V,W} = \left\{(X,\ Y) : X \in V \subset \mathcal{H}(\mathbb{C}^4),\ Y \in W \subset \mathrm{Sym}(\mathbb{C}^4)\right\},$$

where V can be any real subspace of $\mathcal{H}(\mathbb{C}^4)$, the set of all complex Hermitian 4×4 matrices, and W can be any complex subspace of $\mathrm{Sym}(\mathbb{C}^4)$, the set of all complex symmetric 4×4 matrices. In this notation, the subspace $\mathcal{A}$ is described by $V = \{0\}$ and $W = \mathbb{C}\boldsymbol{I}_4$. The condition that the subspace $\Pi_{V,W}$ is an SO(2)-invariant Jordan $\mathcal{A}$-multialgebra is equivalent to the conditions

$$Y^2 + XX^T \in W, \quad YX + XY^H \in V \text{ for all } X \in V,\ Y \in W, \tag{B.2}$$

where $Y^H = \overline{Y}^T$ denotes Hermitian conjugation.[1] Finally, the three-chain condition is equivalent to

$$\mathrm{i}X_1X_2^TX_3 + \left(\mathrm{i}X_1X_2^TX_3\right)^H \in V, \quad \forall\{X_1,\ X_2,\ X_3\} \subset V, \tag{B.3}$$

provided (B.2) holds as well.

The second, quaternionic representation of $\mathcal{T}$ permits us to think of quaternionic multiplication as a real linear transformation on $\mathbb{R}^4$. For $q \in \mathbb{H}$ we denote by $Q(q)$ a linear transformation

$$\mathbb{R}^4 \ni \boldsymbol{h} \mapsto h \in \mathbb{H} \mapsto qh \in \mathbb{H} \mapsto Q(q)\boldsymbol{h} \in \mathbb{R}^4.$$

In this interpretation $Q(q)$ can be written as a 4×4 real matrix. If we now interpret this matrix as a *complex* 4×4 matrix, whose entries just happen to be real, we will obtain the map $\mathfrak{Q}\colon \mathbb{H} \to \mathrm{End}(\mathbb{C}^4) \subset \mathrm{End}(\mathcal{T})$ that allows us to think of quaternions as real linear maps on $\mathcal{T}$. It is easy to see that $\mathfrak{Q}(q) = \mathsf{K}(Q(q), 0)$. We now define

$$V = \left\{\mathrm{i}\mathfrak{Q}(q) : q \in \mathbb{H},\ \mathfrak{Re}(q) = 0\right\} \subset \mathfrak{H}(\mathbb{C}^4), \qquad W = \mathbb{C}\boldsymbol{I}_4 \subset \mathrm{Sym}(\mathbb{C}^4). \tag{B.4}$$

It is not hard to verify that $\Pi_{V,\ W}$ is an SO(2)-invariant Jordan $\mathcal{A}$-multialgebra that does not satisfy (B.3). The corresponding submanifold $\mathbb{M}$, given by (4.4), consists of 'quaternionic materials' with constant determinant:

$$\mathbb{M} = \left\{\mathsf{L} = \begin{bmatrix} \lambda & h \\ \overline{h} & \mu \end{bmatrix} : \lambda > 0,\ \mu > 0,\ h \in \mathbb{H},\ \det_{\mathbb{H}}(\mathsf{L}) = \lambda\mu - |h|^2 = 1\right\}. \tag{B.5}$$

Here we use the quaternionic representation of $\mathcal{T}$ to describe operators in $\mathrm{Sym}(\mathcal{T})$. Of course, $\det_{\mathbb{H}}(\mathsf{L})$ is set to 1 in (B.5) only for simplicity. It can be any positive constant.

Remark 14.1. *Curiously, in three space dimensions we have a positive result, proved in [2], that all L-relations for three-dimensional multifield response polycrystalline composites are exact no matter how many fields are coupled.*

---

[1] We do not use the standard notation Y^* to avoid confusion with our notation for the effective tensor.

Bibliography

[1] Grabovsky Y 2017 From microstructure-independent formulas for composite materials to rank-one convex, non-quasiconvex functions, *Arch. Ration. Mech. Anal.* to be published

[2] Grabovsky Y, Milton G W and Sage D S 2000 Exact relations for effective tensors of polycrystals: necessary conditions and sufficient conditions *Commun. Pure Appl. Math.* **53** 300–53

[3] Milgrom M 1997 Some more exact results concerning multifield moduli of two-phase composites *J. Mech. Phys. Solids* **45** 399–404

[4] Milgrom M and Shtrikman S 1989 Linear response of polycrystals to coupled fields: exact relations among the coefficients *Phys. Rev.* B **40** 5991–4

[5] Milgrom M and Shtrikman S 1989 Linear response of two-phase composites with cross moduli: exact universal relations *Phys. Rev.* A **40** 1568–75

IOP Publishing

Composite Materials
Mathematical theory and exact relations
Yury Grabovsky

Appendix C

Multiplication of SO(3) irreps in real endomorphism algebras

C.1 Irreducible representations (irreps) of SO(3)

Here we give an explicit construction of all irreducible representations of SO(3). We follow the exposition in [2] but provide more details. Since the theory is classical and very well known in quantum mechanics as quantization of angular momentum, we limit ourselves to statements of fact without proof or discussion. These can be found in numerous textbooks on both group representations (for mathematicians) and quantum mechanics (for physicists).

The main idea to deal with complicated Lie groups, like SO(3) is to replace them with geometrically simpler groups, called universal covers. For SO(3), which is topologically a three-dimensional projective space, the universal cover is SU(2)

$$\mathrm{SU}(2) = \left\{ \begin{bmatrix} a & b \\ -\bar{b} & \bar{a} \end{bmatrix} : \{a, b\} \subset \mathbb{C}, |a|^2 + |b|^2 = 1 \right\},$$

which is isomorphic to a three-dimensional unit sphere. The complex SU(2) irreps are spaces $\mathrm{Sym}^m(\mathbb{C}^2)$, where $g \in \mathrm{SU}(2)$ acts on the commutative and associative formal product of any m vectors $\boldsymbol{u}_1 \ldots \boldsymbol{u}_m$ by

$$g \cdot \boldsymbol{u}_1 \ldots \boldsymbol{u}_m = (g\boldsymbol{u}_1) \ldots (g\boldsymbol{u}_m).$$

A basis of $\mathrm{Sym}^m(\mathbb{C}^2)$ is $\{\boldsymbol{e}_1^j \boldsymbol{e}_2^{m-j} : j = 0, \ldots, m\}$. For $\boldsymbol{x} \in \mathbb{R}^3$ we define

$$\rho(\boldsymbol{x}) = \begin{bmatrix} \mathrm{i}x_1 & x_2 + \mathrm{i}x_3 \\ -x_2 + \mathrm{i}x_3 & -\mathrm{i}x_1 \end{bmatrix}.$$

We define $\pi(g) \in \mathrm{SO}(3)$ by its action on $\boldsymbol{x} \in \mathbb{R}^3$:

$$\pi(g)\boldsymbol{x} = \rho^{-1}\left(g\rho(\boldsymbol{x})g^{-1}\right).$$

We note that $\pi : \mathrm{SU}(2) \to \mathrm{SO}(3)$ is a double cover,[1] whose kernel is $\pm \boldsymbol{I}_2$. Thus, any irrep of SU(2) where $-\boldsymbol{I}_2$ acts trivially will also be an irrep of SO(3). Hence SO(3) irreps are $\mathrm{Sym}^m(\mathbb{C}^2)$ for even m.

The subgroup

$$T = \left\{ \begin{bmatrix} e^{i\theta} & 0 \\ 0 & e^{-i\theta} \end{bmatrix} : \theta \in \mathbb{R} \mod 2\pi \right\}$$

is a maximal commutative subgroup of SU(2). It is called a maximal torus. Observe that the 'standard' basis vectors $\boldsymbol{e}_1^j \boldsymbol{e}_2^{m-j}$ are eigenvectors of all $g \in T$. If

$$g_\theta = \begin{bmatrix} e^{i\theta} & 0 \\ 0 & e^{-i\theta} \end{bmatrix}$$

then $g_\theta \boldsymbol{e}_1 = e^{i\theta} \boldsymbol{e}_1$ and $g_\theta \boldsymbol{e}_2 = e^{-i\theta} \boldsymbol{e}_2$. Therefore,

$$g_\theta \cdot \boldsymbol{e}_1^j \boldsymbol{e}_2^{m-j} = e^{(2j-m)\theta} \boldsymbol{e}_1^j \boldsymbol{e}_2^{m-j}.$$

The common eigenvectors of a maximal torus are called the weight vectors. We define the weight of $\boldsymbol{e}_1^j \boldsymbol{e}_2^{m-j}$ to be $s = m/2 - j$, $j = 0, \ldots, m$. In other words, we denote by

$$v_s = \boldsymbol{e}_1^{m/2+s} \boldsymbol{e}_2^{m/2-s}, \qquad s = -m/2, -m/2 + 1, \ldots, m/2 - 1, m/2$$

the weight vectors of weight s. When m is even the weights are integers. From now on we will consider only even m, i.e. we will only deal with SO(3) irreps. It turns out that all SO(3) irreps are of real type. That means that there exists a basis of $\mathrm{Sym}^m(\mathbb{C}^2)$ in which all matrices of operators $\pi(g)$ are real. We denote the real SO(3) irreps by W_k, $k = 0, 1, 2, \ldots$ so that $W_k \otimes \mathbb{C} \cong \mathrm{Sym}^{2k}(\mathbb{C}^2)$. Our goal will be to build a basis of W_k and $W_k \otimes \mathbb{C}$ that is independent of the specific model of irreps. In other words, we assume that the action of SO(3) on W_k is known and that a specific maximal torus $T \subset \mathrm{SO}(3)$ has been chosen. To fix notation we choose this maximal torus to be the group of all rotations around x_1 axis. A natural choice for the basis of $W_k \otimes \mathbb{C}$ would be the eigenvectors of T (weight vectors). However, the eigenvectors are defined up to complex multiples, which will also have to be chosen. We start by considering the zero weight vector w_0: a fixed point of T. This vector can be chosen to lie in W_k, i.e. to be real, since its eigenvalue is 1. Of course, w_0 is still defined up to a real scalar multiple. However, once a specific choice of w_0 is made, the bases of W_k and $W_k \otimes \mathbb{C}$ will be constructed in a canonical way.

C.2 Raising and lowering operators

We recall that every representation of a Lie group[2] defines an action of its Lie algebra as a differentiation. Recall that elements of a Lie algebra are tangent vectors

---

[1] A continuous map, for which the pre-image of any point consists of exactly two points.

[2] A Lie group is a group and a manifold at the same time, where we require that group multiplication be continuous.

$A = \dot{g}(0)$ to curves $g(t) \in \mathrm{SU}(2)$, such that $g(0) = \boldsymbol{I}_2$. A tangent vector A can then be regarded as an operator on W_k defined by its action

$$A \cdot w = \left. \frac{\mathrm{d}(\pi(g(t))w)}{\mathrm{d}t} \right|_{t=0}.$$

We consider the following special curves in SU(2):

$$g_0(t) = \begin{bmatrix} \mathrm{e}^{\mathrm{i}t} & 0 \\ 0 & \mathrm{e}^{-\mathrm{i}t} \end{bmatrix}, \quad g_1(t) = \frac{1}{\sqrt{1+t^2}} \begin{bmatrix} 1 & t \\ -t & 1 \end{bmatrix}, \quad g_2(t) = \frac{1}{\sqrt{1+t^2}} \begin{bmatrix} 1 & \mathrm{i}t \\ \mathrm{i}t & 1 \end{bmatrix}.$$

The corresponding operators A will be called Z_0, Z_1 and Z_2, respectively. We easily compute

$$Z_0 = \dot{g}_0(0) = \begin{bmatrix} \mathrm{i} & 0 \\ 0 & -\mathrm{i} \end{bmatrix}, \quad Z_1 = \dot{g}_1(0) = \begin{bmatrix} 0 & 1 \\ -1 & 0 \end{bmatrix}, \quad Z_2 = \dot{g}_2(0) = \begin{bmatrix} 0 & \mathrm{i} \\ \mathrm{i} & 0 \end{bmatrix}.$$

Returning to our model of $W_k \otimes \mathbb{C}$ we compute (for $A = \dot{g}(0)$)

$$A \cdot v_s = \frac{\mathrm{d}}{\mathrm{d}t}\left[(g(t)e_1)^{k+s}(g(t)e_2)^{k-s} \right] = (k+s)e_1^{k+s-1}e_2^{k-s}(Ae_1) + (k-s)e_1^{k+s}e_2^{k-s-1}(Ae_2).$$

The result is linear in A, and hence this formula will also define the action of any complex linear combination of Z_0, Z_1 and Z_2. We define

$$H = -\mathrm{i}Z_0 = \begin{bmatrix} 1 & 0 \\ 0 & -1 \end{bmatrix}, \quad X_+ = \frac{Z_1 - \mathrm{i}Z_2}{2} = \begin{bmatrix} 0 & 1 \\ 0 & 0 \end{bmatrix}, \quad X_- = \frac{Z_1 + \mathrm{i}Z_2}{2} = \begin{bmatrix} 0 & 0 \\ -1 & 0 \end{bmatrix}.$$

Using the relations

$$He_1 = e_1, \quad He_2 = -e_1, \quad X_+e_1 = 0, \quad X_+e_2 = e_1, \quad X_-e_1 = -e_2, \quad X_-e_2 = 0$$

we obtain

$$\begin{aligned} H \cdot v_s &= (k+s)e_1^{k-s}e_2^{k+s} - (k-s)e_1^{k-s}e_2^{k+s} = 2sv_s, \\ X_+ \cdot v_s &= (k-s)e_1^{k+s+1}e_2^{k-s-1} = (k-s)v_{s+1}, \\ X_- \cdot v_s &= -(k+s)e_1^{k+s-1}e_2^{k-s+1} = -(k+s)v_{s-1}. \end{aligned}$$

For this reason operators X_+ and X_- are called the raising and lowering operators, respectively. Thus, if $w_0 \in W_k$ is the zero weight vector then we can generate the canonical basis $\{v_j : -k \leqslant j \leqslant k\}$ of $W_k \otimes \mathbb{C}$ defining recursively

$$v_{j+1} = \frac{X_+ \cdot v_j}{k-j}, \quad v_{-j-1} = -\frac{X_- \cdot v_{-j}}{k-j}, \quad j = 0, 1, \ldots, k-1,$$

where $v_0 = w_0$. In other words,

$$v_j = \frac{(k-j)!}{k!}X_+^j \cdot w_0, \quad v_{-j} = (-1)^j \frac{(k-j)!}{k!}X_-^j \cdot w_0, \quad j = 1, \ldots, k.$$

Taking complex conjugates we obtain

$$\overline{v_j} = (-1)^j v_{-j}.$$

It follows that $w_j = \mathfrak{Re}(v_j)$ and $w_{-j} = \mathfrak{Im}(v_j)$, $j = 1, \ldots, k$, together with w_0 forms the basis of the real vector space W_k.

It is known that there exists a unique, up to a positive scalar multiple, inner product on $W_k \otimes \mathbb{C}$. Let $\{u, v\} \subset W_k \otimes \mathbb{C}$ and let $g(t)$ be a smooth curve in SU(2) with $g(0) = I_2$. Differentiating $(g(t) \cdot u, g(t) \cdot v) = (u, v)$ at $t = 0$ we obtain

$$(A \cdot u, v) + (u, A \cdot v) = 0.$$

(This means that A is a skew-adjoint operator with respect to such an inner product.) In particular, for $A = Z_0$ we have

$$0 = \left(Z_0 \cdot v_j, v_s\right) + \left(v_j, Z_0 \cdot v_s\right) = \left(2\mathrm{i}jv_j, v_s\right) + \left(v_j, 2\mathrm{i}sv_s\right) = 2\mathrm{i}(j - s)(v_j, v_s).$$

It follows that the weight basis $\{v_j : -k \leqslant j \leqslant k\}$ is orthogonal. We compute

$$Z_1 \cdot v_j = X_+ \cdot v_j + X_- \cdot v_j = (k - j)v_{j+1} - (k + j)v_{j-1}.$$

Similarly,

$$Z_2 \cdot v_j = -\mathrm{i}(X_- - X_+) \cdot v_j = \mathrm{i}(k + j)v_{j-1} + \mathrm{i}(k - j)v_{j+1}.$$

Using these formulas and the skew-adjointness of Z_1 and Z_2 we obtain

$$\|v_j\|^2 = \frac{k + j}{k - j + 1} \|v_{j-1}\|^2,$$

from which we obtain the values of the norms of v_j:

$$\|v_j\|^2 = \frac{\binom{2k}{k}}{\binom{2k}{k + j}} \|w_0\|^2, \quad -k \leqslant j \leqslant k. \tag{C.1}$$

This formula shows, in particular, that $\|v_j\| = \|v_{-j}\|$. It is then easy to check (via formulas)

$$w_j = \frac{v_j + (-1)^j v_{-j}}{2}, \quad w_{-j} = \frac{v_j - (-1)^j v_{-j}}{2\mathrm{i}}, \quad j = 1, \ldots, k,$$

that $\{w_j\}$ is an orthogonal system. We compute

$$\|w_j\|^2 = \|w_{-j}\|^2 = \frac{\|v_j\|^2}{2}, \quad j = 1, \ldots, k.$$

C.3 Homogeneous coordinates of irreps in endomorphism algebras.

From now on we follow the exposition in [1], while giving more details on topics that do not occur in the classical theory of quantization of angular momentum.

Let

$$V = \mathbb{R}^p \otimes (W_0 \oplus W_1 \oplus \cdots \oplus W_n).$$

Then

$$\mathrm{End}(V) = \mathrm{End}(\mathbb{R}^p) \otimes \bigoplus_{j,k=0}^{n} \mathrm{Hom}(W_j, W_k).$$

Recall that $\mathrm{Hom}(W_j, W_k) \cong W_k \otimes W_j$. The decomposition of a tensor product of irreps into a direct sum of irreps is called the Clebsch–Gordan decomposition in representation theory. For SO(3)

$$W_k \otimes W_j \cong \bigoplus_{\alpha=|j-k|}^{j+k} W_\alpha. \tag{C.2}$$

Let $0 \leqslant \alpha \leqslant 2n$. Formula (C.2) shows that for each ordered pair (j, k), such that $|j-k| \leqslant \alpha \leqslant j+k$ there is a unique copy of W_α in $\mathrm{Hom}(W_j, W_k)$, which contains a unique, up to a scalar multiple zero weight vector. Suppose that we have made a choice of a real zero weight vector $\mathsf{E}^{jk\alpha}$ for every such $W_\alpha \subset \mathrm{Hom}(W_j, W_k)$. Then any zero weight vector of a copy of $W_\alpha \subset \mathrm{End}(V)$ can be written as

$$E^\alpha(\mathbb{X}) = \sum_{j,k} X_{jk} \otimes \mathsf{E}^{jk\alpha}$$

for some $p \times p$ matrices $X_{jk} \in \mathrm{End}(\mathbb{R}^p)$. Clearly, $E^\alpha(\mathbb{X})$ and $E^\alpha(\mathbb{Y})$ determine the same $W_\alpha \subset \mathrm{End}(V)$ if and only if $\mathbb{Y}$ is a scalar multiple of $\mathbb{X}$. For this reason the block-matrix $\mathbb{X}$ will be called the matrix of homogeneous coordinates of $W_\alpha \subset \mathrm{End}(V)$. We will denote this copy of W_α by $W_\alpha(\mathbb{X})$.

C.4 Clebsch–Gordan coefficients

In this section we apply the classical theory of preceeding sections and derive the classical decomposition of the tensor product of two irreps into a direct sum of irreps. We give here all the details because we will be applying the same calculation to the matrix product of two irreps.

Let $\{v^j_A : -j \leqslant A \leqslant j\}$ be the canonical basis in $W_j \otimes \mathbb{C}$. It is clear that operators $v^k_{B-A} \otimes v^j_A$ are weight vectors of weight B. Then there are explicitly given real numbers $C^{jk\alpha}_{A,B}$ such that

$$e_B(jk\alpha) = \sum_A C^{jk\alpha}_{A,B} v^k_{B-A} \otimes v^j_A$$

form a canonical basis of $W_\alpha \otimes \mathbb{C} \subset \mathrm{Hom}(W_j, W_k) \otimes \mathbb{C}$.

In order to find the coefficients $C^{jk\alpha}_{A,B}$ (called the Clebsch–Gordan coefficients) we can first consider the highest weight vector

$$e_\alpha(jk\alpha) = \sum_{A=\alpha-k}^{j} C^{jk\alpha}_{A,\alpha} v^k_{\alpha-A} \otimes v^j_A.$$

It has the property that it will be killed by X_+:

$$0 = X_+ \cdot e_\alpha(jk\alpha) = \sum_{A=\alpha-k}^{j} C^{jk\alpha}_{A,\alpha}\left\{(k-\alpha+A)v^k_{\alpha-A+1}\otimes v^j_A + (j-A)v^k_{\alpha-A}\otimes v^j_{A+1}\right\}.$$

Therefore,

$$\sum_{A=\alpha-k+1}^{j}\left\{C^{jk\alpha}_{A-1,\alpha}(j-A+1)+(k-\alpha+A)C^{jk\alpha}_{A,\alpha}\right\}v^k_{\alpha-A+1}\otimes v^j_A = 0.$$

This implies that

$$C^{jk\alpha}_{A,\alpha} = -\frac{j-A+1}{k-\alpha+A}C^{jk\alpha}_{A-1,\alpha}.$$

Hence,

$$C^{jk\alpha}_{A,\alpha} = (-1)^{A+k-\alpha}\frac{(j-A+1)\ldots(j+k-\alpha)}{(k-\alpha+A)!}C^{jk\alpha}_{\alpha-k,\alpha} = (-1)^{A+k-\alpha}\binom{j+k-\alpha}{j-A}C^{jk\alpha}_{\alpha-k,\alpha},$$

and

$$e_\alpha(jk\alpha) = C^{jk\alpha}_{\alpha-k,\,\alpha}\sum_{A=\alpha-k}^{j}(-1)^{A+k-\alpha}\binom{j+k-\alpha}{j-A}v^k_{\alpha-A}\otimes v^j_A.$$

The highest weight vector is defined up to a scalar multiple (dependent on j, k and α), hence, changing index of summation $a = A - \alpha + k$ we obtain

$$e_\alpha(jk\alpha) = \sum_{a=0}^{j+k-\alpha}(-1)^a\binom{j+k-\alpha}{a}v^k_{k-a}\otimes v^j_{\alpha-k+a}.$$

We now apply the lowering operator X_- s times, using the formula

$$X^s_- v^k_A = (-1)^s\binom{k+A}{s}s!\,v^k_{A-s}.$$

$$\begin{aligned} X^s_-\cdot e_\alpha(jk\alpha) &= \sum_{a=0}^{j+k-\alpha}(-1)^a\binom{j+k-\alpha}{a} \\ &\quad\times\sum_{\nu=0}^{s}\binom{s}{\nu}X^\nu_-\cdot v^k_{k-a}\otimes X^{s-\nu}_-\cdot v^j_{\alpha-k+a}. \end{aligned}$$

Thus,

$$\begin{aligned} \binom{2\alpha}{s}e_{\alpha-s}(jk\alpha) &= \sum_{a=0}^{j+k-\alpha}(-1)^a\binom{j+k-\alpha}{a} \\ &\quad\times\sum_{\nu=0}^{s}\binom{2k-a}{\nu}\binom{\alpha+j-k+a}{s-\nu}v^k_{k-a-\nu}\otimes v^j_{\alpha-k+a-s+\nu}. \end{aligned}$$

Introducing notation $B = \alpha - s$ and changing index of summation from ν to $A = k - a - \nu$ we obtain

$$\binom{2\alpha}{\alpha + B} e_B(jk\alpha) = \sum_{a=0}^{j+k-\alpha} (-1)^a \binom{j+k-\alpha}{a} \times \sum_{A=k-a+B-\alpha}^{k-a} \binom{2k-a}{k+A} \binom{\alpha+j-k+a}{j+B-A} v_A^k \otimes v_{B-A}^j.$$

We observe that if $A > k - a$, then $k + A > 2k - a$, so that $\binom{2k-a}{k+A} = 0$, while, if $A < k - a + B - \alpha$ then $j + B - A > j - k + a + \alpha$, so that $\binom{\alpha+j-k+a}{j+B-A} = 0$. In that case we can write

$$e_B(jk\alpha) = \frac{1}{\binom{2\alpha}{\alpha+B}} \sum_{a=0}^{j+k-\alpha} (-1)^a \binom{j+k-\alpha}{a} \times \sum_{A=B-j}^{k} \binom{2k-a}{k+A} \binom{\alpha+j-k+a}{j+B-A} v_A^k \otimes v_{B-A}^j,$$

since $0 \leqslant a \leqslant j + k - \alpha$. Hence,

$$e_B(jk\alpha) = \sum_{A=B-j}^{k} C_{A,B}^{jk\alpha} v_A^k \otimes v_{B-A}^j, \quad |B| \leqslant \alpha, \tag{C.3}$$

where

$$C_{A,B}^{jk\alpha} = \frac{1}{\binom{2\alpha}{\alpha+B}} \sum_{a=0}^{j+k-\alpha} (-1)^a \binom{j+k-\alpha}{a} \binom{2k-a}{k+A} \binom{\alpha+j-k+a}{j+B-A} \tag{C.4}$$

are the Clebsch–Gordan coefficients.[3] From (C.3) we obtain

$$e_0(jk\alpha) = \sum_{A=-\min(j,k)}^{\min(j,k)} C_{A,0}^{jk\alpha} v_A^k \otimes v_{-A}^j.$$

Taking complex conjugate and changing index of summation from A to $-A$ we obtain

$$\overline{e_0(jk\alpha)} = \sum_{A=-\min(j,k)}^{\min(j,k)} C_{-A,0}^{jk\alpha} v_A^k \otimes v_{-A}^j.$$

[3] The Clebsch–Gordan coefficients used in quantum mechanics have different normalization. In a slight abuse of terminology we will call numbers (C.4) the Clebsch–Gordan coefficients.

Changing the index of the other summation from $a' = j + k - \alpha - a$ in (C.4) we obtain

$$C^{jk\alpha}_{-A,0} = (-1)^{j+k-\alpha} C^{kj\alpha}_{A,0}.$$

Finally, we use a symmetry of Clebsch–Gordan coefficients $C^{jk\alpha}_{A,0} = C^{kj\alpha}_{A,0}$, and conclude that zero weight vectors $\mathrm{i}^{j+k-\alpha} e_0(jk\alpha)$ have to be real.

C.5 Choice structure coefficients

Formula (C.3) gives a canonical choice of real zero weight vectors $\mathrm{i}^{j+k-\alpha} e_B(jk\alpha)$ in $W_\alpha \subset \mathrm{Hom}(W_j, W_k)$. However, in applications we usually describe such 'coordinate irreps' in terms that are natural for that specific application. For this reason, we usually have a natural choice of zero weight vectors within these irreps. We denoted these chosen zero weight vectors as $\mathsf{E}^{jk\alpha}$ and used them to define the homogeneous coordinates for specifying an irrep $W_\alpha \subset \mathrm{End}(V)$. Thus, there exist *real* non-zero constants $K_{jk\alpha}$, such that

$$\mathsf{E}^{jk\alpha} = K_{jk\alpha} \mathrm{i}^{j+k-\alpha} e_0(jk\alpha).$$

These constants depend not only on our choice of zero weight vectors $\mathsf{E}^{jk\alpha}$, but also on the choice of real zero weight vectors $w_0^k \in W_k \subset \mathcal{W}$, where

$$\mathcal{W} = W_0 \oplus W_1 \oplus \cdots \oplus W_n,$$

as well as on the specific isomorphisms $\mathrm{Hom}(W_j, W_k) \cong W_k \otimes W_j$. These isomorphisms are determined by a choice of SO(3)-invariant inner products on all $W_k \subset \mathcal{W}$ via the formula

$$(a \otimes b)u = (b, u)_{W_j} a, \qquad a \in W_k,\ \{b, u\} \subset W_j.$$

We call constants $K_{jk\alpha}$ *choice structure coefficients*, since they depend on our choices of zero weight vectors and inner products. In order to compute them we compare the action of our zero weight vectors $\mathsf{E}^{jk\alpha}$ to the action of canonical zero weight vectors $\mathrm{i}^{j+k-\alpha} e_0(jk\alpha)$, given by (C.3). In the case when $j + k - \alpha$ is even the numbers $C^{jk\alpha}_{0,0}$ are non zero and hence,

$$\mathsf{E}^{jk\alpha} w_0^j = K_{jk\alpha} \mathrm{i}^{j+k-\alpha} C^{jk\alpha}_{0,0} \|w_0^j\|^2 w_0^k.$$

Taking an inner product with w_0^k, we obtain

$$\left(\mathsf{E}^{jk\alpha} w_0^j, w_0^k\right)_{W_k} = K_{jk\alpha} \mathrm{i}^{j+k-\alpha} C^{jk\alpha}_{0,0} \|w_0^j\|^2 \|w_0^k\|^2.$$

If $j + k - \alpha$ is odd then we need to apply the raising (or the lowering) operator to w_0^j, obtaining $v_1^j = (X_+ \cdot w_0^j)/j$ and $v_1^k = (X_+ \cdot w_0^k)/k$. Then, using (C.3),

$$e_0(jk\alpha) v_1^j = -C^{jk\alpha}_{1,0} \|v_1^j\|^2 v_1^k = -\frac{j+1}{j} C^{jk\alpha}_{1,0} \|w_0^j\|^2 v_1^k.$$

Hence,

$$\mathsf{E}^{jk\alpha}\left(X_+ \cdot w_0^j\right) = -(j+1)K_{jk\alpha}\mathrm{i}^{j+k-\alpha}\,\|w_0^j\|^2\, C_{1,0}^{jk\alpha} v_1^k.$$

Taking an inner product with v_1^k we obtain the formula

$$\left(\mathsf{E}^{jk\alpha}\left(X_+ \cdot w_0^j\right), \left(X_+ \cdot w_0^k\right)\right)_{W_k} = -(j+1)(k+1)K_{jk\alpha}\mathrm{i}^{j+k-\alpha}\,\|w_0^j\|^2\,\|w_0^k\|^2\, C_{1,0}^{jk\alpha}.$$

Thus, we have the following formulas for computing the choice structure coefficients

$$K_{jk\alpha} = \begin{cases} \dfrac{\left(\mathsf{E}^{jk\alpha} w_0^j, w_0^k\right)}{\mathrm{i}^{j+k-\alpha} C_{0,0}^{jk\alpha}\,\|w_0^j\|^2\,\|w_0^k\|^2}, & j+k+\alpha \text{ is even,} \\[2ex] -\dfrac{\left(\mathsf{E}^{jk\alpha}\left(X_+ \cdot w_0^j\right), \left(X_+ \cdot w_0^k\right)\right)}{(j+1)(k+1)\mathrm{i}^{j+k-\alpha}\,\|w_0^j\|^2\,\|w_0^k\|^2\, C_{1,0}^{jk\alpha}}, & j+k+\alpha \text{ is odd.} \end{cases}$$

These formulas are valid only when $|j-k| \leqslant \alpha \leqslant j+k$. Set $K_{jk\alpha} = 0$ when the inequality $|j-k| \leqslant \alpha \leqslant j+k$ is violated. If we agree to always choose $\mathsf{E}^{jk\alpha}$ to be dual to $\mathsf{E}^{kj\alpha}$, when $k \neq j$, then $K_{jk\alpha} = (-1)^{j+k-\alpha} K_{kj\alpha}$. We note that for computing $K_{jk\alpha}$ we do not need to compute $\{e_B(jk\alpha) : |B| \leqslant \alpha\}$.

For the maximal torus consisting of all rotations around the x_1-axis, the action of the raising operator on $\mathbb{R}^3$ is given by the matrix

$$\boldsymbol{X}_+ = \begin{bmatrix} 0 & \mathrm{i} & 1 \\ -\mathrm{i} & 0 & 0 \\ -1 & 0 & 0 \end{bmatrix}, \qquad X_+ \cdot \boldsymbol{x} = \boldsymbol{X}_+\boldsymbol{x}. \tag{C.5}$$

The action of X_+ on 3×3 matrices is given by the commutator

$$X_+ \cdot \boldsymbol{A} = [\boldsymbol{X}_+, \boldsymbol{A}] = \boldsymbol{X}_+\boldsymbol{A} - \boldsymbol{A}\boldsymbol{X}_+. \tag{C.6}$$

These formulas are sufficient for all physical applications.

C.6 The Racah coefficients

Let us now take $W_\alpha \subset \mathrm{Hom}(W_j, W_k)$ and $W_\beta \subset \mathrm{Hom}(W_k, W_r)$ and observe that $W_\beta W_\alpha \subset \mathrm{Hom}(W_j, W_r)$. Next we also observe that the map

$$\Phi : \mathrm{End}(\mathcal{W}) \otimes \mathrm{End}(\mathcal{W}) \to \mathrm{End}(\mathcal{W}), \qquad \Phi(\mathsf{A} \otimes \mathsf{B}) = \mathsf{AB}$$

is SO(3)-invariant. Then

$$h_0 = \mathrm{i}^{\alpha+\beta-\gamma} \sum_{A=-\min(\alpha,\beta)}^{\min(\alpha,\beta)} C_{A,0}^{\alpha\beta\gamma}\mathrm{i}^{k+r-\beta} e_A(kr\beta)\mathrm{i}^{j+k-\alpha} e_{-A}(jk\alpha)$$

is a real zero weight vector in $W_\gamma \subset \mathrm{Hom}(W_j, W_r)$. Hence, it has to be a real scalar multiple of $\mathrm{i}^{j+r-\gamma} e_0(jr\gamma)$. Simplifying out the powers of ‘i’ we conclude that

$$\hat{h}_0 = \sum_{A=-\min(\alpha,\beta)}^{\min(\alpha,\beta)} C_{A,0}^{\alpha\beta\gamma} e_A(kr\beta)e_{-A}(jk\alpha)$$

must be a real scalar multiple of $e_0(jr\gamma)$. We compute

$$e_A(kr\beta)e_{-A}(jk\alpha) = \left(\sum_b C_{b,A}^{kr\beta} v_b^r \otimes v_{A-b}^k\right)\left(\sum_c C_{c,-A}^{jk\alpha} v_c^k \otimes v_{-A-c}^j\right).$$

Using orthogonality of v_ν^k and (C.1) we obtain

$$e_A(kr\beta)e_{-A}(jk\alpha) = \|w_0^k\|^2 \sum_c (-1)^c \frac{\binom{2k}{k}}{\binom{2k}{k+c}} C_{A+c,\,A}^{kr\beta} C_{c,-A}^{jk\alpha} v_{A+c}^r \otimes v_{-A-c}^j.$$

Hence,

$$\begin{aligned}\hat{h}_0 &= \|w_0^k\|^2 \sum_d \sum_c (-1)^c \frac{\binom{2k}{k}}{\binom{2k}{k+c}} C_{d-c,0}^{\alpha\beta\gamma} C_{d,d-c}^{kr\beta} C_{c,c-d}^{jk\alpha} v_d^r \otimes v_{-d}^j \\ &= \|w_0^k\|^2 R_{jkr}^{\alpha\beta\gamma} e_0(jr\gamma)\end{aligned}$$

for some real numbers $R_{jkr}^{\alpha\beta\gamma}$. Then,

$$R_{jkr}^{\alpha\beta\gamma} C_{d,0}^{jr\gamma} = \sum_c (-1)^c \frac{\binom{2k}{k}}{\binom{2k}{k+c}} C_{d-c,0}^{\alpha\beta\gamma} C_{d,d-c}^{kr\beta} C_{c,c-d}^{jk\alpha}.$$

The absolute constants $R_{jkr}^{\alpha\beta\gamma}$ are called the Racah coefficients in representation theory of SO(3). If W_γ is absent in $W_\beta W_\alpha$ then the Racah coefficient $R_{jkr}^{\alpha\beta\gamma}$ is zero. We can compute the Racah coefficients by means of the formula

$$R_{jkr}^{\alpha\beta\gamma} = \frac{1}{C_{d,0}^{jr\gamma}} \sum_{c=-k}^{k} (-1)^c \frac{\binom{2k}{k}}{\binom{2k}{k+c}} C_{d-c,0}^{\alpha\beta\gamma} C_{d,d-c}^{kr\beta} C_{c,c-d}^{jk\alpha}, \tag{C.7}$$

taking $d = 0$ if $j + r + \gamma$ is even and $d = 1$, if $j + r + \gamma$ is odd. Formula (C.7), as all the subsequent formulas involving division, has the property that the numerator always evaluates to 0 whenever the denominator is 0, in which case we set the value of the fraction to 0. In all of our applications $n \leqslant 2$, so that formulas (C.4), (C.7) can be used directly to compute all the required Clebsch–Gordan and Racah coefficients.

C.7 Multiplication of irreps in End(V)

If we start with our chosen zero weight vectors $\mathsf{E}^{jk\alpha}$ and construct a canonical basis of weight vectors $\mathsf{E}_A^{jk\alpha}$, $|A| \leqslant \alpha$ of $W_\alpha \subset \mathrm{Hom}(W_j, W_k)$, then

$$\mathsf{E}_A^{jk\alpha} = \mathrm{i}^{j+k-\alpha} K_{jk\alpha} e_A(jk\alpha).$$

The canonical basis for $W_\alpha(\mathbb{X})$ is

$$E_B^\alpha(\mathbb{X}) = \sum_{jk} \boldsymbol{X}_{jk} \otimes \mathsf{E}_B^{jk\alpha},$$

where $\boldsymbol{X}_{jk} = 0$, if there is no $W_\alpha \subset \mathrm{Hom}(W_j, W_k)$. Similarly, the canonical basis for $W_\beta(\Upsilon)$ is

$$E_B^\beta(\Upsilon) = \sum_{sr} \boldsymbol{Y}_{sr} \otimes \mathsf{E}_B^{sr\beta}.$$

Then a real zero weight vector $h_0 \in W_\gamma \subset W_\beta(\Upsilon) W_\alpha(\mathbb{X})$ is given by

$$h_0 = \mathrm{i}^{\alpha+\beta-\gamma} \sum_A C_{A,0}^{\alpha\beta\gamma} E_A^\beta(\Upsilon) E_{-A}^\alpha(\mathbb{X}).$$

Putting together the formulas for $E_{-A}^\alpha(\mathbb{X})$, $\mathsf{E}_A^{sr\beta}$ and $e_A(jk\alpha)$ we calculate

$$h_0 = \sum_{jr} \boldsymbol{Z}_{jr} \otimes \mathsf{E}^{jr\gamma},$$

where

$$\boldsymbol{Z}_{jr}^\gamma = \frac{1}{K_{jr\gamma}} \sum_{k=0}^{n} (-1)^k \, \|w_0^k\|^2 \, \boldsymbol{Y}_{kr} \boldsymbol{X}_{jk} R_{jkr}^{\alpha\beta\gamma} K_{kr\beta} K_{jk\alpha}. \tag{C.8}$$

Using our rule about formulas with division (setting $\boldsymbol{Z}_{jr}^\gamma = 0$, whenever $K_{jr\gamma} = 0$), we can extend the range of all Latin indices to $0, \ldots, n$ and all Greek indices to $0, \ldots, 2n$. Looking at formula (C.8) we see that it would be more natural to work with choice structure constants

$$\begin{aligned} \hat{K}_{jk\alpha} &= \|w_0^k\|^2 K_{jk\alpha} \\ &= \begin{cases} \dfrac{\left(\mathsf{E}^{jk\alpha} w_0^j, w_0^k\right)}{\mathrm{i}^{j+k-\alpha} C_{0,0}^{jk\alpha} \|w_0^j\|^2}, & j+k+\alpha \text{ is even}, \\[2ex] -\dfrac{\left(\mathsf{E}^{jk\alpha}\left(X_+ \cdot w_0^j\right), \left(X_+ \cdot w_0^k\right)\right)}{(j+1)(k+1)\mathrm{i}^{j+k-\alpha} \|w_0^j\|^2 C_{1,0}^{jk\alpha}}, & j+k+\alpha \text{ is odd}, \end{cases} \end{aligned} \tag{C.9}$$

while redefining the Racah coefficients

$$\hat{R}_{jkr}^{\alpha\beta\gamma} = (-1)^k R_{jkr}^{\alpha\beta\gamma} = \frac{(-1)^k}{C_{d,0}^{jr\gamma}} \sum_{c=-k}^{k} (-1)^c \frac{\binom{2k}{k}}{\binom{2k}{k+c}} C_{d-c,0}^{\alpha\beta\gamma} C_{d,d-c}^{kr\beta} C_{c,c-d}^{jk\alpha}. \tag{C.10}$$

In this case formula (C.8) simplifies to

$$Z^{\gamma}_{jr} = \frac{1}{\hat{K}_{jr\gamma}} \sum_{k=0}^{n} Y_{kr} X_{jk} \hat{R}^{\alpha\beta\gamma}_{jkr} \hat{K}_{kr\beta} \hat{K}_{jk\alpha}. \tag{C.11}$$

If we choose $\lambda_j w_0^j$ instead of w_0^j for some real non-zero scalars λ_j, while redefining the inner product on $W_j \subset \mathcal{W}$ by $\mu_j(\cdot, \cdot)_{W_j}$ for some positive scalars μ_j, $j = 1, \ldots, n$, then the constants $\hat{K}_{jk\alpha}$ will get multiplied by $\lambda_k\mu_k/\lambda_j\mu_j$, which obviously will not affect the right-hand side in (C.11).

The analysis of products of irreps in endomorphism algebras appears to be a new question in representation theory. We refer to papers by Sage [3–6], who showed that a similar analysis applies to other groups and in more general situations.

C.8 Example: choice structure constants for piezoelectricity

For piezoelectricity

$$V = \mathrm{Sym}(\mathbb{R}^3) \oplus \mathbb{R}^3 \cong W_0 \oplus W_1 \oplus W_2.$$

In this example $p = 1$ and $n = 2$. In order to make our choices we identify

$$W_0 = \{c\boldsymbol{I}_3 : c \in \mathbb{R}\}, \qquad W_2 = \left\{\boldsymbol{A} \in \mathrm{Sym}(\mathbb{R}^3) : \mathrm{Tr}\,\boldsymbol{A} = 0\right\}.$$

We define the following SO(3)-invariant inner products on V:

$$((\boldsymbol{a}, \boldsymbol{A}), (\boldsymbol{b}, \boldsymbol{B})) = \boldsymbol{a} \cdot \boldsymbol{b} + \mathrm{Tr}\,(\boldsymbol{AB}), \qquad \{\boldsymbol{a}, \boldsymbol{b}\} \subset \mathbb{R}^3,\ \{\boldsymbol{A}, \boldsymbol{B}\} \subset \mathrm{Sym}(\mathbb{R}^3).$$

We select the following zero weight vectors in $W_j \subset \mathcal{W} = V$:

$$w_0^0 = \boldsymbol{I}_3 \in W_0, \quad w_0^1 = \boldsymbol{e}_1 \in W_1, \quad w_0^2 = \begin{bmatrix} -2 & 0 & 0 \\ 0 & 1 & 0 \\ 0 & 0 & 1 \end{bmatrix}.$$

These vectors are invariant under rotations around the x_1 axis. We easily compute

$$\|w_0^0\|^2 = \mathrm{Tr}\left(\boldsymbol{I}_3^2\right) = 3, \quad \|w_0^1\|^2 = 1, \quad \|w_0^2\|^2 = (-2)^2 + 1^2 + 1^2 = 6.$$

Next we choose zero weight vectors $\mathsf{E}^{jk\alpha}$.

1. $W_0 \cong \mathrm{Hom}(W_0, W_0)$ is spanned by the map (which we will call E^{000}) that sends $\boldsymbol{I}_3$ into $\boldsymbol{I}_3$, while sending $W_1 \oplus W_2 \subset V$ to zero. We observe that

$$\mathsf{E}^{000} = K_0'\left(\frac{1}{3}\right),$$

where $K'(\kappa)$ is defined in (6.10).
2. The map $\mathsf{E}^{011} \in W_1 \cong \mathrm{Hom}(W_0, W_1)$, sending $w_0^0 = \boldsymbol{I}_3$ to $w_0^1 = \boldsymbol{e}_1$ is obviously a fixed point of rotations around $\boldsymbol{e}_1$ axis, i.e. a zero weight vector. It will be convenient to define $\mathsf{E}^{101} \in W_1 \cong \mathrm{Hom}(W_1, W_0)$ to be the dual of E^{011}. All maps in $\mathrm{Hom}(W_1, W_0)$ have the form $\boldsymbol{A}\boldsymbol{x} = (\boldsymbol{a} \cdot \boldsymbol{x})\boldsymbol{I}_3$ for some $\boldsymbol{a} \in \mathbb{R}^3$. By definition $(\mathsf{E}^{101}\boldsymbol{x}, \boldsymbol{I}_3)_V = (\boldsymbol{x}, \mathsf{E}^{011}\boldsymbol{I}_3)_V = x_1$. Hence, $((\boldsymbol{a} \cdot \boldsymbol{x})\boldsymbol{I}_3, \boldsymbol{I}_3) = x_1$, which

implies that $\boldsymbol{a} = \boldsymbol{e}_1/3$. So, that $\mathsf{E}^{101}\boldsymbol{x} = x_1\boldsymbol{I}_3/3$, while sending $W_0 \oplus W_2 \subset V$ to 0.

3. Identifying $\mathsf{E}^{022} \in W_2 \cong \mathrm{Hom}(W_0, W_2)$ is just as easy: $\mathsf{E}^{022}w_0^0 = w_0^2$. The map $\mathsf{E}^{202} \in \mathrm{Hom}(W_2, W_0)$ is defined by duality: $\mathsf{E}^{202}\boldsymbol{X} = \mathrm{Tr}\,(\boldsymbol{AX})\boldsymbol{I}_3$ and

$$3\mathrm{Tr}\,(\boldsymbol{AX}) = \left(\mathsf{E}^{202}\boldsymbol{X}, \boldsymbol{I}_3\right) = \left(\boldsymbol{X}, \mathsf{E}^{022}\boldsymbol{I}_3\right) = \left(\boldsymbol{X}, w_0^2\right) = \mathrm{Tr}\left(\boldsymbol{X}w_0^2\right).$$

Hence, $\mathsf{E}^{202}\boldsymbol{X} = \mathrm{Tr}\,(w_0^2\boldsymbol{X})\boldsymbol{I}_3/3$. We observe that

$$\mathsf{E}^{022} + \mathsf{E}^{202} = \frac{1}{3}K_2'\left(w_0^2\right),$$

where $K_2'(\boldsymbol{A})$ is defined in (6.11).

4. $\mathrm{Hom}(W_1, W_1) = \mathrm{End}(W_1)$ splits into multiples of the identity map, skew-symmetric operators and symmetric, trace free operators. In the standard basis of $\mathbb{R}^3$ the maps $\mathsf{E}^{11\alpha}$ will be

$$\mathsf{E}^{110}\boldsymbol{x} = \boldsymbol{x}, \qquad \mathsf{E}^{111}\boldsymbol{x} = \boldsymbol{e}_1 \times \boldsymbol{x}, \qquad \mathsf{E}^{112}\boldsymbol{x} = w_0^2\boldsymbol{x} = (-2x_1, x_2, x_3).$$

These maps also send $W_0 \oplus W_2 \subset V$ to 0.

5. Understanding irreps W_1 and W_2 in $\mathrm{Hom}(W_1, W_2)$ is also easy.

$$W_1 = \left\{\boldsymbol{x} \mapsto \boldsymbol{x} \odot \boldsymbol{a} - (1/3)(\boldsymbol{a} \cdot \boldsymbol{x})\boldsymbol{I}_3 : \boldsymbol{a} \in \mathbb{R}^3\right\},$$
$$W_2 = \{\boldsymbol{x} \mapsto [\boldsymbol{A}, \pi(\boldsymbol{x})] : \boldsymbol{A} \in W_2 \subset V\},$$
$$\mathsf{E}^{121}\boldsymbol{x} = \boldsymbol{x} \odot \boldsymbol{e}_1 - (1/3)x_1\boldsymbol{I}_3, \qquad \mathsf{E}^{122}\boldsymbol{x} = \left[w_0^2, \pi(\boldsymbol{x})\right],$$

where $\pi(\boldsymbol{x})\boldsymbol{u} = \boldsymbol{x} \times \boldsymbol{u}$. We define E^{211} and E^{212} by duality: $\mathsf{E}^{211}\boldsymbol{X} = \boldsymbol{X}\boldsymbol{e}_1$, $\mathsf{E}^{212}\boldsymbol{X} = 6\boldsymbol{e}_1 \times \boldsymbol{X}\boldsymbol{e}_1$. It is much more difficult to describe explicitly the irrep $W_3 \subset \mathrm{Hom}(W_1, W_2)$. This, however, is not necessary. We simply declare that we have chosen the zero weight vector E^{123} and its dual $\mathsf{E}^{213} \in W_3 \subset \mathrm{Hom}(W_2, W_1)$. The reason for this is that there is a unique irrep of weight 3 in $\mathrm{Sym}(V)$, and hence, the problem of identifying which $W_3 \subset \mathrm{Sym}(V)$ occurs in $(\Pi_1 \mathcal{A} \Pi_2)_{\mathrm{sym}}$ does not arise. For that same reason we just set $K_{123} = K_{213} = 1$.

6. It is easy to describe $\mathsf{E}^{22\alpha} \in W_\alpha \subset \mathrm{End}(W_2)$ for $\alpha = 0, 1, 2$. $\mathsf{E}^{220}\boldsymbol{X} = \boldsymbol{X}$, $\mathsf{E}^{221}\boldsymbol{X} = [\boldsymbol{X}, \pi(\boldsymbol{e}_1)]$, $\mathsf{E}^{222}\boldsymbol{X} = w_0^2\boldsymbol{X} + \boldsymbol{X}w_0^2 - 2\mathrm{Tr}\,(w_0^2\boldsymbol{X})\boldsymbol{I}_3/3$. We observe that

$$\mathsf{E}^{220} = K_0\left(\frac{1}{2}\right), \qquad \mathsf{E}^{222} = K_2\left(w_0^2\right),$$

where $K_0(\mu)$ and $K_2(\boldsymbol{A})$ are defined in (6.12) and (6.13), respectively. $W_3 \subset \mathrm{End}(W_2)$ consists of antisymmetric operators on V and does not enter into $\mathrm{Sym}(V)$, while there is a unique irrep of weight 4 in $\mathrm{Sym}(V)$. That means that we do not need to describe the maps $\mathsf{E}^{22\alpha} \in W_\alpha \subset \mathrm{End}(W_2)$ for $\alpha = 3, 4$ explicitly. We therefore set $K_{223} = K_{224} = 1$.

Bibliography

[1] Grabovsky Y, Milton G W and Sage D S 2000 Exact relations for effective tensors of polycrystals: necessary conditions and sufficient conditions *Commun. Pure. Appl. Math.* **53** 300–53

[2] Grabovsky Y and Sage D S 1998 Exact relations for effective tensors of polycrystals. II: Applications to elasticity and piezoelectricity *Arch. Ration. Mech. Anal.* **143** 331–56

[3] Kwon N and Sage D S 2008 Subrepresentation semirings and an analog of $6j$-symbols *J. Math. Phys.* **49** 063503

[4] Sage D S 2002 Group actions on central simple algebras *J. Algebra* **250** 18–43

[5] Sage D S 2005 Quantum Racah coefficients and subrepresentation semirings *J. Lie Theory* **15** 321–33

[6] Sage D S 2005 Racah coefficients, subrepresentation semirings, and composite materials *Adv. Appl. Math* **34** 335–57

www.ingramcontent.com/pod-product-compliance
Ingram Content Group UK Ltd.
Pitfield, Milton Keynes, MK11 3LW, UK
UKHW051855100926
8556IPUK00005B/641

9 780750 310499